AF676620

Study on Genotypes and Phenotypes of Neurodegenerative Diseases

Study on Genotypes and Phenotypes of Neurodegenerative Diseases

Editor

Claudia Ricci

MDPI

Basel • Beijing • Wuhan • Barcelona • Belgrade • Novi Sad • Cluj • Manchester

Editor
Claudia Ricci
Department of Medical,
Surgical and
Neurological Sciences
University of Siena
Siena
Italy

Editorial Office
MDPI AG
Grosspeteranlage 5
4052 Basel, Switzerland

This is a reprint of articles from the Special Issue published online in the open access journal *Genes* (ISSN 2073-4425) (available at: https://www.mdpi.com/journal/genes/special_issues/Neurodegenerative_Genotypes_Phenotypes).

For citation purposes, cite each article independently as indicated on the article page online and as indicated below:

Lastname, A.A.; Lastname, B.B. Article Title. *Journal Name* **Year**, *Volume Number*, Page Range.

ISBN 978-3-7258-2531-8 (Hbk)
ISBN 978-3-7258-2532-5 (PDF)
doi.org/10.3390/books978-3-7258-2532-5

© 2024 by the authors. Articles in this book are Open Access and distributed under the Creative Commons Attribution (CC BY) license. The book as a whole is distributed by MDPI under the terms and conditions of the Creative Commons Attribution-NonCommercial-NoDerivs (CC BY-NC-ND) license.

Contents

About the Editor . . . . **vii**

Claudia Ricci
Study on Genotypes and Phenotypes of Neurodegenerative Diseases
Reprinted from: *Genes* **2024**, *15*, 786, doi:10.3390/genes15060786 . . . . **1**

Lea Hentrich, Mered Parnes, Timothy Edward Lotze, Rohini Coorg, Tom J. de Koning, Kha M. Nguyen, et al.
Novel Genetic and Phenotypic Expansion in *GOSR2*-Related Progressive Myoclonus Epilepsy
Reprinted from: *Genes* **2023**, *14*, 1860, doi:10.3390/genes14101860 . . . . **5**

Yury Loika, Elena Loiko, Irina Culminskaya and Alexander M. Kulminski
Exome-Wide Association Study Identified Clusters of Pleiotropic Genetic Associations with Alzheimer's Disease and Thirteen Cardiovascular Traits
Reprinted from: *Genes* **2023**, *14*, 1834, doi:10.3390/genes14101834 . . . . **17**

Anusha Mamidipaka, Isabel Di Rosa, Roy Lee, Yan Zhu, Yineng Chen, Rebecca Salowe, et al.
Factors Associated with Large Cup-to-Disc Ratio and Blindness in the Primary Open-Angle African American Glaucoma Genetics (POAAGG) Study
Reprinted from: *Genes* **2023**, *14*, 1809, doi:10.3390/genes14091809 . . . . **35**

Sangeetha Vishweswaraiah, Ali Yilmaz, Nazia Saiyed, Abdullah Khalid, Purvesh R. Koladiya, Xiaobei Pan, et al.
Integrative Analysis Unveils the Correlation of Aminoacyl-tRNA Biosynthesis Metabolites with the Methylation of the *SEPSECS* Gene in Huntington's Disease Brain Tissue
Reprinted from: *Genes* **2023**, *14*, 1752, doi:10.3390/genes14091752 . . . . **47**

Camille Bouchard, Catherine Gérard, Solange Gni-fiene Yanyabé, Nathalie Majeau, Malek Aloui, Gabrielle Buisson, et al.
Finding an Appropriate Mouse Model to Study the Impact of a Treatment for Friedreich Ataxia on the Behavioral Phenotype
Reprinted from: *Genes* **2023**, *14*, 1654, doi:10.3390/genes14081654 . . . . **61**

Dandan Chen, Jin Li, Hongwei Liu, Yang Xi, Xiaolong Liu, Chenghao Zhang, et al.
Genome-Wide Epistasis Study of Cerebrospinal Fluid Hyperphosphorylated Tau in ADNI Cohort
Reprinted from: *Genes* **2023**, *14*, 1322, doi:10.3390/genes14071322 . . . . **81**

Katarzyna Gaweda-Walerych, Emilia J. Sitek, Małgorzata Borczyk, Ewa Narożańska, Bogna Brockhuis, Michał Korostyński, et al.
A Patient with Corticobasal Syndrome and Progressive Non-Fluent Aphasia (CBS-PNFA), with Variants in *ATP7B*, *SETX*, *SORL1*, and *FOXP1* Genes
Reprinted from: *Genes* **2022**, *13*, 2361, doi:10.3390/genes13122361 . . . . **93**

Marco Salluzzo, Clara Vianello, Sandra Abdullatef, Roberto Rimondini, Giovanni Piccoli and Lucia Carboni
The Role of IgLON Cell Adhesion Molecules in Neurodegenerative Diseases
Reprinted from: *Genes* **2023**, *14*, 1886, doi:10.3390/genes14101886 . . . . **110**

Edward O. Olufunmilayo and R. M. Damian Holsinger
INPP5D/SHIP1: Expression, Regulation and Roles in Alzheimer's Disease Pathophysiology
Reprinted from: *Genes* **2023**, *14*, 1845, doi:10.3390/genes14101845 . . . . **130**

Sue-Faye Siow, Dennis Yeow, Laura I. Rudaks, Fangzhi Jia, Gautam Wali, Carolyn M. Sue and Kishore R. Kumar
Outcome Measures and Biomarkers for Clinical Trials in Hereditary Spastic Paraplegia: A Scoping Review
Reprinted from: *Genes* **2023**, *14*, 1756, doi:10.3390/genes14091756 . . . . **143**

Martin N. Ivanov, Dimo S. Stoyanov, Stoyan P. Pavlov and Anton. B. Tonchev
Distribution, Function, and Expression of the Apelinergic System in the Healthy and Diseased Mammalian Brain
Reprinted from: *Genes* **2022**, *13*, 2172, doi:10.3390/genes13112172 . . . . **165**

About the Editor

Claudia Ricci

Claudia Ricci is a Research Scientist at the Neurogenetics Laboratory of the Department of Medical, Surgical, and Neurological Sciences of the University of Siena. After completing a PhD in Biotechnology, she extended her interest to the genetics of cerebral vascular malformations and motor neuron diseases, particularly amyotrophic lateral sclerosis (ALS), and increased her expertise in the up-to-date techniques of genetic screening and biomarker research. She also deepened her skills in statistical analysis and working with bioinformatics tools. She collaborates with the Italian Consortium for the study of genetics in ALS(ITALSGEN) and the Italy–USA Network and participates in wide studies on the genetic basis of ALS, leading to the identification of several genes related to this disease, including C9orf72, TARDBP, and FUS. Her publication list includes approximately 60 papers in peer-reviewed journals (ORCID 0000-0002-2431-0308). Since 2021, she has been member of the Scientific Committee of the Master's degree programmes "Regulatory Medical Writer" and "Science Popularization" at the University of Siena.

genes

MDPI

Editorial

Study on Genotypes and Phenotypes of Neurodegenerative Diseases

Claudia Ricci

Department of Medical, Surgical and Neurological Sciences, University of Siena, 53100 Siena, Italy; claudia.ricci@unisi.it

Neurodegenerative diseases are a heterogeneous group of age-related disorders that are characterised by the gradual degeneration or death of neurons in the central or peripheral nervous system. The prevalence of these diseases is increasing, partly due to an ageing population; as a result, the economic burden on healthcare systems is also growing. Recent advances in neurobiology and neurogenetics have provided valuable insights into the pathogenesis of neurodegenerative diseases.

Genetic factors play a key role in the pathogenesis of a number of neurodegenerative diseases, acting both as monogenic causes in the inherited forms and as modulating factors in the multifactorial/sporadic forms. The latest advances in cost-effective genetic analysis have significantly increased our knowledge of the genetic basis of several neurodegenerative diseases; in particular, omics studies have provided a better understanding of the mechanisms underlying their pathogenesis.

A common feature of neurodegenerative diseases is that they exhibit a degree of genetic heterogeneity; in other words, the presentation and severity of the disease can vary from person to person. Sometimes, different variants in different genes can even cause the same phenotype. On the other hand, the same mutation can be associated with phenotypic heterogeneity, even within the same family. In some cases, however, a specific variant may be associated with a common phenotype, which may be useful for diagnostic and prognostic purposes. In the era of precision medicine, a better characterisation of genotype–phenotype correlations may help to improve therapeutic approaches, assess individual drug responses, and guide gene-targeted clinical trials.

Several papers in this Special Issue are based on the use of omics technologies to study the basis of neurodegenerative diseases.

In their study, Katarzyna Gaweda-Walerych and colleagues analysed the phenotypic–genetic correlations in a patient diagnosed with early onset corticobasal syndrome with progressive non-fluent aphasia (CBS-PNFA) using a whole-exome sequencing (WES) approach [1]. They identified rare single heterozygous variants in the *ATP7B*, *SORL1*, *SETX*, and *FOXP1* genes, extending the complex clinical spectrum associated with variants in known disease genes and also supporting the hypothesis of oligogenic inheritance. In addition, the study highlights the possible relevance of the *FOXP1* gene in relation to progressive language apraxia in adulthood, whereas previously this gene has only been associated with neurodevelopmental language disorders.

In some cases, the use of large-scale genome-wide association studies (GWASs) has produced inconclusive results. In Alzheimer's disease (AD), for example, the problem of "missing heritability" is still a major challenge. Dandan Chen and colleagues have addressed this issue by focusing on epistasis [2], whereby the effect of a mutation depends on the genetic background in which it occurs. Their study focused on identifying epistasis between two-marker interactions with marginal main effects across the genome, employing cerebrospinal fluid (CSF) P-tau as a quantitative trait. Using this approach, they identified 758 statistically significant SNP pairs and several highly significant SNP–SNP interactions that explained a relatively high amount of variance at the P-tau level. In addition, 331

Citation: Ricci, C. Study on Genotypes and Phenotypes of Neurodegenerative Diseases. *Genes* **2024**, *15*, 786. https://doi.org/10.3390/genes15060786

Received: 27 May 2024
Accepted: 13 June 2024
Published: 14 June 2024

Copyright: © 2024 by the author. Licensee MDPI, Basel, Switzerland. This article is an open access article distributed under the terms and conditions of the Creative Commons Attribution (CC BY) license (https://creativecommons.org/licenses/by/4.0/).

AD-related genes were identified, and 10 gene–gene interaction pairs were replicated in the PPI network, showing associations with AD in terms of neuroinflammation, neurodegeneration, neuronal cell activation, and brain development, leading to a cognitive decline in people with AD. Overall, this study may, therefore, open up new perspectives to complement traditional GWASs and to help explain some of the "missing heritability" of Alzheimer's disease.

AD was also studied in the paper by Yury Loika and colleagues [3]. In this case, the authors aimed to identify clusters of cardiovascular traits that share genetic factors with Alzheimer's disease. They performed a univariate exome-wide association study and a pair-wise pleiotropic analysis focusing on AD and 16 cardiovascular traits (6 diseases and 10 cardio-metabolic risk factors) using 188,260 samples from the UK Biobank (UKB). The analysis identified nine genetic markers in the *APOE* gene region, and four loci mapped to the *CDK11*, *OBP2B*, *TPM1*, and *SMARCA4* genes that showed associations with AD and pleiotropic associations. Using hierarchical cluster analysis, the authors grouped the phenotypes from these pleiotropic associations into seven clusters. Interestingly, all the AD protective alleles were associated with an increased weight and a risk of diabetes mellitus. The overall results of the study suggest that common genetic factors between Alzheimer's disease and cardiovascular traits are influenced in an antagonistic way.

Furthermore, Sangeetha Vishweswaraiah and colleagues investigated the interaction between the metabolome and the epigenome in Huntington's disease (HD) [4]. They performed metabolomic profiling of human post-mortem brain tissue (striatum and frontal lobe) and DNA methylome profiling using the same frontal lobe tissue. Among several perturbed metabolites and differentially methylated loci, they identified aminoacyl-tRNA biosynthesis as being the most significantly perturbed pathway, correlating with two CpGs of the *SEPSECS* gene. This study reveals important links between molecular biomarkers and adds to the knowledge of the metabolic changes that drive the progression of HD.

An important aspect of neurodegenerative diseases is the possible existence of genotype-phenotype correlations. In their article, Lea Hentrich and colleagues describe two patients with progressive myoclonus epilepsy with neurodegeneration that is associated with biallelic variants of the Golgi SNAP receptor complex member 2 gene (*GOSR2*) [5]. The authors report the clinical findings and a genetic characterisation of both patients. The study extends the genotype–phenotype spectrum of *GOSR2*-related disorders and suggests that the *GOSR2* gene should be included in the consideration of the monogenetic causes of dystonia, global developmental delay, and seizures.

The availability of affordable disease models is fundamental to the ongoing search for novel therapeutic approaches. Camille Bouchard and colleagues reported the discovery of a suitable mouse model for Friedreich's Ataxia (FRDA), which is a progressive neurodegenerative disease caused by a triplet repeat in the frataxin gene (*FXN*) that leads to a reduced expression of the frataxin protein [6]. The previously used mouse model (YG8sR mice) showed a milder phenotype compared to human patients, limiting the ability to study the impact of therapy on the phenotype. The authors compared two different mouse models—a modified YG8sR model injected with an AAV encoding an shRNA that targets the human frataxin gene, and the YG8-800, a new mouse model with a human transgene containing 800 GAA repeats. Behavioural tests showed that the mild phenotype of YG8sR mice can be enhanced by injecting them with an AAV expressing an shRNA targeting frataxin, and that the phenotype of YG8-800 mice is comparable to the human ataxic phenotype.

The role of genetic factors in contributing to ethnic differences in health and disease is becoming increasingly recognised. Anusha Mamidipaka and colleagues examined the demographic, optic disc, and genetic risk factors for a large cup-to-disc ratio (LCDR) presentation and its subsequent progression to visual impairment or blindness in a large cohort of individuals of African ancestry that were recruited from the Primary Open-Angle African American Glaucoma Genetics (POAAGG) study [7]. No significant associations were found between genetic variants and LCDR or blindness, while previous glaucoma surgery, increased intraocular pressure, decreased mean deviation, and decreased pattern standard deviation were statistically significant risk factors for blindness in LCDR eyes. These findings highlight the importance of the close monitoring of intraocular pressure and visual function in POAG patients of African descent, especially those with LCDR, in order to preserve visual function.

The importance of risk factors for disease development is also emphasised in one of the four reviews in the Special Issue. Edward O. Olufunmilayo and colleagues [8] review the role of the Src homology 2 (SH2) domain-containing inositol 50 phosphatase 1 (SHIP1) protein, encoded by the *INPP5D* (inositol polyphosphate-5-phosphatase D) gene, in Alzheimer's disease. *INPP5D* expression and SHIP1 activity have been shown to be important risk factors for the development of late-onset Alzheimer's disease, as the various mechanistic actions of SHIP1 in microglia result in the attenuation of signalling and, as a consequence, some of the beneficial functions of these cells. In addition, polymorphisms in the *INPP5D* gene have been found to be associated with a significantly increased risk of AD; *INPP5D*/SHIP1 may, therefore, be an interesting therapeutic target for the disease.

Another review focuses on the role of IgLON cell adhesion molecules in neurodegenerative diseases. Salluzzo and colleagues [9] describe the structure, functional domains, interactions, and activities of the IgLON family, a class of adhesion molecules consisting of five members; they also analyse the available data and suggest a role for IgLONs in neuropsychiatric disorders. The authors highlight the evidence that is already available and the mechanisms that require further in-depth investigation.

Martin N. Ivanov and colleagues discuss the current understanding of the involvement of the apelinergic system in the brain physiology in health and disease [10]. The authors report on the structure and function of the apelin receptor (APLNR) and its ligands—apelin and ELABELA. They also describe the role of the apelinergic system in a typical brain, as well as in several pathological conditions including Parkinson's and Alzheimer's diseases, in which apelin peptides appear to exert protective effects.

Another important issue for neurodegenerative disease research is the identification of outcome measures and biomarkers for clinical trials [11]. Sue-Faye Siow and colleagues conducted a scoping review of outcome measures and biomarkers in hereditary spastic paraplegia (HSP) with the aim of providing recommendations for future studies. They discuss the advantages and limitations of commonly used outcome measures and suggest areas for further research. Overall, the authors report a significant variability and inconsistency in using outcome measures with limited longitudinal data. Given the emergence of several candidate HSP therapies in recent years, they suggest that there is an urgent need to further develop a core set of validated and standardised outcome measures for use in HSP clinical trials to test the efficacy of these therapies.

In conclusion, this Special Issue provides an overview of research into the genotypes and phenotypes of neurodegenerative diseases, ranging from the use of cutting-edge omics techniques to the identification of biomarkers and the characterisation of new experimental models. The picture that emerges is one of ongoing research efforts to more effectively combat these devastating diseases and to identify potential future therapeutic targets.

Conflicts of Interest: The authors declare no conflict of interest.

References

1. Gaweda-Walerych, K.; Sitek, E.J.; Borczyk, M.; Narożańska, E.; Brockhuis, B.; Korostyński, M.; Schinwelski, M.; Siemiński, M.; Sławek, J.; Zekanowski, C. A Patient with Corticobasal Syndrome and Progressive Non-Fluent Aphasia (CBS-PNFA), with Variants in ATP7B, SETX, SORL1, and FOXP1 Genes. *Genes* **2022**, *13*, 2361. [CrossRef] [PubMed]
2. Chen, D.; Li, J.; Liu, H.; Liu, X.; Zhang, C.; Luo, H.; Wei, Y.; Xi, Y.; Liang, H.; Zhang, Q. Genome-Wide Epistasis Study of Cerebrospinal Fluid Hyperphosphorylated Tau in ADNI Cohort. *Genes* **2023**, *14*, 1322. [CrossRef] [PubMed]
3. Loika, Y.; Loiko, E.; Culminskaya, I.; Kulminski, A.M. Exome-Wide Association Study Identified Clusters of Pleiotropic Genetic Associations with Alzheimer's Disease and Thirteen Cardiovascular Traits. *Genes* **2023**, *14*, 1834. [CrossRef] [PubMed]
4. Vishweswaraiah, S.; Yilmaz, A.; Saiyed, N.; Khalid, A.; Koladiya, P.R.; Pan, X.; Macias, S.; Robinson, A.C.; Mann, D.; Green, B.D.; et al. Integrative Analysis Unveils the Correlation of Aminoacyl-tRNA Biosynthesis Metabolites with the Methylation of the SEPSECS Gene in Huntington's Disease Brain Tissue. *Genes* **2023**, *14*, 1752. [CrossRef] [PubMed]

5. Hentrich, L.; Parnes, M.; Lotze, T.E.; Coorg, R.; De Koning, T.J.; Nguyen, K.M.; Yip, C.K.; Jungbluth, H.; Koy, A.; Dafsari, H.S. Novel Genetic and Phenotypic Expansion in GOSR2-Related Progressive Myoclonus Epilepsy. *Genes* **2023**, *14*, 1860. [CrossRef] [PubMed]
6. Bouchard, C.; Gérard, C.; Yanyabé, S.G.; Majeau, N.; Aloui, M.; Buisson, G.; Yameogo, P.; Couture, V.; Tremblay, J.P. Finding an Appropriate Mouse Model to Study the Impact of a Treatment for Friedreich Ataxia on the Behavioral Phenotype. *Genes* **2023**, *14*, 1654. [CrossRef] [PubMed]
7. Mamidipaka, A.; Di Rosa, I.; Lee, R.; Zhu, Y.; Chen, Y.; Salowe, R.; Addis, V.; Sankar, P.; Daniel, E.; Ying, G.-S.; et al. Factors Associated with Large Cup-to-Disc Ratio and Blindness in the Primary Open-Angle African American Glaucoma Genetics (POAAGG) Study. *Genes* **2023**, *14*, 1809. [CrossRef] [PubMed]
8. Olufunmilayo, E.O.; Holsinger, R.M.D. INPP5D/SHIP1: Expression, Regulation and Roles in Alzheimer's Disease Pathophysiology. *Genes* **2023**, *14*, 1845. [CrossRef] [PubMed]
9. Salluzzo, M.; Vianello, C.; Abdullatef, S.; Rimondini, R.; Piccoli, G.; Carboni, L. The Role of IgLON Cell Adhesion Molecules in Neurodegenerative Diseases. *Genes* **2023**, *14*, 1886. [CrossRef] [PubMed]
10. Ivanov, M.N.; Stoyanov, D.S.; Pavlov, S.P.; Tonchev, A.B. Distribution, Function, and Expression of the Apelinergic System in the Healthy and Diseased Mammalian Brain. *Genes* **2022**, *13*, 2172. [CrossRef] [PubMed]
11. Siow, S.-F.; Yeow, D.; Rudaks, L.I.; Jia, F.; Wali, G.; Sue, C.M.; Kumar, K.R. Outcome Measures and Biomarkers for Clinical Trials in Hereditary Spastic Paraplegia: A Scoping Review. *Genes* **2023**, *14*, 1756. [CrossRef] [PubMed]

Disclaimer/Publisher's Note: The statements, opinions and data contained in all publications are solely those of the individual author(s) and contributor(s) and not of MDPI and/or the editor(s). MDPI and/or the editor(s) disclaim responsibility for any injury to people or property resulting from any ideas, methods, instructions or products referred to in the content.

genes

Article

Novel Genetic and Phenotypic Expansion in *GOSR2*-Related Progressive Myoclonus Epilepsy

Lea Hentrich [1,2,3], Mered Parnes [4], Timothy Edward Lotze [4], Rohini Coorg [4], Tom J. de Koning [5,6], Kha M. Nguyen [7], Calvin K. Yip [7], Heinz Jungbluth [8,9], Anne Koy [1,10] and Hormos Salimi Dafsari [1,2,3,8,9,10,*]

1 Department of Pediatrics, Faculty of Medicine and University Hospital Cologne, University of Cologne, 50937 Cologne, Germany; lhentrich@age.mpg.de (L.H.)
2 Max-Planck-Institute for Biology of Ageing, 50931 Cologne, Germany
3 Cologne Excellence Cluster on Cellular Stress Responses in Aging Associated Diseases (CECAD), 50931 Cologne, Germany
4 Division of Pediatric Neurology and Developmental Neuroscience, Department of Pediatrics, Baylor College of Medicine, Houston, TX 77030, USA; parnes@bcm.edu (M.P.); tlotze@bcm.edu (T.E.L.)
5 Department of Genetics, University Medical Center Groningen, University of Groningen, 9713 GZ Groningen, The Netherlands
6 Pediatrics, Department of Clinical Sciences, Lund University, 221 00 Lund, Sweden
7 Department of Biochemistry and Molecular Biology, The University of British Columbia, Vancouver, BC V6T 1Z4, Canada; nguyenkm@student.ubc.ca (K.M.N.); calvin.yip@ubc.ca (C.K.Y.)
8 Department of Paediatric Neurology, Evelina's Children Hospital, Guy's & St. Thomas' Hospital NHS Foundation Trust, London SE1 7EH, UK
9 Randall Division of Cell and Molecular Biophysics, Muscle Signaling Section, King's College London, London WC2R 2LS, UK
10 Center for Rare Diseases, Faculty of Medicine and University Hospital Cologne, University of Cologne, 50937 Cologne, Germany
* Correspondence: hormos.dafsari@uk-koeln.de

Citation: Hentrich, L.; Parnes, M.; Lotze, T.E.; Coorg, R.; de Koning, T.J.; Nguyen, K.M.; Yip, C.K.; Jungbluth, H.; Koy, A.; Dafsari, H.S. Novel Genetic and Phenotypic Expansion in *GOSR2*-Related Progressive Myoclonus Epilepsy. *Genes* **2023**, *14*, 1860. https://doi.org/10.3390/genes14101860

Academic Editor: Claudia Ricci

Received: 3 September 2023
Revised: 18 September 2023
Accepted: 19 September 2023
Published: 25 September 2023

Copyright: © 2023 by the authors. Licensee MDPI, Basel, Switzerland. This article is an open access article distributed under the terms and conditions of the Creative Commons Attribution (CC BY) license (https://creativecommons.org/licenses/by/4.0/).

Abstract: Biallelic variants in the Golgi SNAP receptor complex member 2 gene (*GOSR2*) have been reported in progressive myoclonus epilepsy with neurodegeneration. Typical clinical features include ataxia and areflexia during early childhood, followed by seizures, scoliosis, dysarthria, and myoclonus. Here, we report two novel patients from unrelated families with a *GOSR2*-related disorder and novel genetic and clinical findings. The first patient, a male compound heterozygous for the *GOSR2* splice site variant c.336+1G>A and the novel c.364G>A,p.Glu122Lys missense variant showed global developmental delay and seizures at the age of 2 years, followed by myoclonus at the age of 8 years with partial response to clonazepam. The second patient, a female homozygous for the *GOSR2* founder variant p.Gly144Trp, showed only mild fine motor developmental delay and generalized tonic–clonic seizures triggered by infections during adolescence, with seizure remission on levetiracetam. The associated movement disorder progressed atypically slowly during adolescence compared to its usual speed, from initial intention tremor and myoclonus to ataxia, hyporeflexia, dysmetria, and dystonia. These findings expand the genotype–phenotype spectrum of *GOSR2*-related disorders and suggest that *GOSR2* should be included in the consideration of monogenetic causes of dystonia, global developmental delay, and seizures.

Keywords: *GOSR2*; North Sea progressive myoclonus epilepsy; vesicular trafficking; dystonia; clonazepam

1. Introduction

Biallelic variants in the Golgi SNAP receptor complex member 2 gene (*GOSR2)* have been reported in progressive myoclonus epilepsy (PME, OMIM#614018). Patients usually present with ataxia and areflexia during early childhood, typically followed by seizures and myoclonus of upper limbs during the first decade of life [1]. Dysarthria, creatine kinase elevation, and scoliosis suggestive of a neuromuscular component are also common. Due to common ancestry of patients with a homozygous variant in NM_001012511.3:c.430G>T,p.Gly144Trp in countries in northwestern Europe, the condition was previously termed 'North Sea' PME (NS-PME). Cognitive function is typically spared at disease onset and may be followed by cognitive decline in adulthood. Fever and illness appear to irreversibly worsen NS-PME-associated symptoms, with limited recovery [2]. There are only supportive therapeutic options. *GOSR2* encodes the protein GOSR2/membrin that is involved in intracellular trafficking of COPII vesicles from the endoplasmic reticulum to cis-Golgi [3]. *GOSR2* loss-of-function variants prevent the membrane protein from localizing to cis-Golgi [4], as demonstrated in relevant animal models that demonstrate putative membrin loss of function in neurons and glia [5]. These animal models show similar phenotypes with seizures, particularly if the loss of function affects glial cell localization. In neurons, phenotype alignments are not observed, hinting at an involvement of glia in the pathophysiology of NSPME. However, the precise pathomechanisms and metabolic dysregulation of *GOSR2*-associated disorders remain largely elusive. Here, we report two novel patients with *GOSR2*-related myoclonus epilepsy with an expansion in genotype and phenotype in this ultra-rare disorder.

2. Materials and Methods

2.1. Patient Recruitment

The study was approved by local institutional review boards [6]. All patients and/or their legal guardians gave informed consent, including consent to the anonymous publication of their clinical information. Additional written informed consent was obtained from all participants (and/or their legal guardian) with identifiable photographs.

2.2. Molecular Genetic Investigations

We performed massively parallel sequencing with sequence alignment and variant calling as previously published [7,8]. We performed filtering for rare pathogenic variants below an allele frequency of <0.1% following recessive and dominant inheritance models. In a first step, we filtered for dystonia-associated genes (*GCH1, GCE, TH, SPR, ATP1A3, FTL, PRKRA, TUBB4A, BCAP31, COX20, KIF1C, SLC30A10, ANO3*) [7,9,10].

We performed further in silico and genomic analysis for all variants [11,12], including allele frequency in healthy population databases (gnomAD [13]), and pathogenicity prediction tools such as SpliceAI, GERP++, CADD-Phred, and MutPred2 [14–17]. Investigation of evolutionary conservation was based on NCBI HomoloGene sequences (Supplementary Table S2). All variants were scored on the basis of classification by the standards and guidelines of the American College of Medical Genetics and Genomics—American College of Molecular Pathology (ACMG) for the interpretation of variants [18].

To generate the structural model of the GOSR2/syntaxin-5/Bet1/Sec22b trans-SNARE complex, we used ColabFold and more specifically the AlphaFold-Multimer algorithm with the following protein sequences as input: GOSR2 (107–212), Syntaxin-5 (251–355), Bet1 (1–118), and Sec22b (105–215) [19]. The top-ranked model was subject to energy minimization using PHENIX [20]. The models of the mutants were generated using Pymol mutagenesis function. Figures of the trans-SNARE complex were prepared using Pymol.

3. Results

3.1. Phenotype Expansion

Patient 1 is a male from non-consanguineous Hispanic parents without any notable pregnancy or perinatal history. He showed a mild resting tremor in his hands shortly after falling while running at 18 months of age. On neurological examination at the age of two years his global development was delayed and he presented atypical absence epilepsy with myoclonic jerks exacerbated during a febrile illness. Initial neuroimaging was normal. EEG recordings showed bursts of generalized spike and slow wave activity of 2.5–3.5 Hz lasting up to 5 s. 76 seizures could be recorded within 24 h by EEG monitoring. He developed recurrent atonic seizures with falls at six years of age, and a movement disorder comprising multifocal myoclonus at eight years of age. At 12 years of age, the movement disorder progressed with ataxia after a generalized tonic-clonic status epilepticus. Antiepileptic trials with lamotrigine, zonisamide, clobazam, ethosuximide, and topiramate were ineffective. Valproic acid led to remarkable improvements with seizure reduction to 1–2 per day, lasting 5–10 s, without falls. However, the family reported weakness after long walks in the context of valproic acid. The treatment was ultimately changed to levetiracetam which provided some clinical benefit. Clonazepam led to a partial improvement of the non-epileptic myoclonus. Laboratory investigations revealed an elevated creatine kinase (CK) level of 2900 IU/L. Muscle biopsy showed a mild increase in the number of mitochondria. At the most recent clinical assessment at 15 years of age, he showed mild intellectual disability, walked independently but was unsteady due to a broad based gait, and required assistance due to risk of falling. Current comorbidities include an attention deficit disorder, asthma, obstructive and central sleep apnea, and dysphagia.

Patient 2 is female patient, who was born to non-consanguineous parents of German origin without any notable pregnancy or perinatal history. She presented with fine motor developmental delay with difficulties in handwriting at school age. At 12 years, she developed a generalized tonic–clonic seizure with photosensitivity after a viral infection triggered by flickering light, followed by permanent symptoms of intention tremor and shooting multifocal myoclonus. The movement disorder progressed gradually with increasing ataxia, dysmetria, and dystonia in the hands (Video S1). Hyporeflexia in the upper and lower limbs and scoliosis were noted on examination (Figure 1a). Each generalized tonic—clonic seizure, three in total, led to a further deterioration of the movement disorder. After a urinary tract infection at the age of 17 years, the patient had a focal-onset atonic seizure affecting the lower limb. Interictal EEG recordings revealed multifocal generalized spike and sharp waves with bilateral frontal and occipital accentuation with secondary generalization without any clinical seizure correlations. Brain MRI studies showed left frontal FLAIR hyperintensities and a periventricular cyst (Figure 1b). Serum laboratory analyses revealed a persistent moderate CK elevation (500–1200 IU/L). Tandem mass spectrometry investigations for a range of metabolic diseases were unremarkable. Nerve conduction studies showed reduced sensory conduction velocity of the tibial nerve. An anti-epileptic trial with topiramate was unsuccessful but levetiracetam led to complete seizure remission. Trials with pyrimidone and propranolol did not have any effect on myoclonus and tremor, while carbamazepine only led to a brief improvement in myoclonus. At last follow-up at 20 years of age, the patient was seizure-free but wheelchair-bound. She did not show any signs of cognitive deterioration. The myoclonus of the limbs persisted on voluntary movements. She was unable to write and used a voice-guided computer. The family declined deep brain stimulation.

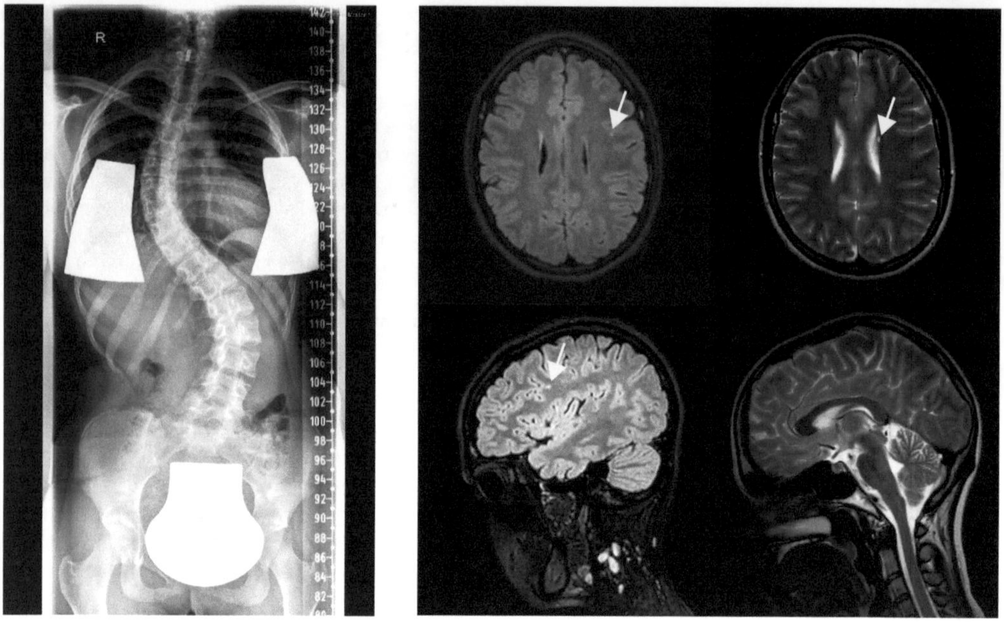

(**a**) (**b**)

Figure 1. (**a**) X-ray showing scoliosis in patient 2 at the age of 17 years. (**b**) Brain MRI in patient 2 at the age of 17 years with left frontal FLAIR hyperintensity (upper left axial and lower left sagittal images), periventricular cyst (upper right axial) but no other abnormalities (lower right sagittal).

3.2. Molecular Genetic Investigations

In patient 1, massively parallel sequencing revealed two novel compound heterozygous variants in *GOSR2*: NM_001012511.3:c.336+1G>A and c.364G>A,p.Glu122Lys. Analyses with SpliceAI showed a delta score at 0.99 for donor loss, predicting deficient protein expression. For the missense variant at p.Glu122Lys, the pathogenicity prediction tools showed a GERP score of 4.87, CADD-Phred score of 23.2, and allele count of 1/152204 in the gnomAD healthy population database [21]. MutPred2 analyses showed a score of 0.951 (loss of N-terminal acetylation $p < 0.01$, altered transmembrane protein $p < 0.05$). The p.Glu122Lys variant has previously been reported as a variant of unclear significance (VUS) with four submissions in ClinVar (ID 589948). On the basis of our clinical, genetic and bioinformatic analyses, we reclassified the variant as likely pathogenic (PM2, PM3, PP3, PP4). Protein models suggest that the p.Glu122Lys mutation would have minimal impact on the SNARE assembly because different rotamers of Lys can be accommodated in this location (Figure 2b,c).

In patient 2, karyotyping revealed 47,XXX without any relevance to the movement disorder. Array CGH did not detect pathogenic copy number alterations. Molecular genetic analysis via panel-based sequencing identified the pathogenic homozygous founder variant in *GOSR2* NM_001012511.3:c.430G>T,p.Gly144Trp, which was confirmed by Sanger sequencing. Pathogenicity prediction tools showed a CADD-Phred score of 32 and a MutPred2 score of 0.766 (altered disordered interface $p < 0.05$, altered transmembrane protein $p < 0.01$). Protein models showed that the p.Gly144Trp mutation introduces steric hindrance with neighboring side chains and potentially destabilizes the SNARE complex assembly (Figure 2b,d).

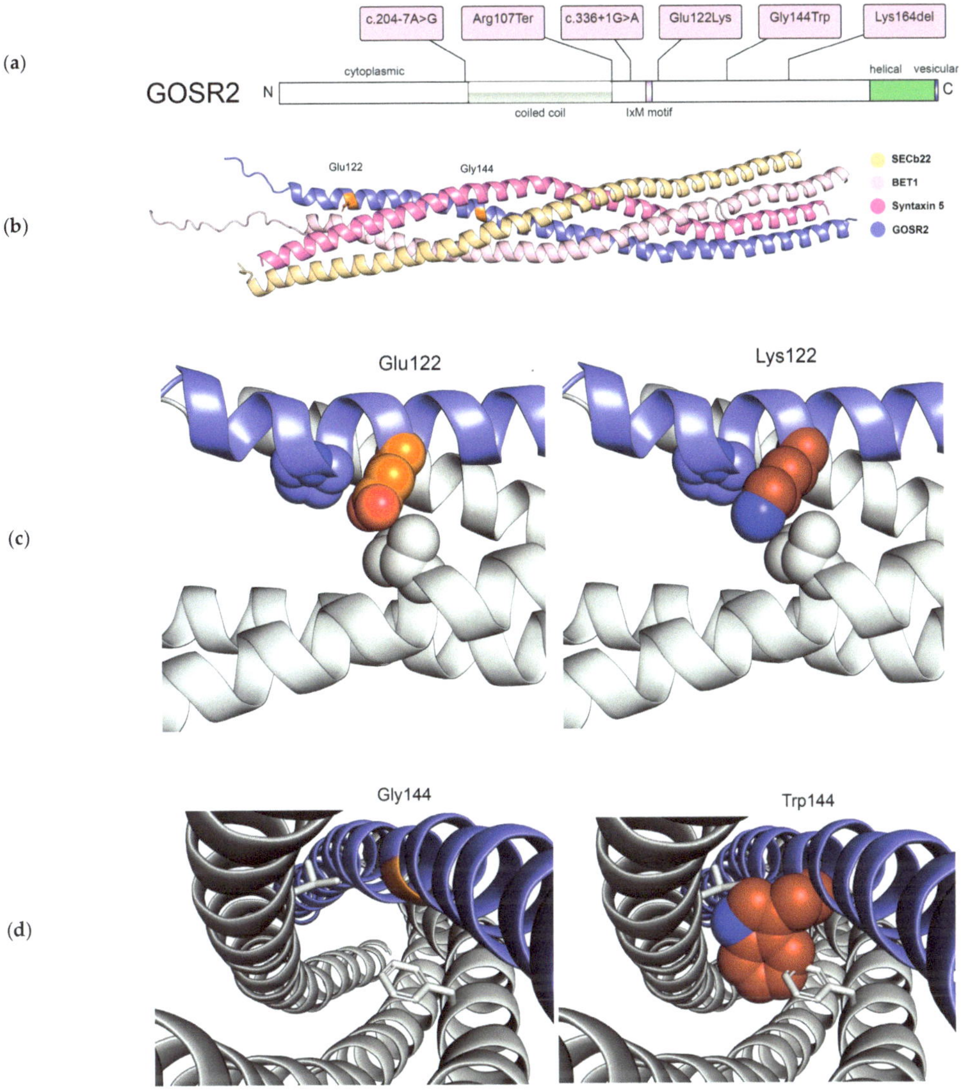

Figure 2. (**a**) Protein overview with all published variants including the p.Gly144Trp variant identified in patient 2 and the novel variants from patient 1 in c.336+1G>A and p.Glu122Lys. (**b**) Energy-minimized AlphaFold model of the SNARE complex composed of SEC22b (105–215), BET1 (1–118), Syntaxin-5 (251–355), and GOSR2 (107–212) colored in orange, pink, lavender, and blue, respectively. The locations of Glu122 and Gly144 are highlighted in orange. (**c**) Zoomed-in view of the structural models of wildtype complex and (left) complex containing GOSR2 p.Glu122Lys mutant (right) with the Glu122 residue colored in orange and Lys122 variant colored in red. These models suggest that the p.Glu122Lys mutation would have minimal impact on the SNARE assembly because different rotamers of Lys can be accommodated in this location. (**d**) Zoomed-in view of the structural models of wildtype complex and (left) complex containing GOSR2 p.Gly144Trp mutant (right) with the Gly144 residue colored in orange and Trp144 variant colored in red. These models show that the p.Gly144Trp mutation introduces steric hindrance with neighboring side chains and potentially destabilizes the SNARE complex assembly.

4. Discussion

PME is mainly characterized by seizures, movement disorders, and progressive neurological decline. The comprehensive genotype–phenotype spectrum in this disorder remains elusive due to limited reports. In *GOSR2*-associated disorders, previous attempts at genotype–phenotype analyses suggested a correlation between residual protein expression and disease severity, ultimately proposing a disease spectrum between PME with missense variants at the milder end and congenital muscular dystrophy with splice site variants at the more severe end [22].

A recent review of 21 patients with *GOSR2*-associated disorders showed the phenotype in the frequently observed p.Gly144Trp founder variant [2], while milder phenotypes with compound heterozygous variants (c.430G>T,c.491_493delAGA) [23] and three patients with α-dystroglycan muscular dystrophy showed a combination of the founder variant and splice site or missense variant that led to an effect in the N-terminal part of the protein (c.430G>T,c.336+1G>A; c.430G>T,c.2T>G; c.430G>T,c.82C>T) [22,24]. Table 1 demonstrates an overview of the core clinical features of *GOSR2*-related disorders in patients from our study and previously published literature. Supplementary Table S1 illustrates additional detailed reports of orthopedic findings and anti-epileptic trials in patients.

Table 1. Overview on core features of patients from this study and the published literature. Abbreviations: y = years; m = months; w = weeks; hom = homozygous; comp het = compound heterozygous; + = present; − = not present; ND = not defined; GTCS = generalized tonic–clonic seizures; ID = intellectual disability; CD = cognitive decline; LD = learning disability; DBS = deep brain stimulation.

Study	Case No. (Age Last Reported)	Genotype—Inheritance Mode, cDNA NM_001012511.3 (Protein NP_004278.2)	Ataxia (Age at Onset)	Myoclonus (Age at Onset)	Seizures (Age at Onset)	Others
Corbett et al., 2011 [1]; Boissé Lomax et al., 2013 [25]	#01 (32 y)	hom, c.430G>T (p.Gly144Trp)	+ (2 y)	+ (8 y)	+ (absence 7y, GTCS with 13 y, drop attacks 13 y)	CD (25 y); pneumonia
	#02 (17 y)	hom, c.430G>T (p.Gly144Trp)	+ (1 y)	+ (6 y)	+ (drop attacks 13 y)	ID
	#03 (32 y)	hom, c.430G>T (p.Gly144Trp)	+ (3 y)	+ (6 y)	+ (GTCS 14 y)	
	#04 (30 y)	hom, c.430G>T (p.Gly144Trp)	+ (2 y)	+ (10 y)	+ (absence 6 y, GTCS 12 y, drop attacks 14 y)	CD (25 y)
	#05 (24 y)	hom, c.430G>T (p.Gly144Trp)	+ (2 y)	+ (6 y)	+ (GTCS 21 y)	CD (30 y)
	#06 (29 y)	hom, c.430G>T (p.Gly144Trp)	+ (2 y)	+ (5 y)	+ (GTCS 24 y, absence 24 y)	
Boissé Lomax et al., 2013 [25]	#07 (10 y)	hom, c.430G>T (p.Gly144Trp)	+ (1 y)	+ (4.5 y)	+ (focal seizures 14 m, GTCS 8.5 y)	
	#08 (29 y)	hom, c.430G>T (p.Gly144Trp)	+ (3 y)	+ (5 y)	+ (drop attacks 2 y, absence 2 y, GTCS 21 y)	LD; febrile seizures
	#09 (27 y)	hom, c.430G>T (p.Gly144Trp)	+ (3.5 y)	+ (4 y)	+ (absence 5 y, GTCS 8 y)	
	#10 (18 y)	hom, c.430G>T (p.Gly144Trp)	+ (2 y)	+ (6 y)	+ (GTCS 10 y, absence)	
	#11 (37 y)	hom, c.430G>T (p.Gly144Trp)	+ (3 y)	+ (12 y)	+ (absence 3 y, GTCS 3 y)	
	#12 (13 y)	hom, c.430G>T (p.Gly144Trp)	+ (2.3 y)	+ (6 y)	+ (GTCS 12 y)	
Tsai et al., 2013 [22]	#13 (36 w)	comp het, c.430G>T (p.Gly144Trp), c.336+1G>A	−	−	−	Muscular dystrophy
van Egmond et al., 2014 [26]	#14 (19 y)	hom, c.430G>T (p.Gly144Trp)	+ (3 y)	+ (5 y)	+ (tonic seizures 9 y)	
	#15 (26 y)	hom, c.430G>T (p.Gly144Trp)	+ (3 y)	+ (6 y)	+ (GTCS 11 y)	Mild LD
	#16 (20 y)	hom, c.430G>T (p.Gly144Trp)	+ (2 y)	+ (6 y)	+ (GTCS 6 y)	Mild LD
	#17 (12 y)	hom, c.430G>T (p.Gly144Trp)	+ (5 y)	+ (8 y)	+ (clonic seizures 3 y)	
	#18 (7 y)	hom, c.430G>T (p.Gly144Trp)	+ (3 y)	+ (6 y)	NA	

Table 1. *Cont.*

Study	Case No. (Age Last Reported)	Genotype—Inheritance Mode, cDNA NM_001012511.3 (Protein NP_004278.2)	Ataxia (Age at Onset)	Myoclonus (Age at Onset)	Seizures (Age at Onset)	Others
Praschberger et al., 2015 [23]	#19 (61 y)	comp het, c.430G>T (p.Gly144Trp), c.491_493delAGA (p.Lys164del)	+ (2 y)	+ (14 y)	+ (14 y)	Mild CD
Anderson et al., 2016 [27]	#20 (10 y)	hom, c.430G>T (p.Gly144Trp)	+ (ND)	+ (5 y)	+ (GTCS 6 y)	DBS
	#21 (30 y)	hom, c.430G>T (p.Gly144Trp)	+ (ND)	+ (ND)	+ (GTCS 8 y)	DBS
	#22 (27 y)	ND	+ (ND)	+ (3 y)	+ (GTCS 5 y)	DBS
Larson et al., 2018 [28]	#23 (died at 5 y)	ND	–	–	+ (absence 2 y)	Muscular dystrophy
	#24 (6 y)	comp het, c.430G>T (p.Gly144Trp), c.2T>G (p.Met1Arg)	–	–	+ (2.5 y)	Muscular dystrophy
Polet et al., 2020 [2]	#25 (5 y)	hom, c.430G>T (p.Gly144Trp)	+ (4 y)	+ (4 y)	NA	
	#26 (8 y)	hom, c.430G>T (p.Gly144Trp)	+ (2 y)	+ (5 y)	+ (7 y)	
	#27 (8 y)	hom, c.430G>T (p.Gly144Trp)	+ (2 y)	+ (3 y)	+ (7 y)	
	#28 (13 y)	hom, c.430G>T (p.Gly144Trp)	+ (3 y)	+ (6 y)	NA	
	#29 (18 y)	hom, c.430G>T (p.Gly144Trp)	+ (5 y)	+ (7 y)	+ (3 y)	DBS
	#30 (19 y)	hom, c.430G>T (p.Gly144Trp)	+ (4 y)	+ (7 y)	+ (9 y)	DBS
	#31 (20 y)	hom, c.430G>T (p.Gly144Trp)	+ (5 y)	+ (5 y)	+ (13 y)	
	#32 (26 y)	hom, c.430G>T (p.Gly144Trp)	+ (3 y)	+ (5 y)	+ (9 y)	
	#33 (26 y)	hom, c.430G>T (p.Gly144Trp)	+ (2 y)	+ (6 y)	+ (8 y)	
	#34 (31 y)	hom, c.430G>T (p.Gly144Trp)	+ (5 y)	+ (8 y)	+ (8 y)	
	#35 (31 y)	hom, c.430G>T (p.Gly144Trp)	+ (8 y)	+ (9 y)	+ (8 y)	
	#36 (32 y)	hom, c.430G>T (p.Gly144Trp)	+ (3 y)	+ (6 y)	+ (11 y)	
	#37 (35 y)	hom, c.430G>T (p.Gly144Trp)	+ (5 y)	+ (ND)	+ (ND)	DBS
	#38 (36 y)	hom, c.430G>T (p.Gly144Trp)	+ (ND)	+ (ND)	+ (6 y)	
	#39 (37 y)	hom, c.430G>T (p.Gly144Trp)	+ (5 y)	+ (ND)	+ (8 y)	DBS
	#40 (41 y)	hom, c.430G>T (p.Gly144Trp)	+ (ND)	+ (5 y)	+ (8 y)	
	#41 (46 y)	hom, c.430G>T (p.Gly144Trp)	+ (3 y)	+ (4 y)	+ (6 y)	
Stemmerik et al., 2021 [29]	#42 (48 y)	comp het, c.319C>T (p.Arg107Ter), c.204-7A>G (p.?)	+ (13 y)	+ (7 y)	NA	Mild CD (46 y)
This study	#43 (15 y)	comp het, c.364G>A (p.Glu122Lys), c.336+1G>A (p.?)	+ (12 y)	+ (8 y)	+ (2 y)	Mild ID
	#44 (21 y)	hom, c.430G>T (p.Gly144Trp)	+ (16 y)	+ (13 y)	+ (GTCS 12y on urinary tract infection)	Gradual progression of movement disorders into dystonia on recurrent viral infections

GOSR2/membrin is a member of the SNARE (soluble NSF (N-ethylmaleimide-sensitive factor) attachment protein receptor) family of proteins that mediate membrane fusion processes involving the ER, the ER–Golgi intermediate compartment, and the Golgi by forming a complex with three other SNAREs (Syntaxin-5, Bet1, and Sec22b) [3,30]. COPII-coated vesicles form on the endoplasmic reticulum (ER) from three subunits (Sar1-GTP, Sec23/24, Sec13/31) to bud vesicles for transport to the Golgi complex [31]. Interaction between the IxM motif in GOSR2 and the subunit Sec24c/d was reported to serve as the primary mechanism for packaging GOSR2/membrin and syntaxin-5 into COPII-coated vesicles for vesicular trafficking under healthy conditions [32].

Through analyses of the variants in our two patients and all published patients, we noted that all of the variants are found in the cytoplasmic part of the protein, some of which are essential for interacting with the SNARE complex (Figure 2a). All of the patient missense and truncating variants are located in regions that are highly conserved throughout evolution (Supplementary Table S2). Inspection of the AlphaFold2-predicted structural model of GOSR2 showed that the two mutation sites map to solvent-accessible regions and do not appear to alter the overall structure of this protein. To assess the effects of the disease mutations in the context of the assembled SNARE complex, we first generated a structural model of this complex composed of only the SNARE domains of GOSR2, syntaxin-5, Bet1, and Sec22b using AlphaFold2, followed by energy minimization refinement, and then constructed the models of the mutant complexes (Figure 2b) [19,20]. According to these models, the variant p.Gly144Trp introduces a large hydrophobic side chain that would sterically repel other side chain residues engaged in the interior of the coiled coil and potentially destabilize the assembly (Figure 2d). On the other hand, multiple lysine rotamers can be accommodated at position 122, which is located in the more solvent-accessible region near the tip of assembly (Figure 2c). Both variants in patient 1, the novel missense variant p.Glu122Lys and splice site variant c.336+1G>A, are located adjacent to the IxM motif. The essential splice variant is predicted to result into donor loss potentially disrupting the IxM motif for packaging GOSR2/membrin and syntaxin-5 into COPII-coated vesicles.

While the splice site variant was previously reported in a patient with muscular dystrophy, our patient 1 presented with a PME phenotype and no muscular hypotonia as in patients with p.Gly144Trp variants without infantile muscular hypotonia. The elevated CK levels with a peak of 2900 IU/L are similar to reports of PME patients in the literature. While this does not automatically preclude the patient from having muscular dystrophy, the muscle biopsy showed only mild increase of mitochondria and there were no other phenotypic signs indicative of muscular dystrophy in this patient, e.g., ocular structural changes, progressive loss of muscle mass, or Gower's sign [7,33]. Indeed, while elevated CK levels may prompt further investigations on future clinical presentation, any CK elevation here was interpreted as likely arising from myoclonus.

Our patient 2 with the founder variant p.Gly144Trp illustrates a milder and later onset of NS-PME with gradual progression of the movement disorder and without cognitive decline. The phenotype expansion included dystonia that may present in other patients with later onsets of movement disorders. While this is not the oldest reported patient to date, the movement disorders began with myoclonus at 13 years and ataxia at 16 years of age. The latest onsets of movement disorders from published reports were at 13 years for ataxia and eight years for myoclonus. Complete seizure remission on levetiracetam monotherapy as in our patient 2 was previously reported in a patient with biallelic *TRAPPC11* variants and congenital muscular dystrophy. TRAPPC11 and GOSR2 are both involved in COPII-related trafficking [34]. Recent work indicated that TRAPPC11 recruits ATG2B–WDR45 to pre-autophagosomal membranes and its Atg9-mediated autophagosome closure during macroautophagy [35], while WDR45 facilitates autophagolysosomal fusion by interacting with the tethering factor EPG5 in neural cells [36]. Monogenic disorders in this pathway include WDR45 deficiency with a bi-phasic disorder from developmental and epileptic–dyskinetic encephalopathy to adult dystonia–Parkinsonism [37]. EPG5 deficiency is associated with a wide spectrum from neurodevelopmental disorders and seizures to Vici syndrome with multi-system features including neurodevelopmental disorders, movement disorders, epilepsy, and immunodeficiency [38,39]. Several anti-epileptic drugs have been included in the management of these disorders of autophagy and intracellular trafficking that may modulate exocytosis or synaptic trafficking, e.g., gabapentin, carbamazepine, or clonazepam [40]. Recent reports of spatial proteomics in *TECPR2*-related hereditary sensory and autonomic neuropathy indeed also implicated these vesicular trafficking pathways, specifically ER-to-Golgi transport via coated vesicles [41].

Notably, phenotype progression in patient 2 occurred on repeated viral infections. It remains elusive how viral infections trigger the disease progression. Recent findings have implicated a viral hijacking of COPII-coated vesicles including flaviviruses and SARS-CoV-2 [42,43]. These viral hijacking routes may be of specific interest as they show a pathomechanism related to COPII vesicle trafficking. For instance, the hepatitis C virus is budded in the endoplasmic reticulum lumen and exits via COPII vesicles by modulating the protein milieu and reorganization at organellar sites such as ER–Golgi intermediate compartments [44]. Future studies may address how viral infection potentially adds on to disorders in COPII-related trafficking or autophagosome biogenesis on the background of biallelic GOSR2 deficiency.

Another important factor in the consideration of viral infections as triggers for disease progression is heat insensitivity itself as it has been reported that patients with GOSR2 deficiency are sensitive to all forms of heat-inducing stressors [5], perhaps indicating a more general mechanism of protein misfolding and ER stress in *GOSR2*-related disorders. Indeed, a *Drosophila melanogaster* model with knockdown of *GOSR2* ortholog membrin indicated seizure-like behavior with hyperthermia. Reduced membrin expression in glia rendered the flies particularly prone to age-dependent seizures. Of note, there was a partial response to treatment with sodium barbital in flies. However, this was not trialed in our patients due to its adverse effects.

It remains unclear which viral pathogens led to the progression in our patient. The additional report of a generalized tonic–clonic seizure after a urinary tract infection in our patient may also suggest a role of bacterial pathogens in disease progression. Recent reports implicated *Legionella pneumophila* in subverting functions of Rab1 and Sec22b for its replication in cells [45]. As Sec24 phosphorylation regulates autophagosome abundance during nutrient deprivation by interacting with Atg9, COPII-related trafficking serves an important mechanism for autophagosome formation [46]. Recently, biallelic variants in *ATG9A* were reported in immune dysregulation with diplopia and inflammatory lesions in brain MRI studies after EBV infection [47]. Other innate errors of autophagy and intracellular trafficking with neurological symptoms after recurrent infections include disorders in phosphatidylinositol 4-kinases which are recruited by the membrin–Arf1 complex in vesicular trafficking [6,48]. Further work is needed to elucidate the role of viral and bacterial infections in the disease progression in patients with GOSR2 deficiency and how pathogen-induced perturbation of vesicular trafficking may eventually trigger neurodegeneration.

5. Conclusions

In conclusion, we report a genotypic and phenotypic expansion of *GOSR2*-associated NS-PME in a patient with early-onset disease and refractory epilepsy likely due to disrupted interaction with the COPII-coated vesicle machinery, and another patient with disease onset during adolescence, seizure remission on levetiracetam, and retained cognitive function during the third decade of life. The presentation with juvenile-onset dystonia suggests that *GOSR2* may also be included in the differential diagnoses of monogenetic dystonia.

Supplementary Materials: The following supporting information can be downloaded at https://www.mdpi.com/article/10.3390/genes14101860/s1: Video S1. Patient with *GOSR2*-associated North Sea progressive myoclonus epilepsy shows tremor and myoclonus of the upper limbs during arm holding test, finger tapping test, and spontaneous movements, at the age of 20 years. The patient shows dystonic posturing of the hands during the arm holding test and the finger tapping test; Table S1. Detailed overview on all patients with GOSR2-related disorders from the published literature and our two novel patients regarding publication, sex, age in years, genetic variant, protein mutation, variant zygosity, first symptoms (with age at onset), ataxia (with age at onset), myoclonus (with age at onset), seizures (with age at onset), orthopedic complications, cognition, mobility at last follow-up, highest recorded CK level (U/L), recurrent infections, management (successful in bold), and others. Abbreviations: LVT, levetiracetam; VPA, valproic acid; CLN, clonazepam; CBZ, clobazam; MDZ, midazolam; PCT, piracetam; PRP, perampanel; PYR: pyrimidone; PRO: propanolol; ESM, ethosuximide; ZNS, zonisamide; ATZ, acetazolamide; LMT, lamotrigine; FBM, felbamate; BCF, baclofen; CRN, carnitine; CBM, carbamazepine; MPD, methylpredisolon; 5HTP, 5-hydroxytryptofaan; CBI, carbidopa; PHB, phenobarbital; DBS, deep brain stimulation; Table S2. Wildtype protein and variants in *Homo sapiens*, as well as illustration of evolutionary conservation of residues with wildtype sequences in *Pan troglodytes*, *Macaca mulatta*, *Canis lupus*, *Bos taurus*, *Gallus gallus*, *Danio rerio*, *Drosophila melanogaster*, *Caenorhabditis elegans*, and *Xenopus tropicalis*.

Author Contributions: Conceptualization, H.S.D.; data curation, H.S.D.; methodology, H.S.D. and C.K.Y.; funding acquisition, H.S.D.; formal analysis, L.H. and H.S.D.; investigation, M.P., C.K.Y. and H.S.D.; visualization, K.M.N., C.K.Y. and H.S.D.; writing—original draft preparation, L.H. and H.S.D.; writing—review and editing, M.P., T.E.L., R.C., T.J.d.K., K.M.N., C.K.Y., H.J., A.K. and H.S.D.; supervision, H.S.D.; project administration, H.S.D. All authors have read and agreed to the published version of the manuscript.

Funding: Lea Hentrich was supported by the Koeln Fortune Program/Faculty of Medicine, University of Cologne (259/2022). Heinz Jungbluth was supported by Action Medical Research, United Kingdom, the GOSH Charity/Sparks, and European Commission H2020-MSCA-ITN-2017. Hormos Salimi Dafsari was supported by the Koeln Fortune Program/Faculty of Medicine, University of Cologne (371/2021 and 243/2022), as well as the Cologne Clinician Scientist Program/Medical Faculty/University of Cologne and German Research Foundation (CCSP, DFG project no. 413543196). We acknowledge support for the Article Processing Charge from the DFG (German Research Foundation, 491454339).

Institutional Review Board Statement: The study was conducted in accordance with the Declaration of Helsinki and approved by the local institutional review boards (Medical Faculty at the University of Cologne, protocol number 20-1711, date of approval 9 June 2021).

Informed Consent Statement: Informed consent was obtained from all subjects involved in the study. We confirm that we read the Journal's position on issues involved in ethical publication and affirm that this work is consistent with those guidelines.

Data Availability Statement: Anonymized data are available upon request from the corresponding author.

Acknowledgments: The authors thank all the patients and families.

Conflicts of Interest: The authors declare no conflict of interest.

References

1. Corbett, M.A.; Schwake, M.; Bahlo, M.; Dibbens, L.M.; Lin, M.; Gandolfo, L.C.; Vears, D.F.; O'Sullivan, J.D.; Robertson, T.; Bayly, M.A.; et al. A Mutation in the Golgi Qb-SNARE Gene GOSR2 Causes Progressive Myoclonus Epilepsy with Early Ataxia. *Am. J. Hum. Genet.* **2011**, *88*, 657–663. [CrossRef] [PubMed]
2. Polet, S.S.; Anderson, D.G.; Koens, L.H.; van Egmond, M.E.; Drost, G.; Brusse, E.; Willemsen, M.A.; Sival, D.A.; Brouwer, O.F.; Kremer, H.P.; et al. A detailed description of the phenotypic spectrum of North Sea Progressive Myoclonus Epilepsy in a large cohort of seventeen patients. *Park. Relat. Disord.* **2020**, *72*, 44–48. [CrossRef] [PubMed]
3. Lowe, S.L.; Peter, F.; Subramaniam, V.N.; Wong, S.H.; Hong, W. A SNARE involved in protein transport through the Golgi apparatus. *Nature* **1997**, *389*, 881–884. [CrossRef] [PubMed]
4. Dibbens, L.M.; Rubboli, G. GOSR2: A progressive myoclonus epilepsy gene. *Epileptic Disord.* **2016**, *18*, S111–S114. [CrossRef]
5. Lambrechts, R.A.; Polet, S.S.; Hernandez-Pichardo, A.; van Ninhuys, L.; Gorter, J.A.; Grzeschik, N.A.; de Koning-Tijssen, M.A.J.; de Koning, T.J.; Sibon, O.C.M. North Sea Progressive Myoclonus Epilepsy is Exacerbated by Heat, A Phenotype Primarily Associated with Affected Glia. *Neuroscience* **2019**, *423*, 1–11. [CrossRef] [PubMed]
6. Dafsari, H.S.; Pemberton, J.G.; Ferrer, E.A.; Yammine, T.; Farra, C.; Mohammadi, M.H.; Ghayoor Karimiani, E.; Hashemi, N.; Souaid, M.; Sabbagh, S.; et al. PI4K2A deficiency causes innate error in intracellular trafficking with developmental and epileptic-dyskinetic encephalopathy. *Ann. Clin. Transl. Neurol.* **2022**, *9*, 1345–1358. [CrossRef]
7. Bayram, N.; Kaçar Bayram, A.; Daimagüler, H.S.; Dafsari, H.S.; Bamborschke, D.; Uyanik, G.; Erdogan, M.; Özsaygılı, C.; Pangal, E.; Yuvaci, İ.; et al. Genotype-phenotype correlations in ocular manifestations of Marinesco-Sjögren syndrome: Case report and literature review. *Eur. J. Ophthalmol.* **2021**, *32*, NP92–NP97. [CrossRef]
8. Dafsari, H.S.; Kawalia, A.; Sprute, R.; Karakaya, M.; Malenica, A.; Herkenrath, P.; Nürnberg, P.; Motameny, S.; Thiele, H.; Cirak, S. Novel mutations in SLC6A5 with benign course in hyperekplexia. *Cold Spring Harb. Mol. Case Stud.* **2019**, *5*, a004465. [CrossRef]

9. Dafsari, H.S.; Sprute, R.; Wunderlich, G.; Daimagüler, H.-S.; Karaca, E.; Contreras, A.; Becker, K.; Schulze-Rhonhof, M.; Kiening, K.; Karakulak, T.; et al. Novel mutations in KMT2B offer pathophysiological insights into childhood-onset progressive dystonia. *J. Hum. Genet.* **2019**, *64*, 803–813. [CrossRef]
10. Saffari, A.; Lau, T.; Tajsharghi, H.; Karimiani, E.G.; Kariminejad, A.; Efthymiou, S.; Zifarelli, G.; Sultan, T.; Toosi, M.B.; Sedighzadeh, S.; et al. The clinical and genetic spectrum of autosomal-recessive TOR1A-related disorders. *Brain* **2023**, *146*, 3273–3288. [CrossRef]
11. Becker, L.L.; Dafsari, H.S.; Schallner, J.; Abdin, D.; Seifert, M.; Petit, F.; Smol, T.; Bok, L.; Rodan, L.; Krapels, I.; et al. The clinical-phenotype continuum in DYNC1H1-related disorders—Genomic profiling and proposal for a novel classification. *J. Hum. Genet.* **2020**, *65*, 1003–1017. [CrossRef]
12. Sprute, R.; Jergas, H.; Ölmez, A.; Alawbathani, S.; Karasoy, H.; Dafsari, H.S.; Becker, K.; Daimagüler, H.S.; Nürnberg, P.; Muntoni, F.; et al. Genotype-phenotype correlation in seven motor neuron disease families with novel ALS2 mutations. *Am. J. Med. Genet. A* **2021**, *185*, 344–354. [CrossRef]
13. Karczewski, K.J.; Francioli, L.C.; Tiao, G.; Cummings, B.B.; Alföldi, J.; Wang, Q.; Collins, R.L.; Laricchia, K.M.; Ganna, A.; Birnbaum, D.P.; et al. The mutational constraint spectrum quantified from variation in 141,456 humans. *Nature* **2020**, *581*, 434–443. [CrossRef] [PubMed]
14. Pejaver, V.; Urresti, J.; Lugo-Martinez, J.; Pagel, K.A.; Lin, G.N.; Nam, H.-J.; Mort, M.; Cooper, D.N.; Sebat, J.; Iakoucheva, L.M.; et al. Inferring the molecular and phenotypic impact of amino acid variants with MutPred2. *Nat. Commun.* **2020**, *11*, 5918. [CrossRef]
15. Rentzsch, P.; Witten, D.; Cooper, G.M.; Shendure, J.; Kircher, M. CADD: Predicting the deleteriousness of variants throughout the human genome. *Nucleic Acids Res.* **2019**, *47*, D886–D894. [CrossRef] [PubMed]
16. de Sainte Agathe, J.M.; Filser, M.; Isidor, B.; Besnard, T.; Gueguen, P.; Perrin, A.; Van Goethem, C.; Verebi, C.; Masingue, M.; Rendu, J.; et al. SpliceAI-visual: A free online tool to improve SpliceAI splicing variant interpretation. *Hum. Genom.* **2023**, *17*, 7. [CrossRef]
17. Davydov, E.V.; Goode, D.L.; Sirota, M.; Cooper, G.M.; Sidow, A.; Batzoglou, S. Identifying a high fraction of the human genome to be under selective constraint using GERP++. *PLoS Comput. Biol.* **2010**, *6*, e1001025. [CrossRef] [PubMed]
18. Richards, S.; Aziz, N.; Bale, S.; Bick, D.; Das, S.; Gastier-Foster, J.; Grody, W.W.; Hegde, M.; Lyon, E.; Spector, E.; et al. Standards and guidelines for the interpretation of sequence variants: A joint consensus recommendation of the American College of Medical Genetics and Genomics and the Association for Molecular Pathology. *Genet. Med.* **2015**, *17*, 405–424. [CrossRef]
19. Jumper, J.; Evans, R.; Pritzel, A.; Green, T.; Figurnov, M.; Ronneberger, O.; Tunyasuvunakool, K.; Bates, R.; Žídek, A.; Potapenko, A.; et al. Highly accurate protein structure prediction with AlphaFold. *Nature* **2021**, *596*, 583–589. [CrossRef]
20. Afonine, P.V.; Grosse-Kunstleve, R.W.; Echols, N.; Headd, J.J.; Moriarty, N.W.; Mustyakimov, M.; Terwilliger, T.C.; Urzhumtsev, A.; Zwart, P.H.; Adams, P.D. Towards automated crystallographic structure refinement with phenix.refine. *Acta Crystallogr. D Biol. Crystallogr.* **2012**, *68*, 352–367. [CrossRef]
21. Dafsari, H.S.; Becker, L.; von der Hagen, M.; Cirak, S. Genomic profiling in neuronal dyneinopathies and updated classifications. *Am. J. Med. Genet. A* **2021**, *185*, 2607–2610. [CrossRef] [PubMed]
22. Tsai, L.; Schwake, M.; Corbett, M.A.; Gecz, J.; Berkovic, S.; Shieh, P.B. P. 1.20 GOSR2: A novel form of Congenital Muscular Dystrophy. *Neuromuscul. Disord.* **2013**, *23*, 748. [CrossRef]
23. Praschberger, R.; Balint, B.; Mencacci, N.E.; Hersheson, J.; Rubio-Agusti, I.; Kullmann, D.M.; Bettencourt, C.; Bhatia, K.; Houlden, H. Expanding the Phenotype and Genetic Defects Associated with the GOSR2 Gene. *Mov. Disord. Clin. Pract.* **2015**, *2*, 271–273. [CrossRef] [PubMed]
24. Henige, H.; Kaur, S.; Pappas, K. Compound heterozygous variants in GOSR2 associated with congenital muscular dystrophy: A case report. *Eur. J. Med. Genet.* **2021**, *64*, 104184. [CrossRef]
25. Boissé Lomax, L.; Bayly, M.A.; Hjalgrim, H.; Møller, R.S.; Vlaar, A.M.; Aaberg, K.M.; Marquardt, I.; Gandolfo, L.C.; Willemsen, M.; Kamsteeg, E.J.; et al. "North Sea" Progressive Myoclonus Epilepsy: Phenotype of Subjects with GOSR2 Mutation. *Brain* **2013**, *136*, 1146–1154. [CrossRef]
26. van Egmond, M.E.; Verschuuren-Bemelmans, C.C.; Nibbeling, E.A.; Elting, J.W.J.; Sival, D.A.; Brouwer, O.F.; de Vries, J.J.; Kremer, H.P.; Sinke, R.J.; Tijssen, M.A.; et al. Ramsay Hunt Syndrome: Clinical Characterization of Progressive Myoclonus Ataxia Caused by GOSR2 Mutation. *Mov. Disord.* **2014**, *29*, 139–143. [CrossRef]
27. Anderson, D.G.; Németh, A.H.; Fawcett, K.A.; Sims, D.; Miller, J.; Krause, A. Deep Brain Stimulation in Three Related Cases of North Sea Progressive Myoclonic Epilepsy from South Africa. *Mov. Disord. Clin. Pract.* **2016**, *4*, 249–253. [CrossRef]
28. Larson, A.A.; Baker, P.R.; Milev, M.P.; Press, C.A.; Sokol, R.J.; Cox, M.O.; Lekostaj, J.K.; Stence, A.A.; Bossler, A.D.; Mueller, J.M.; et al. TRAPPC11 and GOSR2 Mutations Associate with Hypoglycosylation of α-Dystroglycan and Muscular Dystrophy. *Skelet. Muscle* **2018**, *8*, 17. [CrossRef]
29. Stemmerik, M.G.; Borch, J.d.S.; Dunø, M.; Krag, T.; Vissing, J. Myopathy Can Be a Key Phenotype of Membrin (GOSR2) Deficiency. *Hum. Mutat.* **2021**, *42*, 1101–1106. [CrossRef]
30. Hay, J.C.; Klumperman, J.; Oorschot, V.; Steegmaier, M.; Kuo, C.S.; Scheller, R.H. Localization, dynamics, and protein interactions reveal distinct roles for ER and Golgi SNAREs. *J. Cell Biol.* **1998**, *141*, 1489–1502. [CrossRef]
31. Lee, M.C.S.; Miller, E.A.; Goldberg, J.; Orci, L.; Schekman, R. Bi-directional protein transport between the ER and Golgi. *Annu. Rev. Cell Dev. Biol.* **2004**, *20*, 87–123. [CrossRef]

32. Mancias, J.D.; Goldberg, J. Structural basis of cargo membrane protein discrimination by the human COPII coat machinery. *EMBO J.* **2008**, *27*, 2918. [CrossRef]
33. Dafsari, H.S.; Kocaturk, N.M.; Daimagüler, H.-S.; Brunn, A.; Dötsch, J.; Weis, J.; Deckert, M. Bi-allelic mutations in uncoordinated mutant number-45 myosin chaperone B are a cause for congenital myopathy. *Acta Neuropathol. Commun.* **2019**, *7*, 211. [CrossRef]
34. Zhao, S.; Li, C.M.; Luo, X.M.; Siu, G.K.; Gan, W.J.; Zhang, L.; Wu, W.K.; Chan, H.C.; Yu, S. Mammalian TRAPPIII Complex positively modulates the recruitment of Sec13/31 onto COPII vesicles. *Sci. Rep.* **2017**, *7*, 43207. [CrossRef] [PubMed]
35. Stanga, D.; Zhao, Q.; Milev, M.P.; Saint-Dic, D.; Jimenez-Mallebrera, C.; Sacher, M. TRAPPC11 functions in autophagy by recruiting ATG2B-WIPI4/WDR45 to preautophagosomal membranes. *Traffic* **2019**, *20*, 325–345. [CrossRef] [PubMed]
36. Ji, C.; Zhao, H.; Chen, D.; Zhang, H.; Zhao, Y.G. β-propeller proteins WDR45 and WDR45B regulate autophagosome maturation into autolysosomes in neural cells. *Curr. Biol.* **2021**, *31*, 1666–1677.e6. [CrossRef]
37. Saffari, A.; Schröter, J.; Garbade, S.F.; Alecu, J.E.; Ebrahimi-Fakhari, D.; Hoffmann, G.F.; Kölker, S.; Ries, M.; Syrbe, S. Quantitative retrospective natural history modeling of WDR45-related developmental and epileptic encephalopathy—A systematic cross-sectional analysis of 160 published cases. *Autophagy* **2022**, *18*, 1715–1727. [CrossRef]
38. Dafsari, H.S.; Ebrahimi-Fakhari, D.; Saffari, A.; Deneubourg, C.; Fanto, M.; Jungbluth, H. EPG5-Related Disorder. 2022. Available online: http://www.ncbi.nlm.nih.gov/pubmed/29227033 (accessed on 20 September 2023).
39. Byrne, S.; Dionisi-Vici, C.; Smith, L.; Gautel, M.; Jungbluth, H. Vici syndrome: A review. *Orphanet J. Rare Dis.* **2016**, *11*, 21. [CrossRef]
40. Allen, N.M.; Dafsari, H.S.; Wraige, E.; Jungbluth, H. Neck-Tongue Syndrome: An Underrecognized Childhood Onset Cephalalgia. *J. Child. Neurol.* **2018**, *33*, 347–350. [CrossRef] [PubMed]
41. Nalbach, K.; Schifferer, M.; Bhattacharya, D.; Ho-Xuan, H.; Tseng, W.; Williams, L.A.; Stolz, A.; Lichtenthaler, S.F.; Elazar, Z.; Behrends, C. Spatial proteomics reveals secretory pathway disturbances caused by neuropathy-associated TECPR2. *Nat. Commun.* **2023**, *14*. [CrossRef] [PubMed]
42. Hassan, Z.; Kumar, N.D.; Reggiori, F.; Khan, G. How Viruses Hijack and Modify the Secretory Transport Pathway. *Cells* **2021**, *10*, 2535. [CrossRef] [PubMed]
43. Cattin-Ortolá, J.; Welch, L.G.; Maslen, S.L.; Papa, G.; James, L.C.; Munro, S. Sequences in the cytoplasmic tail of SARS-CoV-2 Spike facilitate expression at the cell surface and syncytia formation. *Nat. Commun.* **2021**, *12*, 5333. [CrossRef] [PubMed]
44. Avula, K.; Singh, B.; Samantaray, S.; Syed, G.H. The Early Secretory Pathway Is Crucial for Multiple Aspects of the Hepatitis C Virus Life Cycle. *J. Virol.* **2023**, *97*, e0018023. [CrossRef] [PubMed]
45. Kagan, J.C.; Stein, M.P.; Pypaert, M.; Roy, C.R. Legionella Subvert the Functions of Rab1 and Sec22b to Create a Replicative Organelle. *J. Exp. Med.* **2004**, *199*, 1201. [CrossRef] [PubMed]
46. Davis, S.; Wang, J.; Zhu, M.; Stahmer, K.; Lakshminarayan, R.; Ghassemian, M.; Jiang, Y.; Miller, E.A.; Ferro-Novick, S. Sec24 phosphorylation regulates autophagosome abundance during nutrient deprivation. *eLife* **2016**, *5*, e21167. [CrossRef] [PubMed]
47. Hu, G.; Hauk, P.J.; Zhang, N.; Elsegeiny, W.; Guardia, C.M.; Kullas, A.; Crosby, K.; Deterding, R.R.; Schedel, M.; Reynolds, P.; et al. Autophagy-associated immune dysregulation and hyperplasia in a patient with compound heterozygous mutations in ATG9A. *Autophagy* **2023**, *19*, 678–691. [CrossRef] [PubMed]
48. De Matteis, M.A.; Godi, A. PI-loting membrane traffic. *Nat. Cell Biol.* **2004**, *6*, 487–492. [CrossRef]

Disclaimer/Publisher's Note: The statements, opinions and data contained in all publications are solely those of the individual author(s) and contributor(s) and not of MDPI and/or the editor(s). MDPI and/or the editor(s) disclaim responsibility for any injury to people or property resulting from any ideas, methods, instructions or products referred to in the content.

genes

MDPI

Article

Exome-Wide Association Study Identified Clusters of Pleiotropic Genetic Associations with Alzheimer's Disease and Thirteen Cardiovascular Traits

Yury Loika *, Elena Loiko, Irina Culminskaya and Alexander M. Kulminski

Biodemography of Aging Research Unit, Social Science Research Institute, Duke University, Durham, NC 27708, USA; elena.loiko@duke.edu (E.L.); irina.kulminskaya@duke.edu (I.C.); alexander.kulminski@duke.edu (A.M.K.)
* Correspondence: yury.loika@duke.edu

Abstract: Alzheimer's disease (AD) and cardiovascular traits might share underlying causes. We sought to identify clusters of cardiovascular traits that share genetic factors with AD. We conducted a univariate exome-wide association study and pair-wise pleiotropic analysis focused on AD and 16 cardiovascular traits—6 diseases and 10 cardio-metabolic risk factors—for 188,260 UK biobank participants. Our analysis pinpointed nine genetic markers in the *APOE* gene region and four loci mapped to the *CDK11*, *OBP2B*, *TPM1*, and *SMARCA4* genes, which demonstrated associations with AD at $p \leq 5 \times 10^{-4}$ and pleiotropic associations at $p \leq 5 \times 10^{-8}$. Using hierarchical cluster analysis, we grouped the phenotypes from these pleiotropic associations into seven clusters. Lipids were divided into three clusters: low-density lipoprotein and total cholesterol, high-density lipoprotein cholesterol, and triglycerides. This split might differentiate the lipid-related mechanisms of AD. The clustering of body mass index (BMI) with weight but not height indicates that weight defines BMI-AD pleiotropy. The remaining two clusters included (i) coronary heart disease and myocardial infarction; and (ii) hypertension, diabetes mellitus (DM), systolic and diastolic blood pressure. We found that all AD protective alleles were associated with larger weight and higher DM risk. Three of the four (75%) clusters of traits, which were significantly correlated with AD, demonstrated antagonistic genetic heterogeneity, characterized by different directions of the genetic associations and trait correlations. Our findings suggest that shared genetic factors between AD and cardiovascular traits mostly affect them in an antagonistic manner.

Keywords: exome-wide association study; Alzheimer's disease; pleiotropy; hierarchical cluster analysis; aging; cardio-metabolic traits; coronary heart disease; myocardial infarction; stroke; heart failure

Citation: Loika, Y.; Loiko, E.; Culminskaya, I.; Kulminski, A.M. Exome-Wide Association Study Identified Clusters of Pleiotropic Genetic Associations with Alzheimer's Disease and Thirteen Cardiovascular Traits. *Genes* **2023**, *14*, 1834. https://doi.org/10.3390/genes14101834

Academic Editor: Claudia Ricci

Received: 29 August 2023
Revised: 18 September 2023
Accepted: 20 September 2023
Published: 22 September 2023

Copyright: © 2023 by the authors. Licensee MDPI, Basel, Switzerland. This article is an open access article distributed under the terms and conditions of the Creative Commons Attribution (CC BY) license (https://creativecommons.org/licenses/by/4.0/).

1. Introduction

Alzheimer's disease (AD) is an aging-related neurodegenerative disorder, with few interventions available to ameliorate its symptoms. As the number of AD-affected Americans is expected to increase rapidly, reaching about 12.7 million cases by 2050 [1], identifying the causes of Alzheimer's disease would be beneficial for its prevention and treatment. Several hypotheses have been proposed to explain AD pathology, such as amyloid beta (Aβ) cascade, tau pathology, calcium, inflammatory, cholinergic, and oxidative stress hypotheses [2]. It is likely that many biological processes and pathways related to these hypotheses contribute to the initiation and progression of AD. In addition to the buildup of Aβ plaques and neurofibrillary tangles, the loss of neuronal synapse junctions, neuronal cell death, and brain atrophy can contribute to cognitive impairment underlying AD.

At this time, none of the aforementioned hypotheses alone have succeeded in explaining AD pathology. The amyloid beta hypothesis has been thoroughly investigated in the clinical treatment of AD, with a focus on targeting Aβ plaques. Notably, two drugs, lecanemab and aducanumab, have received approval from the FDA through the accelerated approval mechanism. Additionally, a third drug, donanemab, has shown promising results in clinical trials [3], but there is still no cure for AD at this time. This means that there are other processes accompanying the disease and contributing to its progression and severity. This includes the breakdown of the blood–brain barrier (BBB), the immune system response, reduced cerebral blood flow, altered adult neurogenesis and energy metabolism, cell membranes and the cell cycle, the cytoskeleton, and lipid and protein metabolism [4,5]. These processes can develop independently and/or be initiated by other age-related traits or events. The correlation of AD with these other traits may indicate common biological processes underlying them. Therefore, studying the patterns in the pleiotropic associations of AD and the other traits can facilitate identifying and understanding the biological mechanisms underlying AD pathogenesis.

Genome- and exome-wide association studies (GWAS and EWAS, respectively) are comprehensive tools designed to identify genes and related biological mechanisms associated with different traits. While conventional GWAS and EWAS focus on identifying genetic associations with single traits, pleiotropic GWAS and EWAS take a broader approach. They aim to simultaneously assess genetic associations with multiple traits and identify shared genetic components that contribute to the pleiotropy observed in epidemiological studies. Previous epidemiological studies that considered AD and other related traits helped identify several cardiovascular factors conferring AD risk. For instance, increased levels of LDL-C and TC are considered risk factors for AD [6,7], while increased levels of HDL-C were highlighted as protective [8]. Pleiotropic GWAS/EWAS provided an additional opportunity to identify several genetic components and respective biological mechanisms shared by those multiple traits [9,10].

Certain risk factors can form clusters, i.e., they increase or decrease AD risks simultaneously by acting in additive or synergistic manners. Several studies have already demonstrated the effects of such clusters [11,12]. For example, the clustering of vascular risk factors, such as midlife obesity, high SBP, and high total cholesterol levels, increased the risk of dementia and AD in an additive manner [12]. Clusters of hypertension and heart disease, diabetes, and current smoking also increased AD risk [11].

Many of the previous analyses of clustering of neurodegenerative conditions, including AD, used individual-level data for phenotypes as a distance measure [13–17]. In such types of cluster analyses, strongly correlated phenotypes usually cluster together. This approach is not quite efficient, however, in cases when the underlying biological mechanisms and related genetic components are not major contributors to the phenotypes' correlations. In this study, we applied an approach in which the statistics of univariate and pleiotropic genetic associations were used for identifying clusters of AD and its risk factors. We believe that this approach can more accurately characterize complex roles of genetic and non-genetic factors in AD pathogenesis [18–21], including complex interactions, which can be attributed to different biological mechanisms driving antagonistic genetic heterogeneity [22,23]. Such heterogeneity was shown to be widespread in the genetics of lipid traits [24]. This phenomenon was also observed in pleiotropic associations with AD and some traits related to educational attainment and cardiovascular risk factors [25–27].

The goal of this study was to gain insights into pleiotropic predisposition to cardiovascular and AD risk factors and demonstrate complex clustering patterns of AD with 16 risk factors. We hypothesized that using summary statistics of pleiotropic associations among AD and related phenotypes (risk factors) could capture clusters of phenotypes based on the similarities among the genetic effects, rather than solely on the direct correlations among these phenotypes. This approach may facilitate the discovery of shared biological pathways. An additional goal was to investigate how antagonistic genetic heterogeneity shapes the clustering of AD-related phenotypes. For this purpose, we carried out a pair-wise pleiotropic exome-wide association study (~260 K common genetic variants) on predisposition to AD and on each of the 16 selected traits in a sample of ~190 K individuals from the UK Biobank. The analysis included six qualitative traits (coronary heart disease (CHD), diabetes mellitus (DM), stroke, myocardial infarction (MI), heart failure (HF), and hypertension (HT)), and ten other cardio-metabolic risk factors (blood glucose (BG), BMI, height, weight, four lipid traits

(LDL-C, HDL-C, TG, and TC), and systolic (SBP) and diastolic (DBP) blood pressure). Fisher's method was used to test the pleiotropic associations, and hierarchical cluster analysis was used to identify clusters of AD risk factors with similar genetic components.

2. Materials and Methods

2.1. Accession Numbers

This research has been conducted using data from the UK Biobank, a major biomedical database (http://www.ukbiobank.ac.uk/, accessed on 15 March 2021).

2.2. Study Cohorts

Data from the UK Biobank (UKB) [28,29] on individuals of Caucasian ancestry, men and women combined, were considered in the analyses. Table 1 provides the basic demographic information of the genotyped participants. After quality control of genetic and phenotypic information, 188,260 individuals remained in the subsequent analyses.

Table 1. Basic demographic information of the UK biobank participants with whole exome genotyping information.

AD Status	PAR	N	Women	BC Range	LS	Age	Mortality
	VALUE (SD)	**188,260**	**103,736 (55.10)**	**1936–1970**	**65.21 (8.0)**	**56.70 (8.03)**	**899 (0.48)**
				Quantitative traits			
	AD	DM	HT	CHD	MI	STROKE	HF
Cases	361	53 (15%)	189 (52%)	89 (25%)	37 (10%)	30 (8%)	20 (6%)
Controls	187,899	11,015 (6%)	45,349 (24%)	16,327 (9%)	5834 (3%)	4589 (2%)	3236 (2%)
p-value		1.33×10^{-9}	$<2.2 \times 10^{-16}$	$<2.2 \times 10^{-16}$	3.90×10^{-10}	9.97×10^{-9}	6.867×10^{-6}

Quantitative traits

	BG (mg/dL)	BMI (kg/m^2)	Height (cm)	Weight (kg)	SBP (mmHg)	DBP (mmHg)	HDL-C (mg/dL)	LDL-C (mg/dL)	TC (mg/dL)	TG (mg/dL)
Cases	93.36 (19.93)	27.01 (4.90)	167.18 (9.17)	75.62 (15.40)	146.18 (19.58)	81.79 (10.96)	57.77 (16.30)	133.77 (35.72)	216.97 (47.43)	149.26 (78.77)
Controls	92.13 (21.19)	27.34 (4.73)	168.67 (9.24)	78.03 (15.83)	139.78 (19.58)	82.13 (10.65)	56.49 (14.82)	138.07 (33.44)	221.19 (43.99)	154.26 (89.82)
p-value	5.89×10^{-1}	5.05×10^{-1}	1.61×10^{-1}	1.72×10^{-1}	3.46×10^{-2}	9.04×10^{-1}	4.77×10^{-1}	2.68×10^{-1}	3.96×10^{-1}	5.54×10^{-1}

PAR = the name of characteristics/parameter; SD = standard deviation. N = sample size; Women = number of women (%); BC Range = range of birth years; LS = life span in years; age = mean age measured in years at the selected examination of quantitative traits. p-values for qualitative and quantitative phenotypes were assessed by using the exact Fisher and Wald test, respectively. AD = Alzheimer's disease; DM = diabetes mellitus; HT = hypertension; CHD = coronary heart disease; MI = myocardial infarction; STROKE = stroke; HF = heart failure. BG = blood glucose; BMI = body mass index; SBP = systolic blood pressure; DBP = diastolic blood pressure; HDL-C = high-density lipoprotein cholesterol; LDL-C = low-density lipoprotein cholesterol; TC = total cholesterol; TG = triglycerides.

2.3. Genotypes

Exome sequencing data for ~200 K individuals were available from the UK Biobank Exome Sequencing Consortium [30]. In the analyses, we included directly genotyped markers (no imputation was performed) with minor allele frequency (MAF) greater than 0.5%, which resulted in 258,684 (~250 K) genetic variants. A total of 213,935 SNPs had a missing call rate better than 5%. For these SNPs, all individuals had a missing call rate better than 5%. Only these SNPs were considered for reporting the results of this study. We did not apply the Hardy–Weinberg equilibrium test at this stage because negligible deviation from Hardy–Weinberg equilibrium was still highly significant in the large UKB sample.

2.4. Phenotypes

We considered incident cases of AD and six other diseases, including DM, HT, CHD, MI, stroke, and HF (Table 1). All cases were defined by the UK Biobank based on the ICD-9 and ICD-10 codes. AD status was defined by using codes F00 and G30 according to the ICD-10 classification.

Ten quantitative phenotypes, BG (mg/dL), BMI (kg/m^2), height (m), weight (kg), SBP (mmHg), DBP (mmHg), HDL-C (mg/dL), TG (mg/dL), LDL-C (mg/dL), and TC (mg/dL) were considered. Because longitudinal information (multiple measurements) was available for a small number of participants, we selected measurements based on the first examination for which information was available (Table 1) (see also Section 2.7). We found significant differences between the AD cases and controls in the qualitative phenotypes (assessed using the exact Fisher test) and the mean values for SBP (evaluated using the Wald test).

2.5. Correlations among Phenotypes

Information on the Pearson correlation coefficients between AD and each of the considered quantitative and qualitative phenotypes and their significance is given in Table 2. We observed a small positive significant correlation between AD and each of the six diseases. The largest and most significant correlation was between AD and HT (r = 0.029, $p = 2.41 \times 10^{-23}$). Among the quantitative traits, only weight, height, and SBP demonstrated small significant correlations with AD (Table 2).

Table 2. Pearson correlation coefficient between Alzheimer's disease and each of the considered phenotypes in the UK Biobank based on individual-level data and summary statistics of genetic associations.

	Individual-Level Data		Summary Statistics	
Phenotype	**r**	**P**	**r**	**P**
HT *	0.0290	2.41×10^{-23}	0.0352	$<2.2 \times 10^{-16}$
CHD *	0.0248	1.82×10^{-17}	0.0145	2.15×10^{-11}
MI *	0.0180	6.52×10^{-10}	−0.0169	4.67×10^{-15}
STROKE *	0.0166	1.19×10^{-8}	−0.0174	8.10×10^{-16}
DM *	0.0164	1.76×10^{-8}	0.0365	$<2.2 \times 10^{-16}$
HF *	0.0128	1.08×10^{-5}	−0.0358	$<2.2 \times 10^{-16}$
BG	0.0025	3.82×10^{-1}	−0.0069	1.55×10^{-3}
BMI	−0.0031	2.82×10^{-1}	0.0514	$<2.2 \times 10^{-16}$
Height *	−0.0070	1.61×10^{-2}	−0.0700	$<2.2 \times 10^{-16}$
Weight *	−0.0066	2.30×10^{-2}	0.0092	2.03×10^{-5}
SBP *	0.0143	9.34×10^{-7}	−0.0082	1.41×10^{-4}
DBP	−0.0014	6.35×10^{-1}	0.0164	2.87×10^{-14}
HDL-C	0.0038	1.96×10^{-1}	−0.0074	5.84×10^{-4}
LDL-C	−0.0057	5.17×10^{-2}	−0.0029	1.87×10^{-1}
TC	−0.0042	1.48×10^{-1}	−0.0086	7.62×10^{-5}
TG	−0.0024	4.00×10^{-1}	−0.0078	3.41×10^{-4}

r = Pearson correlation coefficient; P = *p*-values; * = significant correlation based on individual-level data. DM = diabetes mellitus; HT = hypertension; CHD = coronary heart disease; MI = myocardial infarction; STROKE = stroke; HF = heart failure. BG = blood glucose; BMI = body mass index; SBP = systolic blood pressure; DBP = diastolic blood pressure; HDL-C = high-density lipoprotein cholesterol; LDL-C = low-density lipoprotein cholesterol; TC = total cholesterol; TG = triglycerides.

Figure S1 and Table S1 display the Pearson correlation coefficients among the various traits under consideration. Most correlations were statistically significant, with the exception of those involving stroke–height and the ones noted above for AD. Among the observed correlations, the highest were between LDL-C and TC (r = 0.95), CHD and MI (r = 0.58), SBP and DBP (r = 0.67), weight and BMI (r = 0.83), and weight and height (r = 0.54). Conversely, the most pronounced negative correlation was between HDL-C and TG (r = −0.44), as well as HDL-C and weight (r = −0.46).

2.6. Correlations among Summary Statistics

Table 2 also includes information on the Pearson correlation coefficients that were obtained by using summary statistics of the genetic associations with AD and each of the considered quantitative and qualitative phenotypes. Only the correlation coefficient between the summary statistics for AD and LDL-C was statistically non-significant. The correlation coefficients of the summary statistics for the other traits with AD were statistically significant. Moreover, for five traits (MI, stroke, HF, weight, and SBP), the correlation based on individual-level data and on summary statistics demonstrated opposite directions and were statistically significant. This demonstrates that correlations between the phenotypes at an individual level can be shaped by other factors in addition to genetic component(s).

2.7. Statistical Analyses

The univariate unconditional analysis—i.e., the univariate exome-wide association study (EWAS)—of the associations of each SNP with each quantitative phenotype or disease was performed using a linear or logistic regression model, as implemented in *plink 2.0* software. An additive genetic model was considered with the minor allele as an effect allele. All models in all analyses were adjusted for sex and age at the selected examination. Therefore, our EWAS provided 17 summary statistics (effect size, standard error, and p-value) for the estimates of the associations with seven diseases and ten continuous traits for each SNP.

Pleiotropic meta-analysis was performed for the pairs of traits. We used an AD-centric approach when each pair included AD and one of the other traits. Given the small correlations among the traits (Table 2), Fisher's method [31] was used for pleiotropic analysis. It combines p-values across phenotypes, disregarding the effect directions and correlations among them, and addresses the issue of multiple testing by increasing the number of degrees of freedom. Therefore, 16 pleiotropic meta-analyses were conducted.

2.8. Pleiotropic Associations

As Fisher's method corrects pleiotropic p-values for testing multiple phenotypes, the traditional genome-wide (GW) level of significance, $p_{GW} = 5 \times 10^{-8}$, was used to obtain conclusions about the significance of the pleiotropic effects. Here, we report GW significant pleiotropic SNPs when at least one of 16 pleiotropic tests attained GW significance, $p < p_{GW}$, and the EWAS univariate associations with these traits demonstrated significant association at the levels of $p < 5 \times 10^{-4}$ for AD and $p < 5 \times 10^{-2}$ for the other traits. As this analysis was AD-centric, we applied a more stringent threshold for AD associations and a less stringent threshold for each of the other phenotypes.

2.9. Index SNPs and Gene Mapping

The NCBI dbSNP database and variant effect predictor from Ensembl (assembly GRCh38.p13) were used for mapping SNPs to genes. Outside of the *APOE* gene region, one index SNP was selected per locus. Multiple genes were reported when several genes overlapped. The index SNPs were selected based on the most significant association (smallest p-values) within each genetic locus. Furthermore, the selection of index SNPs took into consideration the proximity of the MAF from the UKB exome chip to the MAF found in the UK10 K TWINS and/or 1000 G reference datasets, as the MAF for some SNPs substantially differed. SNPs that met the Hardy–Weinberg threshold p-value = 1×10^{-40}, were retained. We used a small p-value for the Hardy–Weinberg cut-off because even very small deviation from Hardy–Weinberg equilibrium was highly significant in the large UKB sample.

2.10. Cluster Analysis

Hierarchical cluster analysis can be applied if a distance measure is defined. Previous studies [11–17] used individual-level data of phenotype measurements as a distance measure, which assesses the similarity of phenotypes, including their genetic effects, effects of exogenous exposures, and their interaction. In this study, we proposed using genetic associations with phenotypes as a distance measure. This approach is more appropriate for gaining insight into the similarities related to genetic components rather than exogenous exposures, considering that the contributions of genetic effects and exogenous exposures on the correlations among the phenotypes can be substantially different (see Table 2).

In this study, SNP–phenotype clusters were identified, leveraging the results of our univariate and AD-centric pair-wise pleiotropic analyses. First, we identified patterns characterized by the relative directions of the associations of the same alleles with AD and each of the other phenotypes and the significance of the pleiotropic associations. The relative directions were defined by the sign of a product of the effect sizes of the associations of SNPs with AD and the other phenotype in a pair. For instance, if the associations of SNPs with these phenotypes were of the same (opposite) directions, such SNPs had the same (different) pattern.

These patterns can be subdivided further into clusters based on the p-values of the associations of SNPs with phenotypes. To improve the resolution of the cluster analysis, we used p-values from the univariate analyses of non-AD phenotypes and selected pleiotropic SNPs for which the significance of their associations with AD was $p < 5 \times 10^{-4}$. We further categorized minus-log-transformed p-values from the univariate associations of non-AD phenotypes, $-log(p_{uni})$, into five categories, denoted as 0 for $p > 0.05$, 1 for $5 \times 10^{-4} \leq p < 5 \times 10^{-2}$, 2 for $5 \times 10^{-6} \leq p < 5 \times 10^{-4}$, 3 for $5 \times 10^{-8} \leq p < 5 \times 10^{-6}$, and 4 for $p < 5 \times 10^{-8}$. This procedure created an A_{ij} matrix for the ith SNP and jth phenotype, in which the sign of each element was determined by the relative directions of the SNP associations with AD and a respective trait, and its magnitude was defined by the aforementioned categories.

Two approaches were used for cluster analysis. Both approaches utilized hierarchical cluster analysis with Euclidian measure and Ward's method. Under this method, associations are clustered by minimizing the distance measure inside the clusters and maximizing this measure between clusters. Our main approach was applied to matrix A_{ij} as implemented by the *hclust* function in R. This approach estimates clusters based on the similarity of significances and sign-directions of the SNP–phenotype associations. The ad-hoc height cut-off threshold for selecting the clusters was equal to five. This analysis was validated using an approach based on Pearson's correlation as a distance measure. It was implemented using the *pheatmap* function in R package *ecodist*. This approach estimates clusters based on the collinearity of phenotypes as vectors of the associations with selected SNPs (collinearity of columns of matrix A_{ij}) and on the collinearity of the associations of selected SNPs with the considered phenotypes (collinearity of rows of matrix A_{ij}).

3. Results

Figure 1 represents a flowchart of the analyses performed in this study.

3.1. Univariate Associations from EWAS

A univariate EWAS was performed for seven diseases (DM, HT, CHD, MI, HF, stroke, and AD) and ten quantitative phenotypes (BG, BMI, weight, height, SBP, DBP, HDL-C, TG, LDL-C, and TC) in a sample of 188,260 UK biobank participants of European ancestry, for men and women combined (see Section 2).

For the diseases, the only GW significant associations were observed with AD in the *APOE* gene cluster on chromosome 19 for rs112849259 and rs741780 *TOMM40* SNPs and rs440446 and rs429358 *APOE* SNPs (Tables 3 and S2, Supplementary Materials). Also, for SNPs in the *APOE* gene cluster, we identified GW significant associations with at least one of the lipid traits: HDL-C (three SNPs), LDL-C (six SNPs), TG (five SNPs), and TC (six SNPs). In total, this cluster of genes harbored 24 GW significant univariate SNP–trait associations (Table S2).

Table 3. Associations with risk of Alzheimer's disease: SNPs from 4 genetic loci and *APOE* gene region.

N	Gene(s) [1]	SNP [2]	Chr	Location, Base Pairs GRCh38	Ref /Alt	MAF obs (%)	P_{HWE}	Function	Beta	SE	P	Gene GRASP	SNP GRASP	r^2	P_{GRASP}	PMID
1	CDK11B (MMP23B)	rs28688376	1	1,637,577	T/c	26.1	6×10^{-36}	intron	0.298	0.079	1.56×10^{-4}					
2	OBP2B	rs11244035	9	133,205,932	C/t	10.2	2×10^{-30}	missense	0.393	0.109	3.23×10^{-4}	ABO	rs8176694	0.02	1.3×10^{-2}	20061627
3	TPM1	rs4775613	15	63,056,897	A/g	42.8	3×10^{-2}	5′UTR	−0.334	0.078	1.99×10^{-5}	RAB8B	rs10519190	0.00	2.2×10^{-4}	17998437
4	SMARCA4	rs28997580	19	11,013,062	C/t	0.9	3×10^{-2}	synonymous	0.960	0.255	1.64×10^{-4}	LDLR	rs2569540	0.00	1×10^{-9}	35589863
	APOE		19													
5	CBLC	rs80168591	19	44,781,370	G/a	1.4	0.87	splice	0.911	0.205	8.54×10^{-6}	CBLC	rs899087	0.00	5.1×10^{-5}	22832961
6	BCAM	rs28399637	19	44,820,881	G/a	31.5	5×10^{-8}	intron	0.382	0.076	4.71×10^{-7}	BCAM	rs2927480	0.21	5.0×10^{-49}	21460841
7	PVRL2	rs283813	19	44,885,917	T/a	6.7	3×10^{-22}	intron	0.488	0.123	7.16×10^{-5}	PVRL2	rs283813	1	7.6×10^{-28}	33589840
8	TOMM40	rs112849259	19	44,894,050	C/t	2.6	0.73	missense	1.227	0.138	4.59×10^{-19}					
9	TOMM40	rs741780	19	44,901,174	T/c	43.2	7×10^{-2}	intron	−0.454	0.080	1.14×10^{-8}	TOMM40	rs741780	1	1.4×10^{-8}	23565137
10	APOE	rs440446	19	44,905,910	G/c	35.8	2×10^{-2}	missense	−0.561	0.087	1.22×10^{-10}	APOE	rs439401	0.57	1.1×10^{-78}	21460841
11	APOE	rs429358	19	44,908,684	T/c	15.4	0.54	missense	1.371	0.077	4.63×10^{-71}	APOE	rs429358	1	2.7×10^{-78}	21390209
12	APOE	rs7412	19	44,908,822	C/t	8.0	0.66	missense	−0.856	0.200	1.87×10^{-5}	APOE	rs7412	1	5.5×10^{-58}	20885792
13	APOC4; APOC2	rs5167	19	44,945,208	T/g	35.1	0.10	missense	0.315	0.075	2.92×10^{-5}	APOC4; APOC2	rs5167	1	2.8×10^{-9}	21460840

[1] Multiple genes were assigned if the index SNP was within the region of overlapping genes. [2] One single nucleotide polymorphism (**SNP**) per gene locus was retained. **Chr** = chromosome; **Ref/Alt** = reference/alternative allele (majuscule/minuscule letters stay for major/minor allele); **MAF obs** = minor allele frequency **obs**erver in our dataset. **Beta/SE/P** indicates effect size/standard error/*p*-value of the association of the alternative (minor) allele vs. the reference (major) allele. **SNP_{GRASP}** and **$Gene_{GRASP}$** denote SNP and related gene, for which minimum *p*-value has been reported in either the GRASP [32] or GWAS [33] catalog, where **P_{GRASP}** is the reported *p*-value of the association with AD risk and **PMID** is the PubMed index of the respective paper. r^2 represents LD between previously reported SNP and SNP reported here. Empty cells: no index SNPs from the corresponding genetic locus were previously reported neither in the GRASP nor in GWAS catalog for their association with AD risk.

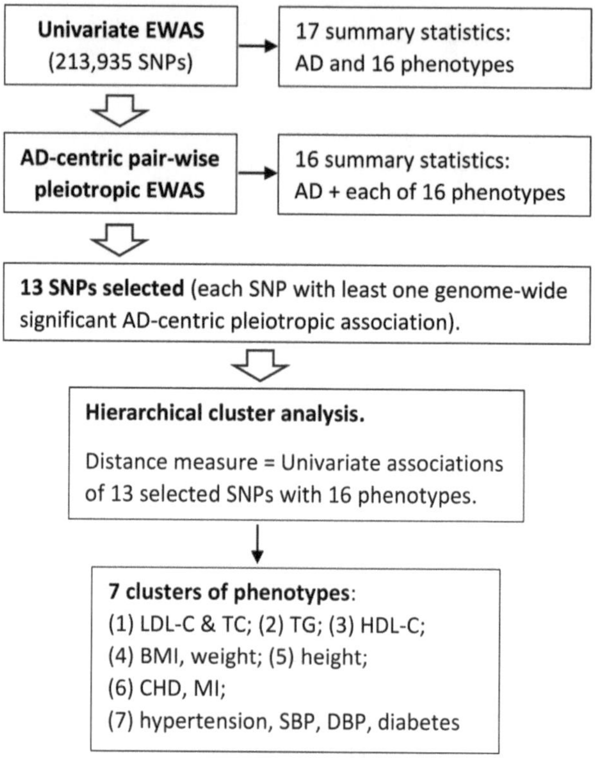

Figure 1. Flowchart of the analyses of this exome-wide association study (EWAS).

Outside of the *APOE* gene cluster, three SNPs demonstrated five GW significant associations with BMI (rs28688376), LDL-C, and TC (rs11244035, rs28997580) (Tables 4 and S2).

Table 4. Univariate and pair-wise pleiotropic associations of SNP, which includes Alzheimer's disease and one of 16 considered phenotypes: 4 genetic loci outside the APOE gene region.

N	Gene(s) [1]	SNP [2]	Chr	Location, Base Pairs GRCh38	Ref /Alt	MAF obs (%)	PH	Beta	SE	P	P_F
1	CDK11B (MMP23B)	rs28688376	1	1,637,577	T/c	26.1	BMI	−0.096	0.017	3.36×10^{-8}	1.42×10^{-10}
2	CDK11B (MMP23B)	rs28688376	1	1,637,577	T/c	26.1	Weight	−0.232	0.052	7.01×10^{-6}	2.37×10^{-8}
3	OBP2B	rs11244035	9	133,205,932	C/t	10.2	LDL-C	1.206	0.187	1.09×10^{-10}	1.13×10^{-12}
4	OBP2B	rs11244035	9	133,205,932	C/t	10.2	TC	1.519	0.243	3.84×10^{-10}	3.81×10^{-12}
5	TPM1	rs4775613	15	63,056,897	A/g	42.8	TC	−0.436	0.146	2.81×10^{-3}	9.91×10^{-7}
6	TPM1	rs4775613	15	63,056,897	A/g	42.8	HDL-C	−0.245	0.048	2.43×10^{-7}	1.31×10^{-10}
7	TPM1	rs4775613	15	63,056,897	A/g	42.8	SBP	0.156	0.062	1.18×10^{-2}	3.84×10^{-6}
8	TPM1	rs4775613	15	63,056,897	A/g	42.8	HF	−0.052	0.026	4.16×10^{-2}	1.24×10^{-5}
9	SMARCA4	rs28997580	19	11,013,062	C/t	0.9	LDL-C	−5.450	0.591	2.81×10^{-20}	$<5 \times 10^{-8}$
10	SMARCA4	rs28997580	19	11,013,062	C/t	0.9	TC	−6.141	0.767	1.20×10^{-15}	$<5 \times 10^{-8}$

[1] Multiple genes were assigned if the index SNP was within the region of overlapping genes. [2] One single nucleotide polymorphism (**SNP**) per gene locus was retained. **Chr** = chromosome; **Ref/Alt** = reference/alternative allele (majuscule/minuscule letters stay for major/minor allele); **MAF obs** = minor allele frequency **obs**erver in our dataset. **Beta/SE/P** indicates effect size/standard error/*p*-value of the association of the alternative (minor) allele vs. the reference (major) allele. $\mathbf{P_F}$ is the *p*-value of the pair-wise pleiotropic analysis obtained by Fisher's method.

In total, 29 GW significant univariate associations were identified.

3.2. Pleiotropic AD-Centric Pair-Wise Associations

We identified 51 GW significant pleiotropic associations (see Section 2) (Tables S2 and S3A). They included 20 associations in the *APOE* gene cluster, which were not driven by the GW significance of the associations with AD. Four SNPs in the *APOE* and *TOMM40* genes, which were associated with AD at the GW level, showed 19 GW significant univariate associations with non-AD phenotypes, totaling 19 GW significant pleiotropic associations. These SNPs were associated with BMI, weight, and DBP (rs112849259 and rs429358); height and TC (rs741780 and rs440446); HDL-C (rs112849259 and rs741780); LDL-C (rs741780); DM (rs112849259, rs440446, and rs429358); HT (rs440446); and CHD and MI (rs429358). Also, the remaining SNPs from the *APOE* gene cluster showed five additional significant pleiotropic associations with AD and BMI, weight, CHD, and MI (rs7412); HDL-C (rs28399637), while these SNPs did not attain GW significance in the univariate analysis. In total, 44 [=20 + 19 + 5] GW significant pleiotropic associations were found for the SNPs mapped to the *APOE* gene cluster. Outside of the *APOE* cluster, we identified seven GW significant pair-wise pleiotropic associations. These included five SNP–trait pairs with GW significant univariate effects (Tables 4 and S2) and two additional pleiotropic associations, including BMI (rs28688376) and HDL-C (rs4775613).

We also identified 11 pleiotropic associations at the suggestive-effect level $5 \times 10^{-8} < p < 5 \times 10^{-6}$. They included associations of rs4775613 with SBP and TC; rs80168591 with HDL-C and stroke; rs28399637 with BMI, weight, and CHD; rs7412 with SBP and HT; as well as rs5167 with height and TC. Thus, in total, we found 62 (=51 + 11) pair-wise pleiotropic associations for 13 SNPs and 14 traits at $p < 5 \times 10^{-6}$ (Tables S2 and S3A). None of the pleiotropic associations attained even a suggestive-effect level for BG and HF, and only one for stroke.

3.3. Clusters of Pleiotropic Associations

For cluster analysis (see Section 2.10), we used 61 pleiotropic associations, excluding unique association with stroke (Table S3A). Matrix A_{ij} for this analysis included information on 13 SNPs and 13 phenotypes (Table S3B). Hierarchical cluster analysis using the Euclidian measure and Ward's method provided two-dimensional dendrogram in the SNP–phenotype domain (Figure 2). Phenotype-specific one-dimensional dendrogram exemplified seven clusters for 13 phenotypes, i.e., (1) (LDL-C, TC), (2) TG, (3) (CHD, MI), (4) height, (5) (HT, SBP, DBP, DM), (6) HDL-C, and (7) (weight, BMI) (Figure 2). Pearson's correlation approach (Figure S2) provided similar results, except that a pair of (HT, SBP) phenotypes was clustered with (CHD, MI) rather than with (DBP, DM). This discrepancy is likely due to diminished power to identify clusters using different distance measures, given the moderate *p*-values from the univariate associations with these six phenotypes (Table S2). Both approaches of cluster analysis aggregated SNPs into the same eight clusters (Figures 2 and S2).

Figure 2 shows that the largest group of GW significant univariate (8 for LDL-C and 8 for TC) and pleiotropic (9 for LDL-C and 10 for TC) associations was observed for the (LDL-C, TC) cluster. The second largest group was observed for TG (five univariate and five pleiotropic GW significant associations), followed by HDL-C (three univariate and seven pleiotropic GW significant associations) and BMI (one univariate and four pleiotropic GW significant associations) + weight (four pleiotropic GW significant associations) clusters. Two distinct clusters emerged from the analysis, namely (LDL-C, TC) and (BMI, weight), which were composed of traits with the largest Pearson correlation coefficients of r = 0.952 and r = 0.833, respectively, evaluated using individual-level data (Figure S1 and Table S1). This arrangement follows the anticipated clustering pattern where correlated traits are grouped together. However, in contrast, height and weight, despite having a substantial significant correlation of r = 0.538, were placed in separate phenotypic clusters.

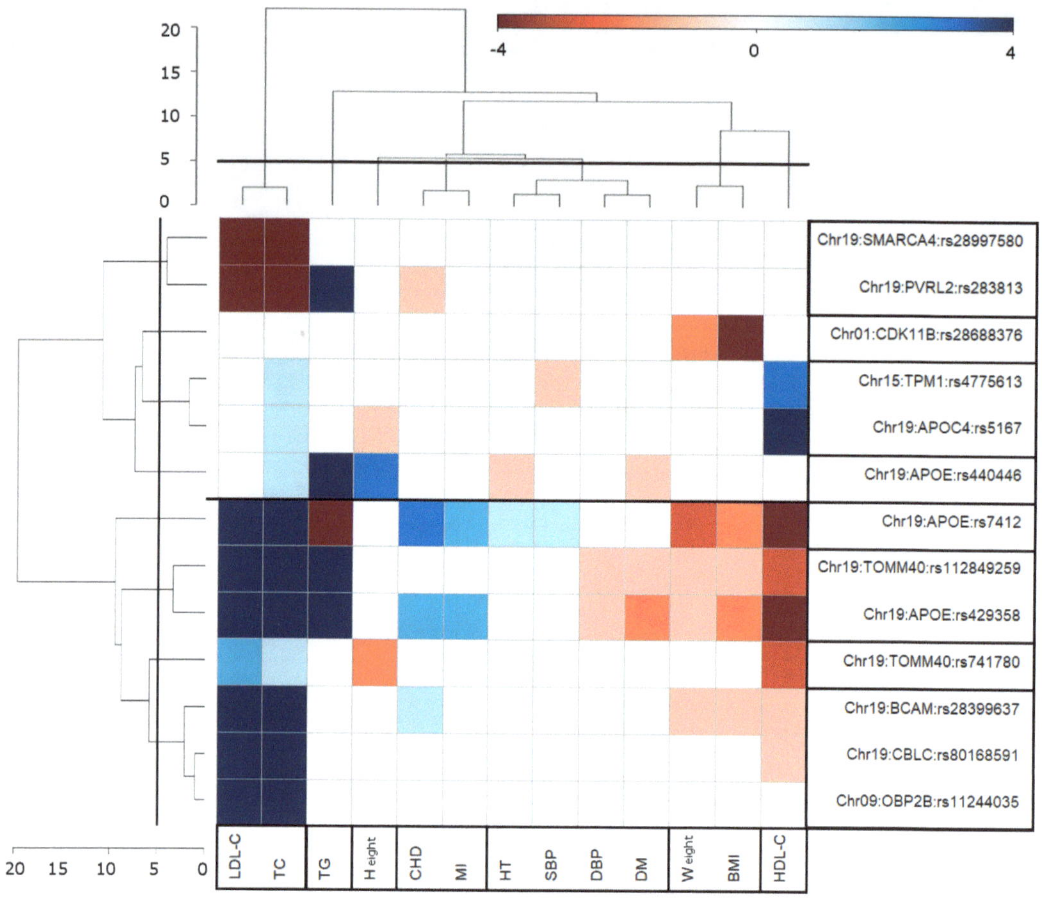

Figure 2. Eight clusters of SNPs and seven clusters of phenotypes based on significant pleiotropic pair-wise associations with AD and each of the other 13 age-related traits. Blue (red) color and its shades shows SNP–phenotype associations with the same (opposite) effect direction compared to the respective SNP–AD associations. Alleles, which demonstrated detrimental association with AD (positive betas), were considered effect alleles. Boxes show SNP and phenotypic clusters of pleiotropic associations, which were selected at a score (height) level of 5, as represented by axes with numbers on the horizontal (phenotypes) and vertical (SNPs) dendrograms. The x-axis shows the phenotypes, while y-axis refers to the SNPs and their respective genes. The horizontal solid line in the center shows the separation of the SNP clusters into two groups based on their significance and the effect directions of their associations with the first phenotypic cluster (LDL-C, TC). Hierarchical cluster algorithm with Euclidian measure and Ward's method as implemented in R function hclust was used for this plot.

3.4. Antagonistic Genetic Heterogeneity

Figure 2 emphasizes the complex forms of genetic heterogeneity when the same alleles from different SNPs can have the same (e.g., ε2-encoding rs7412 SNP) or opposite (e.g., minor allele of rs28997580) directions of the associations with AD and other phenotypes in a pair, e.g., (AD, LDL-C). This pattern, i.e., two sets of SNPs, was observed for each of the eight phenotypes from six phenotypic clusters: LDL-C + TC, TG, HDL-C, height (H), HT + SBP, and CHD.

Figure 3 provides further insight into the phenomenon of antagonistic genetic heterogeneity. It includes the results of the significant pleiotropic associations for the traits that demonstrated significant correlations with AD. Associations with lipid traits, BMI, and DBP were excluded because of their nonsignificant correlations with AD (Table 2). Each cell in Figure 3 displays the sign of a product of effect sizes for SNP–AD and SNP–trait associations and the Pearson correlation coefficient (calculated by using individual-level data) of AD and the corresponding trait. Red (blue) cells, which correspond to the negative (positive) sign of the product mentioned just above, represent pleiotropic associations with antagonistic (non-antagonistic) genetic effects. From this figure, it follows that 33% (7 out of 21) of the pleiotropic associations in 3 of 4 phenotypic clusters (75%) demonstrated antagonistic genetic heterogeneity, i.e., alleles, which demonstrated detrimental association with AD, also demonstrated the opposite sign direction of their association with the considered trait compared to the correlation of that trait with AD. Two traits demonstrated a more homogeneous pattern. Significant associations with DM demonstrated antagonistic pleiotropic effects only, while all significant pleiotropic associations, which included weight, were non-antagonistic ones (Figure 3).

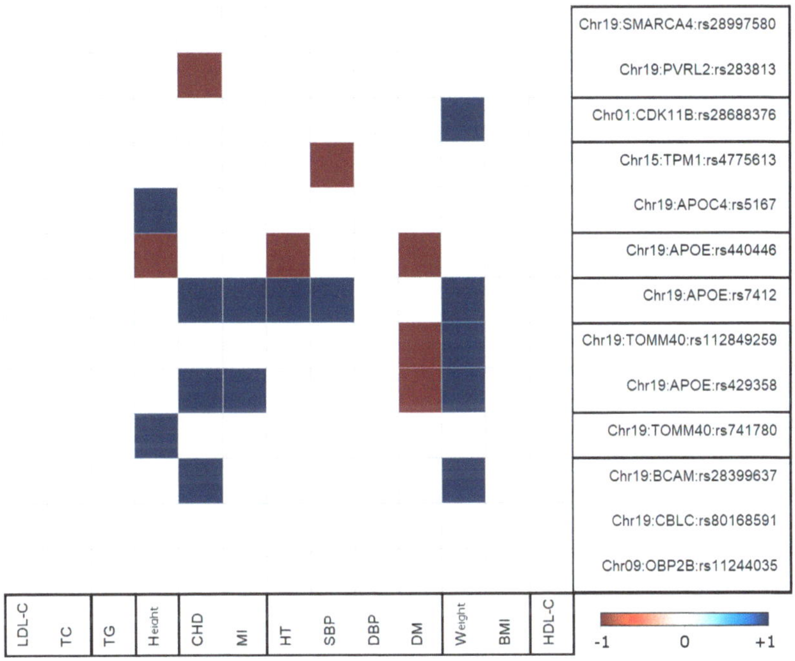

Figure 3. Antagonistic genetic heterogeneity of significant pleiotropic pair-wise associations with AD and each of the other 13 age-related traits. Blue (red) color shows non-antagonistic (antagonistic) genetic heterogeneity. The results are presented for phenotypes that exhibit significant correlations with AD. Therefore, associations involving lipids, BMI, and DBP were excluded from this figure (see Table 2). Refer to Figure 2 caption for other notations.

4. Discussion

In this EWAS, we performed AD-centric pair-wise pleiotropic and cluster analyses of the associations of SNPs with AD and 16 phenotypes of cardiovascular and AD risk factors.

4.1. AD-Centric Pair-Wise Pleiotropic Associations

AD-centric pleiotropic analysis considered the associations of SNPs with AD and one of the 16 phenotypes. This analysis identified 13 SNPs with genome-wide significant pleiotropic associations (Figure 2). Nine SNPs were mapped to the *APOE* gene cluster on chromosome 19 (including *APOE, TOMM40, APOC2/APOC4, PVRL2 (NECTIN2), BCAM,* and *CBCL* genes), which is a well-known genetic risk factor for AD. Four other SNPs were from *CDK11* (rs28688376), *OBP2B* (rs11244035), *TPM1* (rs4775613), and *SMARCA4* (rs28997580) gene loci. These SNPs were associated with AD at $p < 5 \times 10^{-4}$ and attained genome-wide significance ($p < 5 \times 10^{-8}$) in the pleiotropic analysis.

The ***CDK11B*** (cyclin-dependent kinase 11B) gene plays a role in ***cell apoptosis***. Recent studies have demonstrated that *CDK11* showed an altered expression in AD vulnerable neurons, which may be related to APP signaling processes [34,35]. A predominantly increased expression of *CDK11* was observed in the cytoplasm of neuronal cells in AD cases, while it was expressed specifically in the nuclei of post-mitotic neurons in most controls [34]. CDK11 is regulated by checkpoint kinase 2 (CHK2), which phosphorylates tau at an AD-related site, enhancing tau toxicity [36,37], which demonstrates an additional link of the *CDK11B* gene to AD. Altered expression of *CDK11B* can be a cause of neuronal apoptosis and, consequently, of decreased brain weight through this process. This suggestion is consistent with our results that the minor allele of rs28688376 from the *CDK11B* gene was associated with an increased risk of AD and decreased levels of weight and BMI.

TPM1, *TPM3*, and *TPM4* genes encode tropomyosin isoforms in neuronal cells. *TPM1* products were found in the presynaptic compartment of the central nervous system (CNS) neurons [38]. Tropomyosins (Tpm) actin-binding proteins ***stabilize the actin filaments***, which play a key role in the synaptic function of the CNS and mediate processes of memory and learning [39]. Microtubules, neurofilaments, and microfilaments (actin filaments) form the cytoskeleton of neurons. ***Dysregulation*** of tropomyosin and the actin ***cytoskeleton can induce synapse loss***, which takes place in the early stages of AD pathology. Tropomyosin participation in the neurofibrillary pathology of AD was immunochemically demonstrated [40]. A significant increase in tropomyosin-1 abundance in the platelets of AD female patients was recently observed [41], in addition to a proteomics study that showed *TPM1* gene products increasing in the white matter of AD patients when compared to controls [42]. Recently, it has been demonstrated that *TPM1* plays a key role in cardio-genesis and cardiovascular disorders [43], providing a link to the associations of SNPs from this gene with SBP observed in this study.

The protein encoded by the ***SMARCA4*** gene is a member of the SWI/SNF family of proteins, which regulate gene activity by a chromatin remodeling and are involved in repairing damaged DNA, replicating DNA, and controlling the growth, division, and differentiation of cells. *SMARCA4* is in the same gene region as the *LDLR* gene. ***LDLR*** is a key regulator of cholesterol metabolism and encodes the protein involved in receptor-mediated endocytosis of low-density lipoprotein cholesterol [44]. The genome-wide-significant association of rs2569540 from the *LDLR* gene with AD was demonstrated in a recent GWAS study [45].

The ***OBP2B*** gene plays a role in chemosensory behavior and the perception of smell. Impaired sense of smell, or olfactory dysfunction, is often seen as an early indicator of AD pathology in the brain. A recent study indicated that individuals with olfactory dysfunction also exhibit changes in the blood levels of LDL-C and TC [46]. The current study aligns with these findings and adds additional insight, suggesting how blood cholesterol levels and AD-related changes could be intertwined. Specifically, our results suggest that olfactory dysfunction, which is considered as an early manifestation of AD, can be related to altered levels of LDL-C and TC in the blood. This link highlights the need for further investigation to determine whether AD might play a causal role in altering lipid metabolism.

Our findings suggest the involvement of other mechanisms contributing to the disease pathogenesis in addition to the amyloid-beta mechanism. Specifically, they highlight actin filaments, chromatin remodeling, neuronal apoptosis, and lipid abnormalities as potential targets for the development of new AD drugs.

4.2. Clustering of Genome-Wide Significant Pair-Wise AD-Centric Pleiotropic Associations

In previous studies [11–17], the clustering of neurodegenerative conditions, including AD, was considered based on correlations among phenotypes. Our cluster analysis added a pleiotropic genetic component and, therefore, it can help in gaining further insights into the biological mechanisms associated with the pleiotropic effects. Such biological mechanisms may or may not be those which drive correlations among phenotypes. In this study, phenotypes belong to the same cluster if the associations of the SNPs with these phenotypes and AD demonstrated similarity defined by the same effect directions and the level of statistical significance (see Section 2.10). Thus, clustering is considered in a two-dimensional SNP/gene and phenotype domain. Similarity within phenotypic clusters suggests underlying biological mechanisms, which may contribute to AD. Our cluster analysis identified eight clusters of SNPs and seven clusters of phenotypes, which demonstrated similar SNP–phenotype associations (Figure 2). Cluster analysis was harmonized by considering alleles predisposed to AD as effect alleles.

We found that lipids defined three clusters: one cluster was defined by LDL-C and TC, while TG and HDL-C were in two separate clusters. This finding suggests at least partial independence of the biological mechanisms underlying the pleiotropic associations of each of these lipid traits and AD. This result corroborates previous studies [25,47], which reported partially independent mediation of the genetic associations with AD through LDL-C, TG, and HDL-C for SNPs inside and outside of the *APOE* gene region.

A CHD and MI cluster was defined by significant associations of SNPs from the *APOE* gene locus (Figure 2 and Table S2). Interestingly, while most alleles with elevated risks for CHD and MI (rs7412_C: β_{CHD} = 0.12, p = 8.25 $\times$ 10^{-8}; and rs423958_c: β_{CHD} = 0.06, p = 2.90 $\times$ 10^{-4}) were associated with higher risk of AD (rs7412_C: β_{AD} = 0.86, p = 1.87 $\times$ 10^{-5}; and rs423958_c: β_{AD} = 1.37, p = 4.63 $\times$ 10^{-71}), rs283813 (*PVRL2* gene) demonstrated the opposite relationship, i.e., rs283813_a was favorably associated with CHD (β_{CHD} = -0.05, p = 3.01 $\times$ 10^{-2}) but adversely with AD (β_{AD} = 0.49, p = 7.16 $\times$ 10^{-5}). This finding demonstrates that alleles associated with increased risks of CHD can be associated with either a higher or lower risk of AD. This is in line with antagonistic pleiotropic associations identified for AD and other cardiovascular factors, such as DM, that likely reflects contributions of different biological processes [27,48,49].

Four traits (HT, SBP, DBP, DM) defined a cluster with a complex pattern of associations. As they are risk factors for CHD and MI, they are typically considered together with these diseases. Meanwhile, in this study, these phenotypes belong to two different clusters, as there is more similarity within the clusters than among the clusters. All six phenotypes and height would belong to the same cluster if the cut-off of the cluster height level was increased to 7 (Figure 2, see the vertical axis with the numbers in the top left corner). Interestingly, SNPs encoding *APOE* ε2 and ε4 alleles were associated with two different subclusters, (HT, SBP) and (DBP, DM), respectively. This finding emphasizes that the pleiotropic associations of these *APOE* alleles have different etiologies.

Height and (BMI, weight) defined two non-overlapping clusters, indicating a role of independent mechanisms contributing to height and BMI or weight and AD.

Three phenotypes (BG, HF, and stroke) did not define any phenotypic cluster because of the lack of significant genetic associations that could be used for clustering.

4.3. Antagonistic Genetic Heterogeneity Was Observed for AD-Centric Pleiotropic Associations with Five Traits

Antagonistic genetic heterogeneity is characterized by misalignment of the signs of the product of the effects of genetic associations with different phenotypes and the correlations among these phenotypes [22–26]. This heterogeneity in our study was seen in the associations of seven SNPs with five traits in three phenotypic clusters (Figure 3). For each cluster of phenotypes (apart BMI, weight and DBP, DM), there were two sets of SNPs, which were characterized by antagonistic and non-antagonistic genetic effects in the associations with AD and the second phenotype in a pair (Figure 3). For example, rs283813 showed antagonistic genetic heterogeneity because it was adversely associated with AD and favorably with CHD, despite a significant positive correlation between AD and CHD (Table 2). The rs429358 did not show antagonistic heterogeneity with AD and weight because opposite-direction associations of this SNP with these phenotypes were aligned with a negative correlation between AD and weight.

Antagonistic genetic heterogeneity can also be viewed as a source of SNP/gene-phenotype clustering. For example, antagonistic genetic heterogeneity differentiates a cluster of height from that of BMI and weight because the height cluster includes antagonistic heterogeneity whereas the latter does not.

Antagonistic genetic heterogeneity plays a substantial role in the genetic associations with AD and its risk factors. This heterogeneity indicates that an effect allele from an SNP can confer the risk of one trait while being favorably associated with the other traits. Such a relationship likely indicates the role of different biological processes associated with the genes and pathways mapped to such SNPs on AD and its risk factors. For example, for DM, all significant pleiotropic associations were antagonistic. The alleles of SNPs with a significant adverse effect on DM were favorably associated with AD. This finding corroborates previous results [27,48,49] and highlights the complex contribution of genetic components to these diseases.

4.4. Protective Effect of Higher BMI Level against AD Is Related to High Weight

Previous studies demonstrated the protective effect of a high BMI level in late life against AD [20,21]. BMI is defined by two phenotypes, weight and height. Our cluster analysis showed that weight and height belong to different clusters. Moreover, SNPs that demonstrated genome-wide significant pleiotropic associations with BMI and weight were not significantly associated with height and vice versa. This finding suggests that protective effects of a higher BMI level against AD are related to higher weight rather than smaller height. While an increased BMI resulting from weight gain is associated with a higher risk of cardiovascular diseases (such as CHD, MI, and hypertension) and diabetes in midlife, a decreased BMI resulting from weight loss in later life can increase the risk of AD. This risk is attributed to the deteriorated functions of an aging organism, which include impaired metabolism, reduced cholesterol levels, decreased muscle mass, frailty, and sarcopenia, among other factors.

5. Conclusions

This study reports four novel genetic loci showing pleiotropic associations with AD and at least with one of 13 cardiovascular and AD risk factors. Genes harboring these loci are involved in cell apoptosis (*CDK11*), the stabilization of actin filaments, i.e., the cytoskeleton of neurons, (*TPM1*), the regulation of chromatin remodeling (*SMARCA4*), and chemosensory behavior and sensory perception of smell (*OBP2B*). No pleiotropic associations were identified for BG, HF, or stroke. Leveraging information on univariate and pleiotropic genetic associations, we found seven clusters in the domain of 13 SNPs and 13 phenotypes. Nine of thirteen pleiotropic SNPs were mapped to the *APOE* gene cluster (Figure 2).

AD-centric pleiotropic analysis confirmed partially independent mechanisms of pleiotropic associations with AD and lipid traits; although LDL-C and TC defined one phenotypic cluster, TG and HDL-C formed independent clusters because of the different patterns of the genetic associations.

Our analysis identified that 61% (9 of 13) of the considered traits demonstrated antagonistic pleiotropic AD-centric associations. Our cluster analysis found that weight, which was clustered with BMI, but not with height, defined pleiotropy of BMI and AD. SNPs identified by significant pleiotropic associations with AD and DM were from the *APOE* gene cluster, which corroborates previous findings.

Our cluster analysis in a two-dimensional SNP/gene-phenotype domain highlighted a more complex role of genetic and non-genetic factors in AD pathogenesis than the cluster analyses based on phenotype correlations did. Indeed, assuming that the clustering of phenotypes is related to their correlations, it is expected that correlated phenotypes should be in the same cluster. For LDL-C and TC (Pearson's correlation coefficient, r = 0.952), BMI and weight (r = 0.833), and CHD and MI (r = 0.581), this was the case, as each pair of these phenotypes was in its own cluster. This was not the case, however, for the other correlated phenotypes, such as, for example, height and weight (r = 0.538), HDL-C and weight (r = −0.462), and HDL-C and TG (r = −0.439) (see Table S1 and Figure S1) because these phenotypes were in different clusters. Accordingly, our cluster analysis approach was capable of capturing the biological mechanisms relevant to pleiotropy underlying AD and other age-related phenotypes that may be not readily available from analyses relying on information about the correlations of phenotypes. Defining more homogeneous patterns of phenotypes and genetic factors can aid in the identification of more homogenous groups of individuals who are at varying risks of AD. Subsequent research can facilitate the identification of genetic and non-genetic components, contributing to the pleiotropic effects associated with AD. Such insights can be valuable for developing comprehensive treatment approaches that combine drug therapy and lifestyle interventions.

The proposed approach can be applied to published summary statistics of genetic associations with different traits. Such an extension can help identify phenotypes that share genetic components and, therefore, overlapping biological mechanisms between different complex traits.

This study has some limitations. First, the analyses were performed for the entire sample but not separately for groups that are at different risk levels of AD, such as, for instance, within each sex group and/or at different ages. Second, we used a threshold p-value = 5×10^{-4} for associations with AD, which substantially limited the number of significant pleiotropic associations/SNPs. Third, a small number of SNPs (because of the second limitation) were available for hierarchical cluster analysis, which limited the precision of identifying phenotype clustering.

Supplementary Materials: The following supporting information can be downloaded at: https://www.mdpi.com/article/10.3390/genes14101834/s1, Table S1: Pearson's correlation coefficient [%] between 17 considered traits including cardiovascular and Alzheimer's disease risk factors in the UK Biobank; Table S2: Univariate and pleiotropic associations of 13 SNPs in the APOE gene region and other 4 gene loci, which demonstrated genome-wide significance of pair-wise pleiotropic associations with Alzheimer's disease and at least one of 16 traits; Table S3: (A) Patterns/clusters of pleiotropic associations of 13 SNPs in the APOE gene region and the other 4 gene loci, which demonstrated genome-wide significance of pair-wise pleiotropic associations with Alzheimer's disease and at least one of 16 traits. (B) Matrix A that was used in the cluster analysis; Figure S1: Pearson's correlation coefficients between 17 considered traits, which represent cardiovascular and Alzheimer's disease risk factors in the UK Biobank; Figure S2: Clusters of 13 traits and 13 SNPs based on Pearson's correlation coefficients as a distance measure as implemented in R function pheatmap from the R package ecodist.

Author Contributions: Investigation, Data Curation, Validation, Writing—Original Draft Preparation, Writing—Review and Editing: Y.L., E.L., I.C. and A.M.K.; Bioinformatic and Biological Analysis: E.L. and I.C.; Conceptualization, Methodology, and Visualization: Y.L. and A.M.K.; Software and Formal Analysis: Y.L.; Funding Acquisition, Supervision, and Project Administration: A.M.K. All authors have read and agreed to the published version of the manuscript.

Funding: This research was funded by Grants R01AG065477 and R01AG070488 from the National Institute on Aging. The funding source did not participate in the writing of the manuscript or the decision to submit the manuscript for publication.

Institutional Review Board Statement: The study was conducted according to the guidelines of the Declaration of Helsinki, and approved by the Institutional Review Board of Duke University (Pro00105247-INIT-1.0 approved on 26 June 2020; Pro00105346-INIT-1.0 approved on 15 April 2020).

Informed Consent Statement: Informed consent was obtained from all subjects by primary UKB investigators who gathered data.

Data Availability Statement: This manuscript was prepared using limited-access phenotypic and genetic datasets of the UK Biobank, which are available through the following link: http://www.ukbiobank.ac.uk/, accessed on 15 March 2021.

Acknowledgments: This research was conducted using data from UK Biobank, a major biomedical database (http://www.ukbiobank.ac.uk/, accessed on 15 March 2021). See also the Supplementary Materials.

Conflicts of Interest: The authors declare that they have no conflict of interest.

References

1. Gaugler, J.; James, B.; Johnson, T.; Reimer, J.; Solis, M.; Weuve, J. 2022 Alzheimer's disease facts and figures. *Alzheimers Dement.* **2022**, *18*, 700–789. [CrossRef]
2. Malik, R.; Kalra, S.; Bhatia, S.; Harrasi, A.A.; Singh, G.; Mohan, S.; Makeen, H.A.; Albratty, M.; Meraya, A.; Bahar, B.; et al. Overview of therapeutic targets in management of dementia. *Biomed. Pharmacother.* **2022**, *152*, 113168. [CrossRef] [PubMed]
3. Cummings, J. Anti-Amyloid Monoclonal Antibodies are Transformative Treatments that Redefine Alzheimer's Disease Therapeutics. *Drugs* **2023**, *83*, 569–576. [CrossRef]
4. Ferrer, I. Hypothesis review: Alzheimer's overture guidelines. *Brain Pathol.* **2023**, *33*, e13122. [CrossRef]
5. Pons, V.; Rivest, S. Targeting Systemic Innate Immune Cells as a Therapeutic Avenue for Alzheimer Disease. *Pharmacol. Rev.* **2022**, *74*, 1–17. [CrossRef]
6. Tang, Q.; Wang, F.; Yang, J.; Peng, H.; Li, Y.; Li, B.; Wang, S. Revealing a Novel Landscape of the Association Between Blood Lipid Levels and Alzheimer's Disease: A Meta-Analysis of a Case-Control Study. *Front. Aging Neurosci.* **2019**, *11*, 370. [CrossRef] [PubMed]
7. Anstey, K.J.; Ashby-Mitchell, K.; Peters, R. Updating the Evidence on the Association between Serum Cholesterol and Risk of Late-Life Dementia: Review and Meta-Analysis. *J. Alzheimer's Dis.* **2017**, *56*, 215–228. [CrossRef] [PubMed]
8. Zhang, X.; Tong, T.; Chang, A.; Ang, T.F.A.; Tao, Q.; Auerbach, S.; Devine, S.; Qiu, W.Q.; Mez, J.; Massaro, J.; et al. Midlife lipid and glucose levels are associated with Alzheimer's disease. *Alzheimer's Dement.* **2023**, *19*, 181–193. [CrossRef]
9. Bone, W.P.; Siewert, K.M.; Jha, A.; Klarin, D.; Damrauer, S.M.; Program, V.A.M.V.; Chang, K.M.; Tsao, P.S.; Assimes, T.L.; Ritchie, M.D.; et al. Multi-trait association studies discover pleiotropic loci between Alzheimer's disease and cardiometabolic traits. *Alzheimer's Res. Ther.* **2021**, *13*, 34. [CrossRef]
10. Broce, I.J.; Tan, C.H.; Fan, C.C.; Jansen, I.; Savage, J.E.; Witoelar, A.; Wen, N.; Hess, C.P.; Dillon, W.P.; Glastonbury, C.M.; et al. Dissecting the genetic relationship between cardiovascular risk factors and Alzheimer's disease. *Acta Neuropathol.* **2019**, *137*, 209–226. [CrossRef]
11. Luchsinger, J.A.; Reitz, C.; Honig, L.S.; Tang, M.X.; Shea, S.; Mayeux, R. Aggregation of vascular risk factors and risk of incident Alzheimer disease. *Neurology* **2005**, *65*, 545–551. [CrossRef]
12. Kivipelto, M.; Ngandu, T.; Fratiglioni, L.; Viitanen, M.; Kareholt, I.; Winblad, B.; Helkala, E.L.; Tuomilehto, J.; Soininen, H.; Nissinen, A. Obesity and vascular risk factors at midlife and the risk of dementia and Alzheimer disease. *Arch. Neurol.* **2005**, *62*, 1556–1560. [CrossRef]
13. Xiang, Q.; Andersen, S.L.; Sweigart, B.; Gunn, S.; Nygaard, M.; Perls, T.T.; Sebastiani, P. Signatures of Neuropsychological Test Results in the Long Life Family Study: A Cluster Analysis. *J. Alzheimer's Dis.* **2023**, *93*, 1457–1469. [CrossRef]
14. Ramoni, M.F.; Sebastiani, P.; Kohane, I.S. Cluster analysis of gene expression dynamics. *Proc. Natl. Acad. Sci. USA* **2002**, *99*, 9121–9126. [CrossRef] [PubMed]
15. Bondi, M.W.; Edmonds, E.C.; Jak, A.J.; Clark, L.R.; Delano-Wood, L.; McDonald, C.R.; Nation, D.A.; Libon, D.J.; Au, R.; Galasko, D.; et al. Neuropsychological criteria for mild cognitive impairment improves diagnostic precision, biomarker associations, and progression rates. *J. Alzheimer's Dis.* **2014**, *42*, 275–289. [CrossRef]
16. Edmonds, E.C.; McDonald, C.R.; Marshall, A.; Thomas, K.R.; Eppig, J.; Weigand, A.J.; Delano-Wood, L.; Galasko, D.R.; Salmon, D.P.; Bondi, M.W.; et al. Early versus late MCI: Improved MCI staging using a neuropsychological approach. *Alzheimer's Dement.* **2019**, *15*, 699–708. [CrossRef]
17. Jak, A.J.; Preis, S.R.; Beiser, A.S.; Seshadri, S.; Wolf, P.A.; Bondi, M.W.; Au, R. Neuropsychological Criteria for Mild Cognitive Impairment and Dementia Risk in the Framingham Heart Study. *J. Int. Neuropsychol. Soc.* **2016**, *22*, 937–943. [CrossRef] [PubMed]
18. Kulminski, A.M. Genetic Association Studies of Age-Related Traits: New Perspectives. *Adv. Geriatr. Med. Res.* **2021**, *3*, e210003. [CrossRef] [PubMed]
19. Woodward, A.A.; Urbanowicz, R.J.; Naj, A.C.; Moore, J.H. Genetic heterogeneity: Challenges, impacts, and methods through an associative lens. *Genet. Epidemiol.* **2022**, *46*, 555–571. [CrossRef]
20. Pedditzi, E.; Peters, R.; Beckett, N. The risk of overweight/obesity in mid-life and late life for the development of dementia: A systematic review and meta-analysis of longitudinal studies. *Age Ageing* **2016**, *45*, 14–21. [CrossRef]
21. Tolppanen, A.M.; Ngandu, T.; Kareholt, I.; Laatikainen, T.; Rusanen, M.; Soininen, H.; Kivipelto, M. Midlife and late-life body mass index and late-life dementia: Results from a prospective population-based cohort. *J. Alzheimer's Dis.* **2014**, *38*, 201–209. [CrossRef]

22. Kulminski, A.M.; Loika, Y.; Huang, J.; Arbeev, K.G.; Bagley, O.; Ukraintseva, S.; Yashin, A.I.; Culminskaya, I. Pleiotropic Meta-Analysis of Age-Related Phenotypes Addressing Evolutionary Uncertainty in Their Molecular Mechanisms. *Front. Genet.* **2019**, *10*, 433. [CrossRef]
23. Kulminski, A.M.; Huang, J.; Loika, Y.; Arbeev, K.G.; Bagley, O.; Yashkin, A.; Duan, M.; Culminskaya, I. Strong impact of natural-selection-free heterogeneity in genetics of age-related phenotypes. *Aging* **2018**, *10*, 492–514. [CrossRef]
24. Kulminski, A.M.; Loika, Y.; Nazarian, A.; Culminskaya, I. Quantitative and Qualitative Role of Antagonistic Heterogeneity in Genetics of Blood Lipids. *J. Gerontol. Ser. A Biol. Sci. Med. Sci.* **2020**, *75*, 1811–1819. [CrossRef]
25. Loika, Y.; Feng, F.; Loiko, E.; Kulminski, A.M. Mediation of the APOE associations with Alzheimer's and coronary heart diseases through body mass index and lipids. *GeroScience* **2022**, *44*, 1141–1156. [CrossRef] [PubMed]
26. Kulminski, A.M.; Loiko, E.; Loika, Y.; Culminskaya, I. Pleiotropic predisposition to Alzheimer's disease and educational attainment: Insights from the summary statistics analysis. *GeroScience* **2022**, *44*, 265–280. [CrossRef] [PubMed]
27. Kulminski, A.M.; Feng, F.; Loiko, E.; Nazarian, A.; Loika, Y.; Culminskaya, I. Prevailing Antagonistic Risks in Pleiotropic Associations with Alzheimer's Disease and Diabetes. *J. Alzheimer's Dis.* **2023**, *94*, 1121–1132. [CrossRef] [PubMed]
28. Sudlow, C.; Gallacher, J.; Allen, N.; Beral, V.; Burton, P.; Danesh, J.; Downey, P.; Elliott, P.; Green, J.; Landray, M.; et al. UK biobank: An open access resource for identifying the causes of a wide range of complex diseases of middle and old age. *PLoS Med.* **2015**, *12*, e1001779. [CrossRef]
29. Bycroft, C.; Freeman, C.; Petkova, D.; Band, G.; Elliott, L.T.; Sharp, K.; Motyer, A.; Vukcevic, D.; Delaneau, O.; O'Connell, J.; et al. The UK Biobank resource with deep phenotyping and genomic data. *Nature* **2018**, *562*, 203–209. [CrossRef]
30. Szustakowski, J.D.; Balasubramanian, S.; Kvikstad, E.; Khalid, S.; Bronson, P.G.; Sasson, A.; Wong, E.; Liu, D.; Wade Davis, J.; Haefliger, C.; et al. Advancing human genetics research and drug discovery through exome sequencing of the UK Biobank. *Nat. Genet.* **2021**, *53*, 942–948. [CrossRef]
31. Fisher, R.A.S. *Statistical Methods for Research Workers*, 14th ed.; Oliver and Boyd: Edinburgh, UK, 1970.
32. Leslie, R.; O'Donnell, C.J.; Johnson, A.D. GRASP: Analysis of genotype-phenotype results from 1390 genome-wide association studies and corresponding open access database. *Bioinformatics* **2014**, *30*, i185–i194. [CrossRef]
33. Sollis, E.; Mosaku, A.; Abid, A.; Buniello, A.; Cerezo, M.; Gil, L.; Groza, T.; Gunes, O.; Hall, P.; Hayhurst, J.; et al. The NHGRI-EBI GWAS Catalog: Knowledgebase and deposition resource. *Nucleic Acids Res.* **2023**, *51*, D977–D985. [CrossRef] [PubMed]
34. Bajic, V.P.; Su, B.; Lee, H.G.; Kudo, W.; Siedlak, S.L.; Zivkovic, L.; Spremo-Potparevic, B.; Djelic, N.; Milicevic, Z.; Singh, A.K.; et al. Mislocalization of CDK11/PITSLRE, a regulator of the G2/M phase of the cell cycle, in Alzheimer disease. *Cell. Mol. Biol. Lett.* **2011**, *16*, 359–372. [CrossRef] [PubMed]
35. Fredi, B.M.; De Labio, R.W.; Rasmussen, L.T.; Chagas, E.F.B.; Chen, E.S.; Turecki, G.; Smith, M.A.C.; Payao, S.L.M. CDK10, CDK11, FOXO1, and FOXO3 Gene Expression in Alzheimer's Disease Encephalic Samples. *Cell. Mol. Neurobiol.* **2023**, *43*, 2953–2962. [CrossRef]
36. Choi, H.H.; Choi, H.K.; Jung, S.Y.; Hyle, J.; Kim, B.J.; Yoon, K.; Cho, E.J.; Youn, H.D.; Lahti, J.M.; Qin, J.; et al. CHK2 kinase promotes pre-mRNA splicing via phosphorylating CDK11(p110). *Oncogene* **2014**, *33*, 108–115. [CrossRef] [PubMed]
37. Mendoza, J.; Sekiya, M.; Taniguchi, T.; Iijima, K.M.; Wang, R.; Ando, K. Global analysis of phosphorylation of tau by the checkpoint kinases Chk1 and Chk2 in vitro. *J. Proteome Res.* **2013**, *12*, 2654–2665. [CrossRef]
38. Guven, K.; Gunning, P.; Fath, T. TPM3 and TPM4 gene products segregate to the postsynaptic region of central nervous system synapses. *Bioarchitecture* **2011**, *1*, 284–289. [CrossRef]
39. Pelucchi, S.; Stringhi, R.; Marcello, E. Dendritic Spines in Alzheimer's Disease: How the Actin Cytoskeleton Contributes to Synaptic Failure. *Int. J. Mol. Sci.* **2020**, *21*, 908. [CrossRef]
40. Galloway, P.G.; Mulvihill, P.; Siedlak, S.; Mijares, M.; Kawai, M.; Padget, H.; Kim, R.; Perry, G. Immunochemical demonstration of tropomyosin in the neurofibrillary pathology of Alzheimer's disease. *Am. J. Pathol.* **1990**, *137*, 291–300.
41. Reumiller, C.M.; Schmidt, G.J.; Dhrami, I.; Umlauf, E.; Rappold, E.; Zellner, M. Gender-related increase of tropomyosin-1 abundance in platelets of Alzheimer's disease and mild cognitive impairment patients. *J. Proteom.* **2018**, *178*, 73–81. [CrossRef]
42. Castano, E.M.; Maarouf, C.L.; Wu, T.; Leal, M.C.; Whiteside, C.M.; Lue, L.F.; Kokjohn, T.A.; Sabbagh, M.N.; Beach, T.G.; Roher, A.E. Alzheimer disease periventricular white matter lesions exhibit specific proteomic profile alterations. *Neurochem. Int.* **2013**, *62*, 145–156. [CrossRef] [PubMed]
43. England, J.; Granados-Riveron, J.; Polo-Parada, L.; Kuriakose, D.; Moore, C.; Brook, J.D.; Rutland, C.S.; Setchfield, K.; Gell, C.; Ghosh, T.K.; et al. Tropomyosin 1: Multiple roles in the developing heart and in the formation of congenital heart defects. *J. Mol. Cell. Cardiol.* **2017**, *106*, 1–13. [CrossRef] [PubMed]
44. Brown, M.S.; Goldstein, J.L. Receptor-mediated endocytosis: Insights from the lipoprotein receptor system. *Proc. Natl. Acad. Sci. USA* **1979**, *76*, 3330–3337. [CrossRef]
45. Gouveia, C.; Gibbons, E.; Dehghani, N.; Eapen, J.; Guerreiro, R.; Bras, J. Genome-wide association of polygenic risk extremes for Alzheimer's disease in the UK Biobank. *Sci. Rep.* **2022**, *12*, 8404. [CrossRef]
46. Gallo, S.; Byham-Gray, L.; Duffy, V.B.; Hoffman, H.J.; Hayes, J.E.; Rawal, S. Associations of olfactory dysfunction with anthropometric and cardiometabolic measures: Findings from the 2013–2014 national health and nutrition examination survey (NHANES). *Physiol. Behav.* **2020**, *215*, 112702. [CrossRef] [PubMed]

47. Loika, Y.; Loiko, E.; Feng, F.; Stallard, E.; Yashin, A.I.; Arbeev, K.; Kuipers, A.L.; Feitosa, M.F.; Province, M.A.; Kulminski, A.M. Exogenous exposures shape genetic predisposition to lipids, Alzheimer's, and coronary heart disease in the MLXIPL gene locus. *Aging* **2023**, *15*, 3249–3272. [CrossRef]
48. Hao, K.; Di Narzo, A.F.; Ho, L.; Luo, W.; Li, S.; Chen, R.; Li, T.; Dubner, L.; Pasinetti, G.M. Shared genetic etiology underlying Alzheimer's disease and type 2 diabetes. *Mol. Asp. Med.* **2015**, *43–44*, 66–76. [CrossRef]
49. Hardy, J.; de Strooper, B.; Escott-Price, V. Diabetes and Alzheimer's disease: Shared genetic susceptibility? *Lancet Neurol.* **2022**, *21*, 962–964. [CrossRef]

Disclaimer/Publisher's Note: The statements, opinions and data contained in all publications are solely those of the individual author(s) and contributor(s) and not of MDPI and/or the editor(s). MDPI and/or the editor(s) disclaim responsibility for any injury to people or property resulting from any ideas, methods, instructions or products referred to in the content.

genes

MDPI

Article

Factors Associated with Large Cup-to-Disc Ratio and Blindness in the Primary Open-Angle African American Glaucoma Genetics (POAAGG) Study

Anusha Mamidipaka, Isabel Di Rosa, Roy Lee, Yan Zhu, Yineng Chen, Rebecca Salowe, Victoria Addis, Prithvi Sankar, Ebenezer Daniel, Gui-Shuang Ying and Joan M. O'Brien *

Department of Ophthalmology, Scheie Eye Institute, University of Pennsylvania, Philadelphia, PA 19104, USA; anusha.mamidipaka@pennmedicine.upenn.edu (A.M.); isabel.dirosa@pennmedicine.upenn.edu (I.D.R.); roy.lee@pennmedicine.upenn.edu (R.L.); yan.zhu1@pennmedicine.upenn.edu (Y.Z.); yineng.chen@pennmedicine.upenn.edu (Y.C.); rebecca.salowe@pennmedicine.upenn.edu (R.S.); victoria.addis@pennmedicine.upenn.edu (V.A.); prithvi.sankar@pennmedicine.upenn.edu (P.S.); ebdaniel@pennmedicine.upenn.edu (E.D.); gs.ying@pennmedicine.upenn.edu (G.-S.Y.)

* Correspondence: joan.o'brien@pennmedicine.upenn.edu

Abstract: Background/Aims: Primary open-angle glaucoma (POAG) disproportionately affects individuals of African ancestry. In these patients' eyes, a large cup-to-disc ratio (LCDR > 0.90) suggests greater retinal ganglion cell loss, though these patients often display varied visual ability. This study investigated the prevalence and risk factors associated with LCDR in African ancestry individuals with POAG and explored the differences between blind (>20/200) and not blind (≤20/200) LCDR eyes. Methods: A case–control methodology was used to investigate the demographic, optic disc, and genetic risk factors of subjects in the Primary Open-Angle African American Glaucoma Genetics Study. Risk factors were analyzed using univariable and multivariable logistic regression models with inter-eye correlation adjusted using generalized estimating equations. Results: Out of 5605 eyes with POAG, 1440 eyes (25.7%) had LCDR. In the multivariable analysis, LCDR was associated with previous glaucoma surgery (OR = 1.72), increased intraocular pressure (OR = 1.04), decreased mean deviation (OR = 1.08), increased pattern standard deviation (OR = 1.06), thinner retinal nerve fiber layer (OR = 1.05), nasalization of vessels (OR = 2.67), bayonetting of vessels (OR = 1.98), visible pores in the lamina cribrosa (OR = 1.68), and a bean-shaped cup (OR = 2.11). Of LCDR eyes, 30.1% were classified as blind (≤20/200). In the multivariable analysis, the statistically significant risk factors of blindness in LCDR eyes were previous glaucoma surgery (OR = 1.72), increased intraocular pressure (OR = 1.05), decreased mean deviation (OR = 1.04), and decreased pattern standard deviation (OR = 0.90). Conclusions: These findings underscore the importance of close monitoring of intraocular pressure and visual function in African ancestry POAG patients, particularly those with LCDR, to preserve visual function.

Keywords: primary open-angle glaucoma POAG; cup-to-disc ratio; vision loss; African American; neurodegeneration

Citation: Mamidipaka, A.; Di Rosa, I.; Lee, R.; Zhu, Y.; Chen, Y.; Salowe, R.; Addis, V.; Sankar, P.; Daniel, E.; Ying, G.-S.; et al. Factors Associated with Large Cup-to-Disc Ratio and Blindness in the Primary Open-Angle African American Glaucoma Genetics (POAAGG) Study. *Genes* **2023**, *14*, 1809. https://doi.org/10.3390/genes14091809

Academic Editor: Claudia Ricci

Received: 11 August 2023
Revised: 14 September 2023
Accepted: 15 September 2023
Published: 16 September 2023

Copyright: © 2023 by the authors. Licensee MDPI, Basel, Switzerland. This article is an open access article distributed under the terms and conditions of the Creative Commons Attribution (CC BY) license (https://creativecommons.org/licenses/by/4.0/).

1. Introduction

Primary open-angle glaucoma (POAG), the most common form of glaucoma, is an optic nerve degeneration that is the leading cause of irreversible blindness worldwide [1]. POAG disproportionately affects African ancestry individuals; these individuals are over five times more likely to be diagnosed with POAG and up to 15 times more likely to experience glaucomatous vision loss from the disease than individuals of European descent [2]. Cup-to-disc ratio (CDR), defined as the ratio of the optic cup diameter to the optic disc diameter, is a commonly used measurement to assess and track the progression of glaucomatous damage to the optic nerve. In terms of its diagnostic power, CDR has been considered one of "the most valuable optic disc variables for early detection of glaucomatous nerve damage" [3]. Cupping often occurs in POAG due to progressive neuroretinal rim narrowing [4] and mechanical stress on the lamina cribrosa caused by elevated intraocular pressure (IOP) [5]. Vertical CDR (VCDR), which is measured on the vertical meridian, has been shown to have a strong genetic correlation with POAG susceptibility [6]. A large VCDR (LCDR) suggests greater ganglion cell loss due to displacement of the lamina cribrosa and loss of axon fibers [4,7,8]. This axonal compromise, in addition to the mechanical failure of connective tissue in the lamina cribrosa, peripapillary sclera, and scleral canal wall, is a central pathophysiology that underlies glaucomatous damage [5].

Current studies on LCDR are limited, despite its evaluative importance in POAG and other ophthalmic diseases. Existing studies in European and Asian ancestry populations have found that risk factors for LCDR include higher IOP [6], greater lamina cribrosa depth [9], older age, and male sex [10]. While several studies have identified a significant correlation between LCDR and visual field loss in POAG subjects [11–15], others have found no such correlation, suggesting that intrasubject asymmetry in central corneal thickness (CCT) explains the varying levels of visual field damage in glaucoma patients [16]. Another study found no significant difference in CDR among POAG groups with visual field progression and without progression [17]. Interestingly, this suggests that, despite having axonal loss and displacement of the lamina cribrosa due to cupping, at least a subset of patients with LCDR maintain normal visual outcomes that do not progress to blindness. However, there is a lack of literature regarding why some eyes with LCDR maintain normal vision while others progress to blindness, especially in the overaffected African ancestry population.

In this paper, we examined the risk factors for presentation with LCDR and subsequent progression to visual impairment or blindness in a large cohort of African ancestry individuals. These individuals were recruited from the Primary Open-Angle African American Glaucoma Genetics (POAAGG) study. The goal of this paper was to understand the demographic, optic disc, phenotypic, and genetic characteristics that predispose these individuals to extreme cupping, and, among those with LCDR, to determine the risk factors for blindness. A better understanding of these risk factors can help to inform decisions on treatment escalation among individuals presenting with LCDR.

2. Materials and Methods

2.1. Study Design and Population

A case–control methodology is used in the POAAGG study. The POAAGG study cohort (details described elsewhere) consists of 10,255 individuals of African descent recruited from the greater Philadelphia area [18]. Each enrolled subject was classified by a glaucoma specialist or ophthalmologist as a case, control, or suspect based on detailed criteria. Each subject signed an informed consent form and provided a DNA sample. The degree of each subject's African ancestry was quantified using the fastSTRUCTURE program, described elsewhere [19]. The Declaration of Helsinki was followed, and research was carried out with Institutional Review Board clearance from the University of Pennsylvania.

2.2. Demographic Characteristics

Based on previously recognized criteria, a glaucoma expert categorized participants as POAG cases, POAG suspicions, or controls [18]. Participants who also had an eye injury history, non-glaucomatous optic disc neuropathy, inflammatory eye disorders, Grave's disease with ocular symptoms, or any type of glaucoma other than POAG were excluded. At enrollment, Clinical Research Coordinators performed standardized interviews to gather data on systemic illness, behavior, and demographics. Demographic data included age, gender, self-described race, usage of cigarettes and alcohol, hypertension, and diabetes. Electronic medical records were evaluated to determine previous glaucoma surgery history to supplement the onsite examination data.

2.3. Quantitative Phenotypic Characteristics

At enrollment, ocular phenotypic data for subjects were collected from both eyes, including IOP, CDR, CCT, visual acuity (VA), retinal nerve fiber layer (RNFL) thickness, mean deviation (MD), and pattern standard deviation (PSD). Eyes with very low vision (less than 20/200) were assigned the following logMAR values: 2.2 for counting fingers (CF), 2.3 for hand motion (HM), 2.4 for light perception (LP), and 2.5 for no light perception (NLP).

2.4. Qualitative Phenotypic Characteristics

Baseline 30-degree color stereo disc photographs of POAAGG subjects were collected using the Topcon TRC 50EX retinal camera (Topcon Corp. of America, Paramus, NJ, USA) and were uploaded onto a secure server at the Scheie Image Reading Center. Three non-physician graders were trained by glaucoma specialists to grade digital stereo color images of the optic disc using a stereo viewer (Screen-Vu stereoscope) [20]. Using a standardized grading form, two readers independently graded quantitative and qualitative features of the optic disc, and differences were adjudicated by the ophthalmologist director of the Reading Center. The qualitative optic disc characteristics measured were the presence of beta peripapillary atrophy, stereoscopically identified disc tilt, disc hemorrhage, arteriole narrowing, venule narrowing, baring of the lamina cribrosa, nasalization of the vessels, cilioretinal vessels, grey crescent, disc pallor, and visible pores in the lamina cribrosa. Additionally, graders characterized the disc shape as round, oval, or other and the cup shape as conical, cylindrical, or bean pot.

2.5. Definition of LCDR

This study included eyes from POAG cases. Eyes with severe optic nerve cupping (VCDR: 0.90–1) were defined as having an LCDR. The maximum VCDR value was determined during clinical eye examinations by glaucoma specialists. A typical retinal fundus photograph of the patients with LCDRs and those of the healthy control is provided in Figure 1.

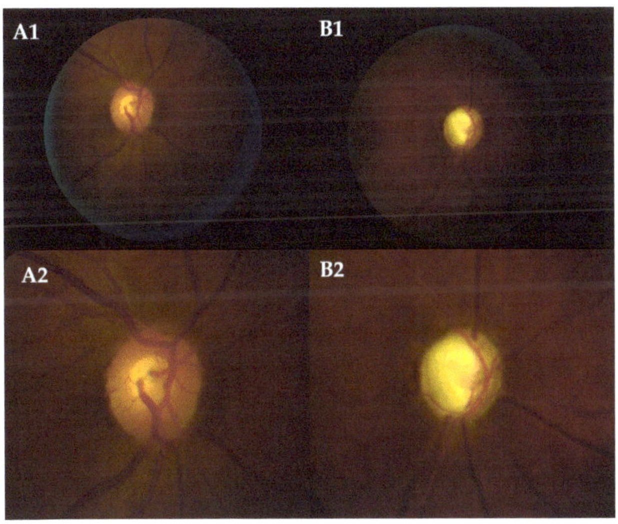

Figure 1. Funduscopic color images of the retinal with examples of a healthy control and LCDR subject. (**A1**) Right eye shows the phenotypic features of the retina of a healthy control subject with a normal optic disc CDR of 0.35. (**A2**) Close-up image of (**A1**) demonstrates the distinction between optic cup and disc use to calculate a CDR of 0.35. (**B1**) Right eye shows the phenotypic features of a retina of a case subject with an optic disc LCDR of 1. (**B2**) Close-up image of the central features of (**B1**) used to calculate a CDR of 1.

2.6. Genotype

We previously conducted a genome-wide association study (GWAS) for POAG in 11,275 individuals of African ancestry (6003 cases, 5272 controls), including the African Descent and Glaucoma Evaluation Study (n = 1999) [21], the Genetics of Glaucoma in People of African Descent (GGLAD) consortium (n = 2952) [22], and the POAAGG study (n = 6324). We discovered 46 risk loci associated with POAG at genome-wide significance in a discovery mega-analysis, with replication analyses, trait colocalization analyses, and functional studies implicating three likely causal loci: rs1666698 mapping to *DBF4P2*, rs34957764 mapping to *ROCK1P1*, and rs11824032 mapping to *ARHGEF12*. We selected these three SNPs and tested their association with LCDR and with blindness in this study. Each variant was stratified into three groups based on the number of risk alleles (0, 1, and 2) present. Generalized estimating equations accounted for the inter-eye correlation.

2.7. Statistical Analysis

We performed a descriptive analysis for risk factors (demographic, phenotypic, and optic disc) using mean (SD) for continuous factors and using proportion for categorical factors. We performed the risk factor analysis for LCDR and blindness using univariable and multivariable logistic regression models. Generalized estimating equations accounted for the inter-eye correlation. The initial univariate analysis included all risk factors with a p-value < 0.2, and backward variable selection was used to reach the final multivariate analysis that retained only risk factors with $p < 0.05$. All statistical analyses were performed in R version 4.03 and SAS v9.4 (SAS Institute Inc., Cary, NC, USA).

3. Results

A total of 5605 eyes from POAAGG cases met eligibility criteria for this study, with 1440 eyes (25.7%) meeting the criteria for LCDR. Univariate analyses of demographic and clinical (Table S1), optic disc (Table S2), and phenotypic (Table S3) characteristics of LCDR eyes were conducted.

A multivariate analysis of risk factors for the development of LCDRs was also conducted (Table 1). Out of 5605 eyes, 1890 eyes were included in the final multivariable model due to missing data in the univariate variables: previous glaucoma surgery (n = 350), IOP (n = 11), MD (n = 1392), PSD (n = 1384), RNFL (n = 1335), VA (n = 674), shape of cup (n = 2322), vessel bayonetting (n = 2310), nasalization of the vessels (n = 2203), and visible pores in lamina cribrosa (n = 2306).

Table 1. Multivariable analysis for association of LCDR ratio with demographic, phenotypic, and optic disc characteristics (n = 1890 eyes).

Source	Number of Eyes	LCDR, n (%)	Odds Ratio [OR] (95% CI)	p-Value
Previous glaucoma surgery				
No	1412	257(18.2%)	Ref	0.002
Yes	478	177 (37.0%)	1.72 (1.23–2.42)	
IOP (per 1 mmHg increase)	1890		1.04 (1.02–1.06)	0.0006
MD (per 1 unit decrease)	1890		1.08 (1.06–1.11)	<0.0001
PSD (per 1 unit increase)	1890		1.06 (1.01–1.12)	0.02
VA (per 1 logMAR unit increase)	1890		1.88 (1.32–2.65)	0.0004
RNFL (per 1 μm decrease)	1890		1.05 (1.04–1.07)	<0.0001
Visible pores in lamina cribrosa				
No	671	84 (12.52%)	Ref	0.004
Yes	1219	350 (28.7%)	1.68 (1.18–2.40)	

Table 1. *Cont.*

Source	Number of Eyes	LCDR, *n* (%)	Odds Ratio [OR] (95% CI)	*p*-Value
Vessel bayonetting				
No	1234	195 (15.8%)	Ref	<0.0001
Yes	656	239 (36.4%)	1.98 (1.44–2.74)	
Nasalization of the vessels				
No	1181	161 (13.6%)	Ref	<0.0001
Yes	709	273 (38.5%)	2.67 (1.92–3.73)	
Shape of cup				
Conical	770	148 (19.2%)	Ref	<0.0001
Cylindrical	863	148 (17.2%)	0.68 (0.48–0.97)	
Bean Pot	248	136 (54.8%)	2.11 (1.35–3.31)	
Others	9	2 (22.2%)	1.55 (0.34–6.96)	

LCDR was significantly associated with previous glaucoma surgery (odds ratio (OR) 1.72, 95% CI 1.23–2.42, $p = 0.002$), 1 mmHg increase in IOP (OR 1.04, 95% CI 1.02–1.06, $p = 0.0006$), vessel bayonetting (OR 1.98, 95% CI 1.44–2.74, $p < 0.0001$), vessel nasalization (OR 2.67, 95% CI 1.92–3.73, $p < 0.0001$), visible pores in the lamina cribrosa (OR 1.68, 95% CI 1.18–2.40, $p < 0.0001$), bean = pot-shaped cup (OR 2.11, 95% CI 1.35–3.31, $p < 0.0001$), 1 unit decrease in MD (OR 1.08, 95% CI 1.06–1.11, $p < 0.0001$), 1 unit increase in PSD (OR 1.06, 95% CI 1.01–1.12, $p = 0.02$), worse VA (1 logMAR unit increase, OR 1.88, 95% CI 1.32–2.65, $p = 0.0004$), and 1 µm decrease in RNFL thickness (OR 1.05, 95% CI 1.04–1.07, $p < 0.0001$).

The univariable analysis of genetic variants and LCDR in glaucoma cases (Table 2) did not unveil significant associations between the studied SNPs and LCDR.

Table 2. Univariable analysis for association of three genetic variants and LCDR among POAAGG glaucoma cases ($n = 3242$ eyes).

Variants	No. of ALT	LCDR Eyes	Non-LCDR Eyes	OR (95% CI)	*p*-Value
rs1666698_ALT	0	24 (16%)	124 (84%)	Ref	-
	1	254 (19%)	1074 (81%)	1.22 (0.78, 1.92)	0.383
	2	370 (21%)	1396 (79%)	1.36 (0.88, 2.14)	0.166
rs11824032_ALT	0	355 (19%)	1488 (81%)	Ref	-
	1	262 (21%)	968 (79%)	1.13 (0.95, 1.35)	0.154
	2	31 (18%)	138 (82%)	0.94 (0.63, 1.42)	0.773
rs34957764_ALT	0	585 (20%)	2386 (80%)	Ref	-
	1	63 (23%)	208 (77%)	1.24 (0.93, 1.65)	0.152
	2	0	0	-	-

To further investigate why a subset of eyes with LCDR progress to blindness while others maintain normal VA, we conducted additional analyses to compare the two groups. Among the 1440 LCDR eyes, 186 eyes were excluded due to missing VA values. Of the 1254 eyes analyzed, 448 had normal VA (with worst VA $\leq 20/40$), 429 had impaired VA (VA between 20/40–20/200), and 377 were blind (VA > 20/200) (Figure 2). To simplify the univariate and multivariate analyses of risk factors for blindness, the eyes of the normal VA and impaired VA groups were combined into one group ($n = 877$) and compared to the blind group ($n = 377$).

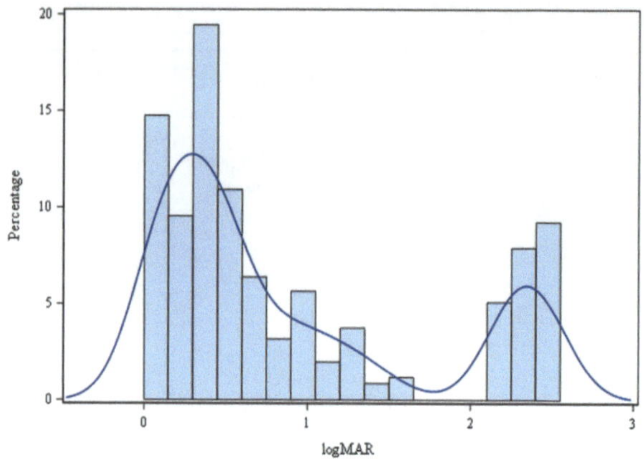

Figure 2. Distribution of VA in LCDR eyes (n = 1254). LogMAR = logarithm of the minimum angle of resolution.

A univariate analysis of demographic characteristics (Table S4), optic disc (Table S5), and phenotypic (Table S6) characteristics of LCDR eyes predicting blindness was conducted. A multivariate analysis of risk factors for the development of blindness among eyes with LCDR was also conducted. Out of 1254 eyes, 440 eyes were excluded due to missing data in previous glaucoma surgery (n = 79), MD (n = 366), and PSD (n = 366). In this multivariate analysis, previous glaucoma surgery (OR 1.72, 95% CI 1.17–2.53, p = 0.0062), a 1 mmHg increase in IOP (OR 1.05, 95% CI 1.03–1.06, p < 0.0001), 1 unit decrease in MD (OR 1.04, 95% CI 1.02–1.07, p < 0.0001), and 1 unit increase in PSD (OR 0.90, 95% CI 0.85–0.95, p = 0.002) were significantly associated with blindness (Table 3).

Table 3. Multivariate analysis for associations of blindness with phenotypic, demographic, and optic disc characteristics in eyes with LCDR (n = 814 eyes).

Source	Number of Eyes	Blind, n (%)	Odds Ratio	p-Value
IOP (per 1 mmHg increase)	814	384 (59.6%)	1.05 (1.03–1.06)	<0.0001
MD (per 1 unit decrease)	814	384 (59.6%)	1.04 (1.02–1.07)	<0.0001
PSD (per 1 unit increase)	814	384 (59.6%)	0.90 (0.85–0.95)	0.0002
Previous glaucoma surgery				
No	437	75 (17.2%)	Ref	0.0062
Yes	377	109 (28.9%)	1.72 (1.17–2.53)	

The univariable genetic analysis (Table 4) did not reveal significant associations between the studied genetic variants and blindness among LCDR eyes.

Table 4. Univariable analysis for association of three genetic variants and blind (LCDR eyes with impaired vision) vs. non-blind (LCDR eyes with normal vision) among POAAGG glaucoma cases (n = 1050 eyes).

Variants	No. of ALT	Blind	Non-Blind	OR (95% CI)	p-Value
rs1666698_ALT	0	11 (22%)	38 (78%)	Ref	-
	1	97 (23%)	316 (77%)	1.06 (0.53, 2.13)	0.870
	2	153 (26%)	435 (74%)	1.22 (0.62, 2.40)	0.573

Table 4. *Cont.*

Variants	No. of ALT	Blind	Non-Blind	OR (95% CI)	p-Value
rs11824032_ALT	0	148 (25%)	437 (75%)	Ref	-
	1	96 (23%)	313 (77%)	0.91 (0.69, 1.19)	0.478
	2	17 (30%)	39 (70%)	1.29 (0.72, 2.31)	0.399
rs34957764_ALT	0	238 (25%)	710 (75%)	Ref	-
	1	23 (23%)	79 (77%)	0.87 (0.54, 1.39)	0.557
	2	0	0	-	-

4. Discussion

This study aimed to characterize risk factors associated with the development of LCDR and to assess predictors of blindness among African ancestry individuals with POAG. Among the African ancestry cohort, we found 26.5% of eyes met the criteria for LCDR (CDR > 0.9). LCDR was significantly correlated with worse overall visual field presentations, worse VA, higher IOP, previous glaucoma surgery, thinner RNFL, presence of visible pores in the lamina cribrosa, and vessel bayonetting and nasalization. In total, 377 (30%) of the 1254 LCDR eyes were blind (VA < 20/200). Among LCDR eyes, blindness was associated with previous glaucoma surgery, higher IOP, and worse visual field presentation characterized by higher MD.

The prevalence and risk factors associated with LCDR differ among individuals of African, Asian, and European ancestries. According to studies, non-Hispanic White individuals with glaucoma had smaller CDRs, thicker CCT, and less severe visual field defects compared to other ethnicities [23,24]. Conversely, on average, African Americans have a larger CDR compared to White populations [25,26]. The African Descent and Glaucoma Evaluation Study found that African-descent individuals have greater optic disc areas than European-descent individuals [27]. In fact, studies have found optic discs to be significantly smaller in White and Hispanic populations compared to Asian and African-ancestry ones [28,29]. In studies on Asian populations, the risk factors for LCDR were found to be higher IOP, lower diastolic and higher systolic blood pressure, higher AST/ALT ratio, longer axial length, lower BMI, older age, male sex, and previous cataract surgery [30,31]. Notably, higher IOP emerged as the most significant determinant of a LCDR, similar to our findings in an African ancestry population [31].

In our study, eyes with LCDR were more likely to experience elevated IOP, which is a major risk factor for POAG [32,33]. Studies have consistently found a significant correlation between higher IOP and LCDR, an indication of more severe glaucomatous damage [6,30,34]. Furthermore, research has shown that both IOP and VCDR have significant heritability and are genetically correlated with each other, as well as with the risk of POAG [6,34,35]. Studies in monkey models revealed that artificially increasing IOP, quantified as the Pressure Insult, subsequently caused an increase in CDR [36,37]. Values of Pressure Insult above 11 mmHg·Days/Day were associated with significant cupping, emphasizing the role of elevated IOP in CDR changes [37]. The study also showed that trabeculectomy to lower Pressure Insult below this threshold was correlated with slowing the rate of CDR progression [37].

In addition to studying IOP and other variables, we examined the relationship between age and LCDR in the POAG population. Although the univariate analysis initially suggested a correlation between older age and LCDR eyes with African ancestry, our multivariate analysis could not confirm this independent association when considering other factors. Thinning of the RNFL with age can influence CDR, as it contributes to structural changes in the optic nerve [38]. Additionally, the development of LCDR can be affected by age-related changes in ocular biomechanics, such as progressive thinning of the outer plexiform layer and increased collagen production leading to tissue stiffness and elevated IOP, as revealed by glaucoma mouse models [39–41]. While some studies suggest an age-related decline in neuroretinal rim areas [42], the Rotterdam Study [43] and Singapore Malay Study [31] found no progressive age-related decline, similar to the findings in our study. These nonpathological aging-related changes that affect CDR may contribute to the complexity of interpreting the sole effects of glaucomatous aging-related changes on CDR.

In our study, as in prior studies in different populations, LCDR was also associated with lower MD, higher PSD, and RNFL thinning [35,44]. Visible pores on lamina cribrosa, vessel bayonetting, and nasalization of vessels were also associated with LCDR in our multivariate analysis. Visible pores on the lamina cribrosa may be related to biomechanics-driven lamina cribrosa remodeling and displacement in response to elevated IOP [9,45]. Examinations of glaucomatous laminar pores in both humans and non-human primates have identified the replacement of RGC axons with irregular deposits of extracellular collagen and basement membrane components, clinically observable as optic cupping [46–48]. This association between lamina cribrosa pore abnormalities and LCDR highlights the significance of structural changes in the lamina cribrosa in understanding glaucoma progression.

As the optic disc becomes increasingly excavated, the lamina cribrosa may become more susceptible to deformation and displacement, leading to compression and distortion of vessels passing through the optic nerve head (ONH), resulting in vessel bayonetting and nasalization of vessels [49]. A study found that nasalization of the central retinal vessel trunk by 60% or more was significantly associated with glaucoma conversion, indicating progression of disease [49]. In a monkey model of chronic open-angle glaucoma, glaucomatous damage was simulated by increasing IOP. As cupping increased, retinal blood vessels shifted position, particularly moving from far from the ONH edge onto ONH in response to glaucoma damage, possibly further contributing to LCDR [50].

Past studies have shown that while some eyes with LCDRs maintain normal vision, others progress to blindness [17,51,52]. As expected, our multivariate analysis revealed that elevated IOP was strongly predictive of blindness in LCDR eyes. Although individuals with initially normal-range IOPs (between 10 and 20 mmHg) can still develop glaucoma, the Collaborative Normal Tension Glaucoma Trial revealed that lowering IOP by 30% through medical or surgical treatment resulted in reduced rates of glaucoma progression [36]. In African ancestry populations, higher IOP may have more damaging effects on the ONH due to larger optic disc diameters associated with increased vulnerability to pressure-induced deformation [53]. One study found that chronically elevated IOP demonstrated significant axonal dephosphorylation in the retina, ONH, and even the optic chiasm in animal models [36]. This dephosphorylation parallels glaucoma-induced axonal damage, revealing an increase in CDR in all eyes studied and potentially contributing to vision impairment [36]. Additionally, elevated IOP induces cellular stress, leading to retinal astrocyte activation and hypertrophy [54,55]. These hypertrophied astrocytes, observed in experimental rat models, potentially cause hypoxia, which impede the optic axons' function and contribute to LCDR and optic nerve damage [54].

In our study, LCDR eyes with poor VA were also more likely to have lower (worse) MD, which is associated with faster deterioration rates [56,57]. Paradoxically, lower PSD was associated with blindness. A possible explanation is that, despite the initial gradual elevation of PSD values due to uneven loss of sensitivity in the visual field, advanced glaucoma leads to a general decrease in sensitivity, resulting in lower PSD values overall [58]. Therefore, patients with high IOP and poor visual field measures should be closely monitored for disease progression and considered for more aggressive treatment to prevent visual disability.

Interestingly, previous glaucoma surgery was associated with poor VA in LCDR eyes. However, it should be noted that surgical options are typically only considered in severe glaucoma cases when medical and laser treatments fail to achieve IOP reduction, with the goal of preventing further vision loss [59]. Additionally, surgical interventions can result in complications such as choroidal effusion, shallow anterior chamber, and persistent corneal edema with frequent need for post-operative interventions [60]. Although permanent vision loss, known as the "snuff-out phenomenon", after trabeculectomy typically occurs in only 2% of patients, various degrees of transient vision loss are reported in up to 50% of patients, and it can take up to two years to recover completely [61].

The genetic analysis for LCDR did not unveil significant associations between the three studied genetic variants and LCDR or blindness among LCDR eyes, indicating that these specific genetic variants might not substantially contribute to the LCDR trait or blindness in POAG. Further studies with larger sample sizes or explorations of other potential risk genes are warranted to better elucidate the genetic underpinnings of VA in LCDR POAG eyes.

This study has some limitations that should be considered. The sample consisted of cross-sectional African American patients from the Philadelphia area, which may limit the generalizability of the findings to the national African American population due to referral bias and geographic isolation. While optic nerve parameters were determined by trained graders with a robust adjudication process, there is a possibility of subjectivity in grading. Moreover, patients with unavailable or low-quality images were excluded from the analysis, introducing potential for selection bias. This study also did not consider the interaction between social, demographic, and biological factors affecting health outcomes. These confounding factors could contribute to elevated risk of poor VA.

In conclusion, in our study, 30% of African ancestry eyes with LCDR progressed to blindness. The strongest risk factors for blindness in African ancestry eyes with LCDR were previous glaucoma surgery, higher IOP, and worse visual field presentation. African ancestry individuals are more likely to experience glaucomatous visual impairment compared to European ancestry individuals, and although management of POAG includes various therapies to reduce IOP, many patients experience substantial limitations in visual function and quality of life despite all treatments. Understanding genetic, demographic, and optic disc phenotypic risk factors for developing poor VA, particularly in patients with LCDRs, is crucial in preventing irreversible blindness.

Supplementary Materials: The following supporting information can be downloaded at https://www.mdpi.com/article/10.3390/genes14091809/s1, Supplemental Table S1: Univariate analysis of demographic and clinical characteristics of eyes with and without LCDR; Supplemental Table S2: Univariate analysis of optic disc characteristics of eyes with and without LCDR; Supplemental Table S3: Univariate analysis of phenotypic characteristics between eyes with and without LCDR; Supplemental Table S4: Univariable analysis for comparison of demographic characteristics between non-blind eyes vs. blind eyes among LCDR eyes; Supplemental Table S5: Univariable analysis for comparison of optic disc characteristics between non-blind eyes vs. blind eyes among LCDR eyes; Supplemental Table S6: Univariable analysis for comparison of phenotypic characteristics between non-blind eyes vs. blind eyes among LCDR eyes.

Author Contributions: Conceptualization, E.D., A.M. and I.D.R.; Methodology, J.M.O., V.A., P.S., G.-S.Y., E.D. and R.L.; Software, R.L.; Validation, J.M.O.; Formal Analysis, R.L., Y.Z., Y.C. and G.-S.Y.; Investigation, A.M., I.D.R. and R.S.; Resources, J.M.O.; Data Curation, R.L. and Y.Z.; Writing—Original Draft Preparation, A.M. and I.D.R.; Writing—Review and Editing, A.M., I.D.R. and R.S.; Visualization, Y.C.; Supervision, R.S. and J.M.O.; Project Administration, J.M.O. All authors have read and agreed to the published version of the manuscript.

Funding: The Primary Open-Angle African American Glaucoma Genetics (POAAGG) study was supported by the National Eye Institute, Bethesda, Maryland (grant #1R01EY023557-01), and Vision Research Core Grant (P30 EY001583). Funds also come from the F.M. Kirby Foundation (N/A), Research to Prevent Blindness (N/A), The UPenn Hospital Board of Women Visitors (N/A), and The Paul and Evanina Bell Mackall Foundation Trust (N/A). The sponsor or funding organization had no role in the design or conduct of this research.

Institutional Review Board Statement: The study was conducted in accordance with the Declaration of Helsinki, and approved by the Institutional Review Board of the University of Pennsylvania (protocol code #812036, continuing review approved on 12 March 2021).

Informed Consent Statement: The study protocol and the consent statement were approved by the University of Pennsylvania IRB (IRB protocol #812036), and the research adhered to the tenets of the Declaration of Helsinki. Informed consent was obtained from all subjects involved in the study.

Data Availability Statement: The data presented in this study are available on request from the corresponding author. The data are not publicly available due to their containing information that could compromise the privacy of research participants.

Conflicts of Interest: There are no competing interest for any author.

References

1. Bourne, R.; Steinmetz, J.D.; Flaxman, S.; Briant, P.S.; Taylor, H.R.; Resnikoff, S.; Casson, R.J.; Abdoli, A.; Abu-Gharbieh, E.; Afshin, A.; et al. Trends in Prevalence of Blindness and Distance and near Vision Impairment over 30 Years: An Analysis for the Global Burden of Disease Study. *Lancet Glob. Health* **2021**, *9*, e130–e143. [CrossRef] [PubMed]
2. Muñoz, B. Causes of Blindness and Visual Impairment in a Population of Older Americans: The Salisbury Eye Evaluation Study. *Arch. Ophthalmol.* **2000**, *118*, 819. [CrossRef] [PubMed]
3. Jonas, J.B.; Bergua, A.; Schmitz-Valckenberg, P.; Papastathopoulos, K.I.; Budde, W.M. Ranking of Optic Disc Variables for Detection of Glaucomatous Optic Nerve Damage. *Investig. Ophthalmol. Vis. Sci.* **2000**, *41*, 1764–1773.
4. Weinreb, R.N.; Leung, C.K.S.; Crowston, J.G.; Medeiros, F.A.; Friedman, D.S.; Wiggs, J.L.; Martin, K.R. Primary Open-Angle Glaucoma. *Nat. Rev. Dis. Primers* **2016**, *2*, 16067. [CrossRef]
5. Burgoyne, C.F.; Crawford Downs, J.; Bellezza, A.J.; Francis Suh, J.-K.; Hart, R.T. The Optic Nerve Head as a Biomechanical Structure: A New Paradigm for Understanding the Role of IOP-Related Stress and Strain in the Pathophysiology of Glaucomatous Optic Nerve Head Damage. *Prog. Retin. Eye Res.* **2005**, *24*, 39–73. [CrossRef]
6. Charlesworth, J.; Kramer, P.L.; Dyer, T.; Diego, V.; Samples, J.R.; Craig, J.E.; Mackey, D.A.; Hewitt, A.W.; Blangero, J.; Wirtz, M.K. The Path to Open-Angle Glaucoma Gene Discovery: Endophenotypic Status of Intraocular Pressure, Cup-to-Disc Ratio, and Central Corneal Thickness. *Investig. Ophthalmol. Vis. Sci.* **2010**, *51*, 3509. [CrossRef]
7. Quigley, H.A.; Green, W.R. The Histology of Human Glaucoma Cupping and Optic Nerve Damage: Clinicopathologic Correlation in 21 Eyes. *Ophthalmology* **2020**, *127*, S45–S69. [CrossRef]
8. Waisberg, E.; Micieli, J.A. Neuro-Ophthalmological Optic Nerve Cupping: An Overview. *Eye Brain* **2021**, *13*, 255–268. [CrossRef]
9. Jung, K.I.; Jeon, S.; Park, C.K. Lamina Cribrosa Depth Is Associated with the Cup-to-Disc Ratio in Eyes with Large Optic Disc Cupping and Cup-to-Disc Ratio Asymmetry. *J. Glaucoma* **2016**, *25*, e536–e545. [CrossRef]
10. Kim, Y.J.; Kim, J.M.; Shim, S.H.; Bae, J.H.; Park, K.H. Epidemiologic Survey Committee of the Korean Ophthalmological Society Associations between Optic Cup-to-Disc Ratio and Systemic Factors in the Healthy Korean Population. *Korean J. Ophthalmol.* **2015**, *29*, 336–343. [CrossRef]
11. Jonas, J.B.; Gusek, G.C.; Naumann, G.O.H. Optic Disc Morphometry in Chronic Primary Open-Angle Glaucoma: II. Correlation of the Intrapapillary Morphometric Data to Visual Field Indices. *Graefe's Arch. Clin. Exp. Ophthalmol.* **1988**, *226*, 531–538. [CrossRef] [PubMed]
12. Cullinane, A.B. Optic Disc Cup Slope and Visual Field Indices in Normal, Ocular Hypertensive and Early Glaucomatous Eyes. *Br. J. Ophthalmol.* **2002**, *86*, 555–559. [CrossRef] [PubMed]
13. Araie, M.; Sekine, M.; Suzuki, Y.; Koseki, N. Factors Contributing to the Progression of Visual Field Damage in Eyes with Normaltension Glaucoma. *Ophthalmology* **1994**, *101*, 1440–1444. [CrossRef] [PubMed]
14. Siesky, B.; Wentz, S.M.; Januleviciene, I.; Kim, D.H.; Burgett, K.M.; Verticchio Vercellin, A.C.; Rowe, L.W.; Eckert, G.J.; Harris, A. Baseline Structural Characteristics of the Optic Nerve Head and Retinal Nerve Fiber Layer Are Associated with Progressive Visual Field Loss in Patients with Open-Angle Glaucoma. *PLoS ONE* **2020**, *15*, e0236819. [CrossRef] [PubMed]
15. Lee, Y.-A.; Shih, Y.-F.; Lin, L.L.-K.; Huang, J.-Y.; Wang, T.-H. Association between High Myopia and Progression of Visual Field Loss in Primary Open-Angle Glaucoma. *J. Formos. Med. Assoc.* **2008**, *107*, 952–957. [CrossRef] [PubMed]
16. Yazdani, S.; Doozandeh, A.; Haghighat, M.; Akbarian, S.; Pakravan, M.; Yaseri, M. Intrasubject Difference in CCT among POAG versus Normal Individuals. *Optom. Vis. Sci.* **2015**, *92*, 879–883. [CrossRef]
17. Kitaoka, Y.; Tanito, M.; Yokoyama, Y.; Nitta, K.; Katai, M.; Omodaka, K.; Nakazawa, T. A Small Disc Area Is a Risk Factor for Visual Field Loss Progression in Primary Open-Angle Glaucoma: The Glaucoma Stereo Analysis Study. *J. Ophthalmol.* **2018**, *2018*, 8941489. [CrossRef]
18. Charlson, E.S.; Sankar, P.S.; Miller-Ellis, E.; Regina, M.; Fertig, R.; Salinas, J.; Pistilli, M.; Salowe, R.J.; Rhodes, A.L.; Merritt, W.T.; et al. The Primary Open-Angle African American Glaucoma Genetics Study. *Ophthalmology* **2015**, *122*, 711–720. [CrossRef]
19. Cole, B.S.; Gudiseva, H.V.; Pistilli, M.; Salowe, R.; McHugh, C.P.; Zody, M.C.; Chavali, V.R.M.; Ying, G.S.; Moore, J.H.; O'Brien, J.M. The Role of Genetic Ancestry as a Risk Factor for Primary Open-Angle Glaucoma in African Americans. *Investig. Ophthalmol. Vis. Sci.* **2021**, *62*, 28. [CrossRef]
20. Addis, V.; Oyeniran, E.; Daniel, E.; Salowe, R.; Zorger, R.; Lee, R.; Pistilli, M.; Maguire, M.; Cui, Q.; Miller-Ellis, E.; et al. Non-Physician Grader Reliability in Measuring Morphological Features of the Optic Nerve Head in Stereo Digital Images. *Eye* **2019**, *33*, 838–844. [CrossRef]

21. Zangwill, L.M.; Ayyagari, R.; Liebmann, J.M.; Girkin, C.A.; Feldman, R.; Dubiner, H.; Dirkes, K.A.; Holmann, M.; Williams-Steppe, E.; Hammel, N.; et al. The African Descent and Glaucoma Evaluation Study (ADAGES) III. *Ophthalmology* **2019**, *126*, 156–170. [CrossRef] [PubMed]
22. Taylor, K.D.; Guo, X.; Zangwill, L.M.; Liebmann, J.M.; Girkin, C.A.; Feldman, R.M.; Dubiner, H.; Hai, Y.; Samuels, B.C.; Panarelli, J.F.; et al. Genetic Architecture of Primary Open-Angle Glaucoma in Individuals of African Descent. *Ophthalmology* **2019**, *126*, 38–48. [CrossRef] [PubMed]
23. Kooner, K.S.; Joseph, A.; Shar, A.; Marquardt, F.A.; AlBdoor, M. Dallas Glaucoma Registry: Preliminary Results. *J. Clin. Exp. Ophthalmol.* **2011**, *2*, 164. [CrossRef] [PubMed]
24. European Glaucoma Prevention Study (EGPS) Group. Predictive Factors for Open-Angle Glaucoma among Patients with Ocular Hypertension in the European Glaucoma Prevention Study. *Ophthalmology* **2007**, *114*, 3–9. [CrossRef] [PubMed]
25. Beck, R.W.; Messner, D.K.; Musch, D.C.; Martonyi, C.L.; Lichter, P.R. Is There a Racial Difference in Physiologic Cup Size? *Ophthalmology* **1985**, *92*, 873–876. [CrossRef]
26. Chi, T. Racial Differences in Optic Nerve Head Parameters. *Arch. Ophthalmol.* **1989**, *107*, 836. [CrossRef]
27. Girkin, C.A. African Descent and Glaucoma Evaluation Study (ADAGES): II. Ancestry Differences in Optic Disc, Retinal Nerve Fiber Layer, and Macular Structure in Healthy Subjects. *Arch. Ophthalmol.* **2010**, *128*, 541. [CrossRef]
28. Doss, E.L.; Doss, L.; Han, Y.; Huang, S.; Porco, T.; Pekmezci, M.; Lin, S. Risk Factors for Glaucoma Suspicion in Healthy Young Asian and Caucasian Americans. *J. Ophthalmol.* **2014**, *2014*, 726760. [CrossRef]
29. Mansour, A.M. Racial Variation of Optic Disc Size. *Ophthalmic Res.* **1991**, *23*, 67–72. [CrossRef]
30. Lee, M.H.; Kim, H.K.; Kim, S.S. Risk Factors Associated With a Large Vertical Cup-to-Disc Ratio: Korean National Health and Nutritional Examination Survey. *J. Glaucoma* **2023**, *32*, 221–226. [CrossRef]
31. Amerasinghe, N. Determinants of the Optic Cup to Disc Ratio in an Asian Population: The Singapore Malay Eye Study (SiMES). *Arch. Ophthalmol.* **2008**, *126*, 1101. [CrossRef] [PubMed]
32. Allingham, R.R.; Liu, Y.; Rhee, D.J. The Genetics of Primary Open-Angle Glaucoma: A Review. *Exp. Eye Res.* **2009**, *88*, 837–844. [CrossRef] [PubMed]
33. Le, A.; Mukesh, B.N.; McCarty, C.A.; Taylor, H.R. Risk Factors Associated with the Incidence of Open-Angle Glaucoma: The Visual Impairment Project. *Investig. Ophthalmol. Vis. Sci.* **2003**, *44*, 3783. [CrossRef] [PubMed]
34. Chang, T.C.; Congdon, N.G.; Wojciechowski, R.; Muñoz, B.; Gilbert, D.; Chen, P.; Friedman, D.S.; West, S.K. Determinants and Heritability of Intraocular Pressure and Cup-to-Disc Ratio in a Defined Older Population. *Ophthalmology* **2005**, *112*, 1186–1191. [CrossRef]
35. Forchheimer, I.; De Moraes, C.G.; Teng, C.C.; Folgar, F.; Tello, C.; Ritch, R.; Liebmann, J.M. Baseline Mean Deviation and Rates of Visual Field Change in Treated Glaucoma Patients. *Eye* **2011**, *25*, 626–632. [CrossRef] [PubMed]
36. Kashiwagi, K.; Ou, B.; Nakamura, S.; Tanaka, Y.; Suzuki, M.; Tsukahara, S. Increase in Dephosphorylation of the Heavy Neurofilament Subunit in the Monkey Chronic Glaucoma Model. *Investig. Ophthalmol. Vis. Sci.* **2003**, *44*, 154. [CrossRef]
37. Nickells, R.W.; Schlamp, C.L.; Li, Y.; Kaufman, P.L.; Heatley, G.; Peterson, J.C.; Faha, B.; Ver Hoeve, J.N. Surgical Lowering of Elevated Intraocular Pressure in Monkeys Prevents Progression of Glaucomatous Disease. *Exp. Eye Res.* **2007**, *84*, 729–736. [CrossRef]
38. Harwerth, R.S.; Wheat, J.L.; Rangaswamy, N.V. Age-Related Losses of Retinal Ganglion Cells and Axons. *Investig. Ophthalmol. Vis. Sci.* **2008**, *49*, 4437. [CrossRef]
39. Palko, J.R.; Morris, H.J.; Pan, X.; Harman, C.D.; Koehl, K.L.; Gelatt, K.N.; Plummer, C.E.; Komáromy, A.M.; Liu, J. Influence of Age on Ocular Biomechanical Properties in a Canine Glaucoma Model with ADAMTS10 Mutation. *PLoS ONE* **2016**, *11*, e0156466. [CrossRef]
40. Fuchs, M.; Scholz, M.; Sendelbeck, A.; Atorf, J.; Schlegel, C.; Enz, R.; Brandstätter, J.H. Rod Photoreceptor Ribbon Synapses in DBA/2J Mice Show Progressive Age-Related Structural Changes. *PLoS ONE* **2012**, *7*, e44645. [CrossRef]
41. Ke, T.-L.; Clark, A.F.; Gracy, R.W. Age-Related Permeability Changes in Rabbit Corneas. *J. Ocul. Pharmacol. Ther.* **1999**, *15*, 513–523. [CrossRef] [PubMed]
42. Garway-Heath, D.F.; Wollstein, G.; Hitchings, R.A. Aging Changes of the Optic Nerve Head in Relation to Open Angle Glaucoma. *Br. J. Ophthalmol.* **1997**, *81*, 840–845. [CrossRef] [PubMed]
43. Ramrattan, R.S.; Wolfs, R.C.W.; Jonas, J.B.; Hofman, A.; De Jong, P.T.V.M. Determinants of Optic Disc Characteristics in a General Population: The Rotterdam Study. *Ophthalmology* **1999**, *106*, 1588–1596. [CrossRef] [PubMed]
44. Sommer, A. Clinically Detectable Nerve Fiber Atrophy Precedes the Onset of Glaucomatous Field Loss. *Arch. Ophthalmol.* **1991**, *109*, 77. [CrossRef] [PubMed]
45. Downs, J.C.; Girkin, C.A. Lamina Cribrosa in Glaucoma. *Curr. Opin. Ophthalmol.* **2017**, *28*, 113–119. [CrossRef]
46. Wallace, D.M.; O'Brien, C.J. The Role of Lamina Cribrosa Cells in Optic Nerve Head Fibrosis in Glaucoma. *Exp. Eye Res.* **2016**, *142*, 102–109. [CrossRef]
47. Hernandez, M.R. The Optic Nerve Head in Glaucoma: Role of Astrocytes in Tissue Remodeling. *Prog. Retin. Eye Res.* **2000**, *19*, 297–321. [CrossRef]
48. Morrison, J.C.; Johnson, E.C.; Cepurna, W.; Jia, L. Understanding Mechanisms of Pressure-Induced Optic Nerve Damage. *Prog. Retin. Eye Res.* **2005**, *24*, 217–240. [CrossRef]

49. Kim, Y.W.; Yun, Y.I.; Choi, H.J. Screening Fundus Photography Predicts and Reveals Risk Factors for Glaucoma Conversion in Eyes with Large Optic Disc Cupping. *Sci. Rep.* **2023**, *13*, 81. [CrossRef]
50. Kuroda, A.; Enomoto, N.; Ishida, K.; Shimazawa, M.; Noguchi, T.; Horai, N.; Onoe, H.; Hara, H.; Tomita, G. Movement of Retinal Vessels toward the Optic Nerve Head after Increasing Intraocular Pressure in Monkey Eyes with Experimental Glaucoma. *Exp. Eye Res.* **2017**, *162*, 110–115. [CrossRef]
51. Spaeth, G.L.; Lopes, J.F.; Junk, A.K.; Grigorian, A.P.; Henderer, J. Systems for Staging the Amount of Optic Nerve Damage in Glaucoma: A Critical Review and New Material. *Surv. Ophthalmol.* **2006**, *51*, 293–315. [CrossRef] [PubMed]
52. Caprioli, J. Correlation of Visual Function with Optic Nerve and Nerve Fiber Layer Structure in Glaucoma. *Surv. Ophthalmol.* **1989**, *33*, 319–330. [PubMed]
53. Bonnemaijer, P.W.M.; Lo Faro, V.; Sanyiwa, A.J.; Hassan, H.G.; Cook, C.; Van De Laar, S.; Lemij, H.G.; Klaver, C.C.W.; Jansonius, N.M.; et al.; GIGA study group. Differences in Clinical Presentation of Primary Open-angle Glaucoma between African and European Populations. *Acta Ophthalmol.* **2021**, *99*, e1118–e1126. [CrossRef]
54. Macanian, J.; Sharma, S.C. Pathogenesis of Glaucoma. *Encyclopedia* **2022**, *2*, 1803–1810. [CrossRef]
55. Burgoyne, C.F. A Biomechanical Paradigm for Axonal Insult within the Optic Nerve Head in Aging and Glaucoma. *Exp. Eye Res.* **2011**, *93*, 120–132. [CrossRef] [PubMed]
56. Peters, D.; Bengtsson, B.; Heijl, A. Factors Associated with Lifetime Risk of Open-angle Glaucoma Blindness. *Acta Ophthalmol.* **2014**, *92*, 421–425. [CrossRef] [PubMed]
57. Grant, W.M.; Burke, J.F. Why Do Some People Go Blind from Glaucoma? *Ophthalmology* **1982**, *89*, 991–998. [CrossRef] [PubMed]
58. Heo, D.W.; Kim, K.N.; Lee, M.W.; Lee, S.B.; Kim, C. Properties of Pattern Standard Deviation in Open-Angle Glaucoma Patients with Hemi-Optic Neuropathy and Bi-Optic Neuropathy. *PLoS ONE* **2017**, *12*, e0171960. [CrossRef]
59. Wagner, I.V.; Stewart, M.W.; Dorairaj, S.K. Updates on the Diagnosis and Management of Glaucoma. *Mayo Clin. Proc. Innov. Qual. Outcomes* **2022**, *6*, 618–635. [CrossRef]
60. Gedde, S.J.; Herndon, L.W.; Brandt, J.D.; Budenz, D.L.; Feuer, W.J.; Schiffman, J.C. Postoperative Complications in the Tube Versus Trabeculectomy (TVT) Study during Five Years of Follow-Up. *Am. J. Ophthalmol.* **2012**, *153*, 804–814.e1. [CrossRef]
61. Francis, B.A. Vision Loss and Recovery after Trabeculectomy: Risk and Associated Risk Factors. *Arch. Ophthalmol.* **2011**, *129*, 1011. [CrossRef] [PubMed]

Disclaimer/Publisher's Note: The statements, opinions and data contained in all publications are solely those of the individual author(s) and contributor(s) and not of MDPI and/or the editor(s). MDPI and/or the editor(s) disclaim responsibility for any injury to people or property resulting from any ideas, methods, instructions or products referred to in the content.

genes

MDPI

Article

Integrative Analysis Unveils the Correlation of Aminoacyl-tRNA Biosynthesis Metabolites with the Methylation of the *SEPSECS* Gene in Huntington's Disease Brain Tissue

Sangeetha Vishweswaraiah [1], Ali Yilmaz [2], Nazia Saiyed [2], Abdullah Khalid [2], Purvesh R. Koladiya [2], Xiaobei Pan [3], Shirin Macias [3], Andrew C. Robinson [4], David Mann [4], Brian D. Green [3], Ieva Kerševičiūte [5], Juozas Gordevičius [5], Uppala Radhakrishna [1] and Stewart F. Graham [1,2,6,*]

[1] Department of Obstetrics and Gynecology, Corewell Health William Beaumont University Hospital, 3601 W. 13 Mile Road, Royal Oak, MI 48073, USA; sangeetha.vishweswaraiah@corewellhealth.org (S.V.); uppalar99@gmail.com (U.R.)

[2] Metabolomics Department, Corewell Health Research Institute, 3811 W. 13 Mile Road, Royal Oak, MI 48073, USA; ali.yilmaz@corewellhealth.org (A.Y.); nazia.saiyed@corewellhealth.org (N.S.); abdullah.khalid@corewellhealth.org (A.K.); purveshrohitkumar.koladiya@corewellhealth.org (P.R.K.)

[3] Advanced Asset Technology Centre, Institute for Global Food Security, Queen's University Belfast, Belfast BT9 5DL, UK; x.pan@qub.ac.uk (X.P.); smacias01@qub.ac.uk (S.M.); b.green@qub.ac.uk (B.D.G.)

[4] Faculty of Biology, Medicine and Health, School of Biological Sciences, Division of Neuroscience, The University of Manchester, Salford Royal Hospital, Salford M6 8HD, UK; andrew.c.robinson@manchester.ac.uk (A.C.R.); davidmann1948@hotmail.com (D.M.)

[5] VUGENE, LLC, 625 Kenmoor Ave Suite 301 PMB 96578, Grand Rapids, MI 49546, USA; ieva@vugene.us (I.K.); juozas@vugene.us (J.G.)

[6] Department of Obstetrics and Gynecology, Oakland University-William Beaumont School of Medicine, Rochester, MI 48309, USA

* Correspondence: stewart.graham@corewellhealth.org

Citation: Vishweswaraiah, S.; Yilmaz, A.; Saiyed, N.; Khalid, A.; Koladiya, P.R.; Pan, X.; Macias, S.; Robinson, A.C.; Mann, D.; Green, B.D.; et al. Integrative Analysis Unveils the Correlation of Aminoacyl-tRNA Biosynthesis Metabolites with the Methylation of the *SEPSECS* Gene in Huntington's Disease Brain Tissue. *Genes* **2023**, *14*, 1752. https://doi.org/10.3390/genes14091752

Academic Editor: Claudia Ricci

Received: 2 August 2023
Revised: 28 August 2023
Accepted: 30 August 2023
Published: 2 September 2023

Copyright: © 2023 by the authors. Licensee MDPI, Basel, Switzerland. This article is an open access article distributed under the terms and conditions of the Creative Commons Attribution (CC BY) license (https://creativecommons.org/licenses/by/4.0/).

Abstract: The impact of environmental factors on epigenetic changes is well established, and cellular function is determined not only by the genome but also by interacting partners such as metabolites. Given the significant impact of metabolism on disease progression, exploring the interaction between the metabolome and epigenome may offer new insights into Huntington's disease (HD) diagnosis and treatment. Using fourteen post-mortem HD cases and fourteen control subjects, we performed metabolomic profiling of human postmortem brain tissue (striatum and frontal lobe), and we performed DNA methylome profiling using the same frontal lobe tissue. Along with finding several perturbed metabolites and differentially methylated loci, Aminoacyl-tRNA biosynthesis (adj p-value = 0.0098) was the most significantly perturbed metabolic pathway with which two CpGs of the *SEPSECS* gene were correlated. This study improves our understanding of molecular biomarker connections and, importantly, increases our knowledge of metabolic alterations driving HD progression.

Keywords: metabolomics; epigenetics; integrative omics; epi-metabolomics; Huntington's disease; brain

1. Introduction

Huntington's disease (HD) is a neurodegenerative disorder that results in a decline of neurological function, characterized by worsening motor skills, cognitive abilities, and psychological symptoms [1]. The root cause of HD is the expansion of a CAG trinucleotide repeat in the huntingtin (*HTT*) gene, producing a toxic protein that adversely affects the brain's cells [2]. The average age for symptom onset in HD is between 30 and 50 years, although, in certain instances, symptoms may begin before the age of 20. The early indications of HD include irritability, depression, small uncontrolled movements, impaired coordination, and difficulty with learning or decision-making. In advanced stages, many individuals with HD experience involuntary twitching movements that become more severe over time. Walking, speaking, and swallowing may also become difficult [3,4]. The global prevalence of HD is estimated to be around 5–10 cases per 100,000 individuals, and it is inherited in an autosomal dominant pattern [5]. The onset and severity of HD symptoms can vary significantly among individuals, even within families with identical genetic mutations. Presently, no cure exists for HD, and available treatments only provide relief for symptoms associated with the disease [6].

HD involves neurodegeneration, mainly in the basal ganglia, particularly the striatum, impacting movement and reward-related neural activity [7]. The neocortex, which connects to the striatum, shrinks during HD, affecting other brain areas later [8]. Disease development involves complex mechanisms like transcription dysfunction and abnormal trafficking [9]. Mood and behavior changes related to frontal lobe dysfunction serve as early HD indicators [10–13].

In the pursuit of comprehending complex biological systems, multi-omics studies hold significant importance. Such studies involve the integration and analysis of data from various "-omics" fields, such as genomics, transcriptomics, proteomics, and metabolomics, to obtain a holistic view of biological processes. By exploring the interactions between different molecular levels, multi-omics studies can reveal novel pathways and mechanisms that contribute to complex diseases such as neurodegenerative diseases [14,15]. Multi-omics studies can aid in identifying new drug targets and developing more effective therapeutic strategies [16].

DNA methylation is an epigenetic mechanism that plays a crucial role in regulating gene expression, and numerous studies have revealed the involvement of altered DNA methylation in the pathogenesis of neurodegenerative disorders, including HD [17,18]. In recent years, DNA methylation has emerged as a promising biomarker for HD diagnosis and prognosis [18]. For instance, a study has indicated that differences in DNA methylation levels of the *HTT* gene may contribute to the tissue-specific variation of its expression. Additionally, the mutant *HTT* gene appears to impact the epigenetic age of individuals with HD [18]. Another study reported that DNA methylation levels in certain genes correlated with the advancement and severity of the disease [19]. Furthermore, some studies have explored the therapeutic potential of targeting DNA methylation in HD [18,20,21].

Metabolomics possesses the capability to reveal the underlying mechanisms of disease progression that have the potential to function as prognostic biomarkers for risk assessment [22]. The discovery of metabolic biomarkers and their corresponding pathways linked to the progression of HD can enhance comprehension of the disease's pathophysiological mechanisms. Moreover, this can stimulate the development of efficacious treatments by providing precise assessments of the disease advancement [23,24]. Several metabolomics investigations involving human subjects and other models have produced intriguing outcomes, indicating significant alterations in metabolic pathways associated with HD [23,25]. Several -omics studies, including some in the field of metabolomics, have showcased metabolic modifications, encompassing alterations in protein, carbohydrate, cholesterol, lipid, and amino acid metabolism [26–29].

Given the potential impact of both DNA methylation and metabolomics on the pathogenesis of HD, and the potential for interplay between these two -omics technologies, we proposed that investigating them in tandem could reveal potential signaling mechanisms for therapeutic targeting. To this end, we conducted DNA methylation profiling of the frontal lobe, as well as metabolite profiling in both the striatum and frontal lobe of postmortem individuals with HD. The implications of dysregulated pathways are presented and discussed.

2. Methods

2.1. Study Samples

This research was carried out by utilizing brain tissue samples collected posthumously from HD-affected patients and from normal controls. The Manchester Brain Bank provided the brain tissue [30]. The use of post-mortem human brain samples was approved by the Manchester Brain Bank Management Committee working under conditions agreed upon with the Newcastle and North Tyneside 1 Research Ethics Committee (REC Reference# 09/H0906/52 + 5). Further ethical approval was granted by the Beaumont Institutional Review Board (IRB# 2014–353). Fourteen post-mortem HD cases and fourteen controls were used for metabolomics profiling, and the same fourteen cases, and thirteen controls, were considered for the methylation analysis. Control subjects had no evidence of Huntington's pathology and there was no recent history of dementia-associated medication use among them. Both striatum and frontal lobe tissues were metabolically profiled, whereas only frontal lobe tissue was available for epigenetic analysis. The HD cases all exhibited a moderate to severely atrophied corpus striatum consistent with grades 2 or 3. The exact CAG repeat numbers were not available, but the diagnosis of HD was confirmed using genetic testing except for in two cases, where the diagnosis was based on the presence of ubiquitinated/p62 positive intra-nuclear inclusions within cortical and striatal neurons, as observed in all other HD cases but not in control cases. The demographic characteristics of the study group are provided in Table 1.

Table 1. A comparison of demographic features between individuals diagnosed with Huntington's disease and control subjects without the condition.

	HD Patients	**Controls**	***p*-Value**
Number of subjects	14	14	n/a
Age, Mean (SD)	54.64 (12.39)	78.5 (13.46)	<0.0001
Individual age in years:			
Patient/control–1	70	84	
Patient/control–2	57	84	
Patient/control–3	51	81	
Patient/control–4	52	87	
Patient/control–5	67	90	
Patient/control–6	51	89	
Patient/control–7	33	89	
Patient/control–8	47	54	
Patient/control–9	48	53	
Patient/control–10	na	84	
Patient/control–11	50	60	
Patient/control–12	68	89	
Patient/control–13	72	83	
Patient/control–14	75	90	
Sex			
Males	8 (57.1)	8 (57.1)	0.45
Females	6 (42.8)	6 (42.8)	
Postmortem delay (PMD)-Minutes			
Mean (SD)	77.35 (71.63)	69.28 (38.09)	0.65

2.2. ^{1}H NMR Analysis

The samples were first stored at −80 °C before being prepared. To minimize the amount of heat produced, the samples were subsequently lyophilized and milled into a fine powder under liquid nitrogen. To analyze the samples using ^{1}H NMR, previously optimized methods were employed [31]. The 50 mg samples of the lyophilized and milled tissue were extracted in 50% methanol/water (1 mg per 10 μL; 0.1 g/mL) in a sterile 2 mL Eppendorf tube. The samples were mixed and sonicated for 20 min at a frequency of 50–60 Hz to achieve homogenization, cell disruption, and compound extraction, and centrifuged at 13,000× *g* at 4 °C for 30 min to remove any macro molecules that may affect the NMR signal. After collecting the supernatants, they were dried under vacuum using a Savant DNA Speedvac (Thermo Scientific, USA) and reconstituted in 285 μL of 50 mM potassium phosphate buffer (pH 7.0), 30 μL of Sodium 2,2-dimethyl-2-silapentane-5-sulfonate (DSS), and 35 μL of D_2O. A volume of 200 μL of the reconstituted sample was then transferred to a 3 mm Bruker NMR tube for analysis. All samples were stored at 4 °C in a thermostatically controlled SampleJet autosampler (Bruker-Biospin, Billerica, MA, USA) and heated to room temperature over 3 min prior to NMR analysis.

To record all 1D ^{1}H NMR data, a randomized running order was utilized on a Bruker ASCEND HD 600 MHz spectrometer (Bruker-Biospin, Billerica, MA, USA) coupled with a 5 mm TCI cryoprobe at a temperature of 300 (±0.5) K. For each sample, 256 transients were collected as 64k data points with a spectral width of 12 kHz (20 ppm) using a pulse sequence called CPP WaterSupp (Bruker pulse program: pusenoesypr1d), which was developed by Mercier et al. (2011) [32], with an inter-pulse delay of 9.65 s. The data collection protocol involved a 180 s temperature equilibration period, fast 3D shimming utilizing the z-axis profile of the ^{2}H NMR solvent signal, receiver gain adjustment, and acquisition. The free induction decay signal was zero-filled to 128 k points and exponentially multiplied with a 0.1 Hz line broadening factor. After Fourier transformation, the zero and first-order phase constants were manually optimized, and a polynomial baseline correction of the FID (degree 5) was applied for precise quantitation. Chenomx NMR Suite (ver. 8.1, Chenomx, Edmonton, AB, Canada) was used for processing and analyzing all spectra.

2.3. Direct Injection/Liquid Chromatography–Mass Spectral Analysis (DI/LC-MS/MS)

Post-mortem brain samples were prepared and analyzed using a method previously reported by Urban et al. and our group [33,34]. Post-mortem brain samples were subjected to freeze drying using a Christ α 1-4LD Plus freeze dryer (IMA Life) and milled to a fine powder under liquid nitrogen using a 6870-freezer mill (SpexSamplePrep, Metuchen, NJ, USA). Subsequently, 10 mg of tissue was mixed with 300 μL of 85% ethanol and 15% phosphate-buffered saline. Samples were sonicated for 5 min in ice water, mixed for 30 s, and then centrifuged at 10,000× *g* for 5 min at 4 °C.

A total of 10 μL of the subsequent supernatant was analyzed using the targeted, commercially available quantitative DI/LC-MS/MS AbsoluteIDQ p180 kit (Biocrates, Innsbruck, Austria), as previously described [35], and all sample preparatory steps were completed as detailed by the manufacturer. Metabolites were separated and quantified using a reverse phase column (Waters ACQUITY UPLC BEH C18 2.1 × 50 mm, 1.7 μm; Wexford, Ireland) on a UPLC system (I-Class, Waters Corporation, Budapest, Hungary) coupled to a triple quadrupole mass spectrometer (Xevo TQ-S, Waters Corporation) operating in the multiple reaction monitoring (MRM) mode. All remaining compounds (acylcarnitines, hexoses, glycerophospholipids, and sphingolipids) were quantified using the same mass spectrometer using flow injection analysis, operating in multiple reaction monitoring mode. Metabolite concentrations were calculated and expressed in micromolar (μM).

2.4. Genome-Wide DNA Methylation Assay

The QIAamp DNA Mini Kit (Qiagen) was used to extract genomic DNA from the lyophilized and milled brain tissue for the genome-wide DNA methylation assay. The Illumina Infinium Epic HumanMethylation850K arrays were employed to profile DNA methylation (Illumina, Inc., San Diego, CA, USA), which covers >850,000 methylation sites per sample, including translation start sites, enhancers, gene bodies, promoters, intronic regions, and also the CpG islands all over the genome at a single-nucleotide resolution. A total of 500 ng of genomic DNA was utilized from each sample that was subjected to bisulfite conversion using the EZ DNA Methylation-Direct Kit (Zymo Research, Orange, CA, USA).

The BeadChips were handled in accordance with the manufacturer's guidelines, and the Illumina iScan (Illumina, USA) was used to image the fluorescently stained BeadChips.

2.5. Data Analysis

Metabolomic profiling: The raw metabolomic data underwent sum normalization, autoscaling, and log transformation. Principal Component Analysis (PCA) was employed to detect outliers; any sample deviating by more than 3 standard deviations away from the center of the first three principal components was considered an outlier (i.e., 99.7% confidence). No outliers were identified. To estimate sample diagnosis, a logistic regression model was applied, using the first five principal components, as well as sample age, sex, and post-mortem interval as inputs. To identify differentially expressed metabolites, a robust linear regression model was fitted to each metabolite, with sample diagnosis, age, sex, and PMI as covariates. *p*-values for diagnosis were calculated using empirical Bayes treatment of fitted models. Metabolites with FDR $q < 0.05$ were deemed significantly differentially expressed in the disease. To estimate metabolite pathway enrichment, the metabolites were ranked based on their log-transformed *p* value multiplied by the sign of fold change. This ranking method ensured that metabolites with a significant increase in abundance were at the top of the list, while those with the most significant decrease in abundance were at the bottom. Any metabolites without an HMDB ID were removed from the ranked list. Metabolomic pathways were downloaded from the KEGG database using the R package multiGSEA, and then metabolite set enrichment was calculated with 10,000 permutations using the fgsea package (version 1.16.0).

DNA methylation profiling: The raw *idat files were read using *minfi* package of bioconductor, which was also considered to mark the failed methylation probes. To identify outliers in the data, several quality control measures were performed. First, samples with more than 20% of failed probes were considered outliers based on the proportion of failed probes. The median probe intensity of U and M probes was then calculated, and samples in the lower-left corner of the plot were marked as outliers. Gender prediction was performed by comparing the median signal in X and Y chromosomes, and samples with a mismatch between the predicted and known gender were marked as outliers. Principal component analysis of the centered, unnormalized β value matrix was used to identify samples deviating by more than 2 standard deviations from the mean of the first three principal components and marked as outliers. Noob normalization was used to normalize the samples, and methylation β values were extracted for the remaining probes. To adjust for the effect of sample position on the EPIC array, the empiricalBayesLM function from the WGCNA package was used, and the effect was modeled as a second-degree polynomial. The proportion of neuronal cells in each sample was estimated using the flow-sorted PFC samples and the estimateCellCount function from the minfi package. The sample DNA methylation age was estimated using the ENmix package. Age acceleration was defined as the residuals of a linear model, where DNAmAge was the dependent variable, and sample chronological age, sex, postmortem interval, and proportion of neuronal cells were the independent variables. To identify differentially methylated CpGs, robust linear regression was employed using the R package limma with sample diagnosis, age, sex, postmortem interval, and proportion of neuronal nuclei as covariates. *p*-value estimates for diagnosis were obtained after the empirical Bayes treatment of the fitted models. Cytosines with a false discovery rate (FDR) q-value of less than 0.05 were considered significantly altered in HD individuals.

To conduct an epigenome enrichment analysis, cytosines were mapped to gene names using the UCSC_RefGene_Name column specified in the Bioconductor EPIC array annotation package IlluminaHumanMethylationEPICanno.ilm10b4.hg19. When a CpG locus was annotated with multiple genes multiple times, the one that was most frequently associated with the locus was chosen. The genes were ranked based on the significance of the affected cytosines multiplied by the sign of fold change. For genes that mapped to multiple cytosines, the one with the smallest p-value was selected. To conduct the pathway enrichment analysis, KEGG pathways were downloaded and the fgsea function was used.

2.6. Epigenome–Metabolome Interactions

To establish epigenome–metabolome interactions within each pathway, metabolites and transcripts pertaining to the pathway were selected. Pathway–protein associations were obtained from SMPDB 63 (https://smpdb.ca). Only CpGs that were associated with any of the transcripts in the pathway were further considered. The concordance of each metabolite–CpG pair was established by fitting a robust linear regression model without an intercept, where the standardized methylation value was the response variable, and standardized metabolite abundance, as well as diagnosis, age, and sex were the independent variables. Concordance p values were adjusted using FDR, and those with FDR $q < 0.2$ were reported.

2.7. Diagnostic Models

To evaluate the diagnostic models based on metabolome, the normalized data were used, and no other additional data preprocessing was performed. Two basic diagnostic models were trained: *glmnet* (LASSO), and Random Forest (rf). In addition, an ensemble of those models was built using either linear (ensemble linear) or glmnet-weighted (ensemble glmnet) combination of basic predictors. The area under the curve (AUC) metric was used as the training objective. The data were randomly split into ten training and validation sets. Within each training set, bootstrapping was performed to train and select the best models. The models were then evaluated on the validation set.

For the methylome, we transformed the normalized β values into M values and applied batch correction to remove Sentrix array ID and Position effects from the data. We randomly split the data into a training (N = 20) and validation (N = 4) sets. For each such split, we used the training data to filter the loci by applying a t-test and selecting top 50.000 loci for further use in training the models. We used glmnet, random forest, and SVM with linear and radial kernel algorithms to build and evaluate the models. In addition, we combined these models into ensembles using either linear weighting of predictors (ensemble linear) or glmnet inferred weights of predictors (ensemble glmnet). Finally, we trained a classifier using each of the above algorithms on the full dataset.

3. Results

3.1. HD Brain Metabolomic Profile

To study any underlying metabolic alterations in the brain of people who died from HD, we profiled the striatum and the frontal lobe of these individuals and compared them with controls. In both the striatum and frontal lobes, the variance inflation analysis showed that the control subjects were more likely to be younger, which correlates with the chronological age. However, age was controlled for in all linear models. Variance inflation for the striatum and frontal lobe are shown in Supplementary Figures S1 and S2, respectively. Normalized intensities using the striatum and frontal lobe data are provided in Supplementary Figures S3 and S4, respectively.

Targeted metabolomic profiling revealed that, of the 166 metabolites, 4 were significantly upregulated in the striatum of HD brains when compared to controls (FDR adjusted q-value < 0.05). This included taurine, SM.C16.0, PC.aa.C40.3, and SM.C18.1 (Supplementary Table S1). In the frontal lobe, we identified 14 metabolites that were significantly different between HD and control brains (FDR adjusted q-value < 0.05). Eight were upregulated and the remaining six were downregulated in HD brains. The top four were PC.aa.C38.4, lysoPC.a.C18.2, C18.1.OH, and C2 (Supplementary Table S2).

We also performed metabolomic set enrichment analysis for both the striatum (Figure 1) and the frontal lobe (Figure 2). Interestingly, both analyses highlighted the same two pathways as being significantly perturbed. These include aminoacyl-tRNA biosynthesis and the biosynthesis of amino acids (FDR adjusted q-value less than 0.05).

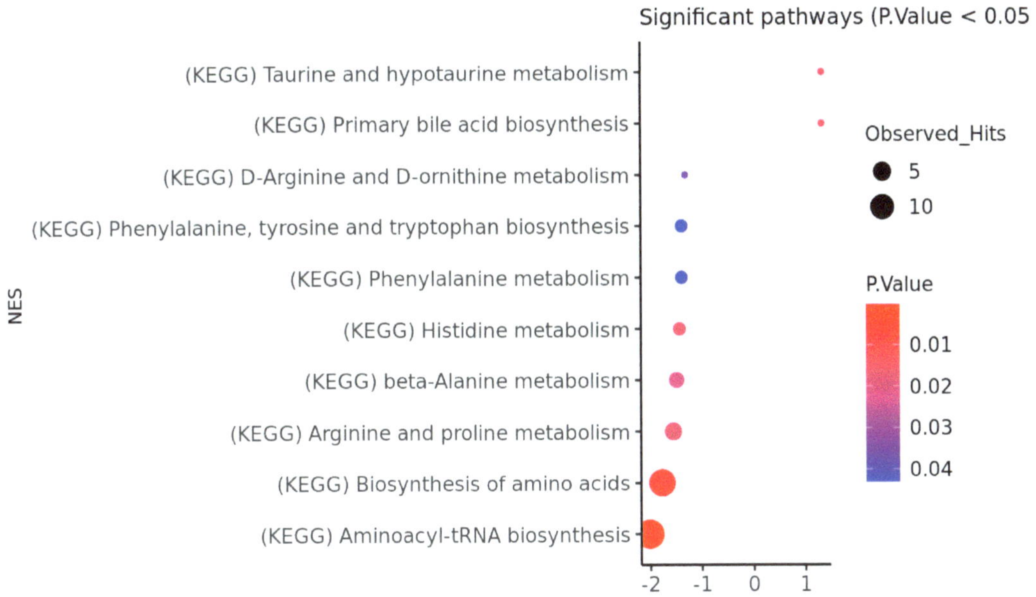

Figure 1. The metabolic pathways that are significantly enriched with metabolites have been identified using the striatum region of HD patients. The *x*-axis represents the hyper-methylated (+) and hypo-methylated (−) status of pathways, whereas the *y*-axis provides the pathways that were perturbed. The node size indicates the number of observed hits that are associated with significant pathways. The node color indicates the significance level based on the *p*-value (<0.05).

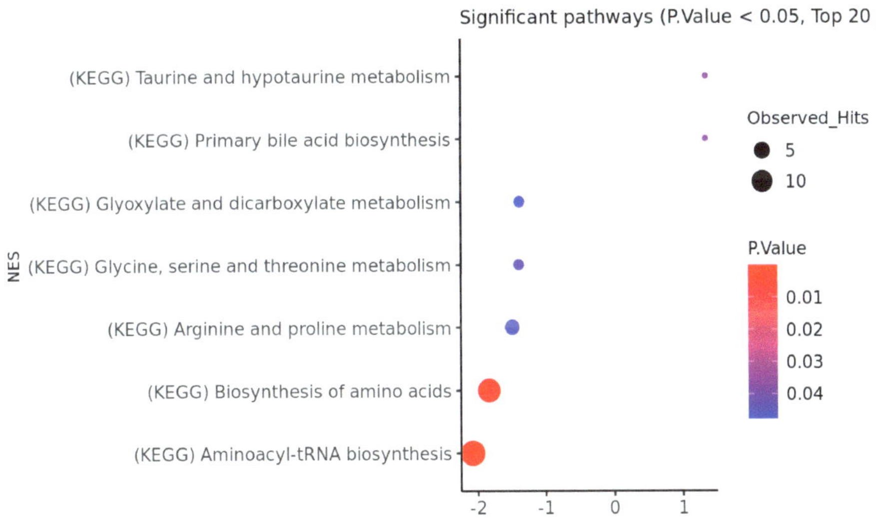

Figure 2. The metabolic pathways that are significantly enriched with metabolites have been identified using the frontal lobe region of HD patients.

3.2. HD Brain Methylation Profile

The Sentrix ID of the sample wells was found to be inflated and adjusted (Supplementary Figure S5). Further, the enrichment of genomic regions by methylation markers using the frontal lobe of HD patients vs. the control group revealed enrichment of hypomethylated markers in the intron and 3′ UTRs. CGI shores are enriched with hyper-methylated loci, and the open sea areas are enriched with hypo-methylated loci. In terms of genomic features, hypo-methylated loci tend to overlap introns and 3′UTR regions of genes. The enrichment of genomic regions is depicted in Supplementary Figure S6, and the details are provided in Supplementary Table S3. There were 11,955 significantly differentially methylated CpGs identified (FDR adjusted q-value = 0.05) when the frontal lobes of HD patients were compared with those of control subjects. Among them, 11,292 CpGs were hypomethylated, and the remaining 663 were significantly hypermethylated (Supplementary Table S4).

3.3. Epigenome and Metabolome Correlation

This study found correlations between DNA methylation and the metabolome, specifically identifying several genes (Trafficking Protein Particle Complex Subunit 10 [*TRAPPC10*], CUB And Sushi Multiple Domains 3 [*CSMD3*], 5-Methyltetrahydrofolate-Homocysteine Methyltransferase Reductase [*MTRR*], Pecanex 1 [*PCNX*], PC-Esterase Domain Containing 1B [*PCED1B*], Glucose-6-Phosphatase Catalytic Subunit 2 [*G6PC2*], Protocadherin 7 [*PCDH7*], Glypican 6 [*GPC6*], Ribosomal Protein S25 [*RPS25*], and Casein Kinase 1 γ 3 [*CSNK1G3*]) that are correlated with the metabolites, phenylalanine, and methionine. Further details are available in Supplementary Table S5.

The linear model interaction network between CpGs and metabolites was established. The most significantly affected pathway in this study was the Aminoacyl-tRNA biosynthesis pathway. Within this pathway, two metabolites, valine and phenylalanine, were found to interact significantly with two CpGs of the same gene, *SEPSECS*. Specifically, HMDB0000159 (phenylalanine) interacted with cg17068512 (p-value = 0.0005), HMDB0000159 interacted with cg14837557 (p-value = 0.004), and HMDB0000883 (valine) interacted with cg17068512 (p-value = 0.007). These CpGs were located on different transcripts of the *SEPSECS* gene body. Interestingly, both metabolites were found to be negatively correlated with the methylation of sites belonging to the *SEPSECS* gene (see Figure 3 and Supplementary Table S6).

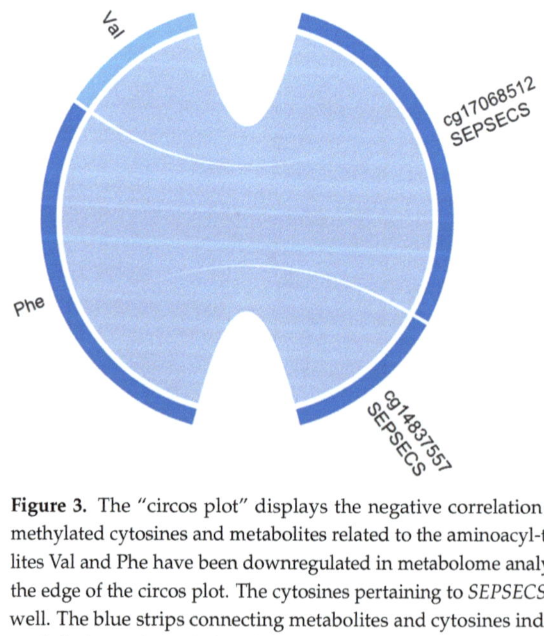

Figure 3. The "circos plot" displays the negative correlation between significantly differentially methylated cytosines and metabolites related to the aminoacyl-tRNA biosynthesis pathway. Metabolites Val and Phe have been downregulated in metabolome analysis, as indicated by the blue color on the edge of the circos plot. The cytosines pertaining to *SEPSECS* gene have been hypo-methylated as well. The blue strips connecting metabolites and cytosines indicate a negative correlation between methylation and metabolite abundance when adjusted for age, sex, postmortem delay, and condition.

3.4. Diagnostic Models

Using metabolomics data acquired from striatal tissue (Supplementary Figure S7), we were able to develop a predictive model with an AUC of 0.82 on glmnet. This was followed by ensemble Linear AUC with 0.79, rf AUC with 0.75, and ensembleGlmnet AUC with 0.74. Using frontal lobe metabolomics data (Supplementary Figure S8), we developed a predictive model with an AUC of 0.84 on glmnet, while both ensembleLinear and ensembleGlmnet had an AUC of 0.8. The rf analysis achieved an AUC of 0.74.

Based on the methylome data, we evaluated the performance of several models. Our results indicate that the svmRadial model achieved the highest AUC value of 0.95. Additionally, both ensembleGlmnet and ensembleLinear models performed well, with an AUC value of 0.9. The glmnet, random forest, and svmLinear models achieved an AUC of 0.85 (Supplementary Figure S9).

4. Discussion

Earlier studies [36] have shown that alterations in DNA methylation can contribute to the development of Huntington's disease (HD). However, the relationship between methylation and the metabolome in HD has not been extensively studied. The goal of our study was to investigate the potential interaction between DNA methylation and the metabolome in the human brain of HD individuals compared to controls. We analyzed the metabolome in two brain regions, the striatum and frontal lobe, and performed methylome profiling only in the frontal lobe.

Environmental factors can lead to epigenetic changes, and metabolites can act as interacting partners that determine cellular function. Even minor and short-term changes in nutrition can have significant and long-lasting effects on gene expression, possibly by interacting with epigenetic factors. This "memory" of past metabolic conditions may also contribute to the development of certain diseases over time [37]. We hypothesized that there exists a robust connection between DNA methylation patterns and metabolic measures. We aimed to discover diagnostic indicators and understand the methylome–metabolome interplay in HD. Our study reveals the complex relationship between the epigenome and metabolome, potentially impacting HD development.

Our study highlights the complex interplay between the epigenome and metabolome, which may play a role in the development of HD.

Analysis of the acquired data showed that there was a significant correlation between phenylalanine and methionine, and the CpGs of *GPC6*, *CSNK1G3*, and *PCNX* genes. While phenylalanine is not directly involved in the development of HD or the formation of protein aggregates, it can influence the levels of neurotransmitters, which are critical in the pathogenesis of HD [38,39]. Similarly, while methionine is not directly implicated in the development of HD, it may play a role in the neurodegeneration [40,41] and the pathogenesis of HD through its involvement in the generation of reactive oxygen species [42] and impaired homocysteine metabolism [43,44]. According to previously reported research, GPC6 is a significant protein involved in synapse formation [45]. *CSNK1G3* is a member of the casein kinase family of genes, which play a critical role in neuronal and synaptic network functions. It has been studied in relation to neurodegenerative disorders, including HD [46]. Meanwhile, *PCNX* is a member of the pecanex family of genes, which act as maternal-effect neurogenic genes studied in Drosophila. This gene is thought to potentially contribute to nervous system development [47]. Additionally, genes including TRAPPC10, CSMD3, MTRR, G6PC2, PCDH7, PCED1B, RPS25, and TRAPPC4 were found to be highly correlated with phenylalanine. Specifically, CSMD3 has been linked to synaptogenesis and neurogenesis in the context of neurodevelopmental disorders. This gene may also play a role in regulating a network of genes involved in synaptic organization and immune activity [48]. The *MTRR* gene encodes for 5-methyltetrahydrofolate-homocysteine methyltransferase reductase, which is necessary for the normal function of the *MTHFR* gene. Dysregulation of S-adenosyl methionine, methionine, and homocysteine, which are all involved with the *MTHFR* gene, can lead to neurotoxicity [18,49]. *PCDH7* has been shown to play a role in regulating dendritic spine morphology and synaptic function [50]. *RPS25* participates in nucleotide repeat expansions through the regulation of an unconventional translation process that is commonly associated with the pathogenesis of Huntington's disease [51,52]. Our study identified methylation on the promoter region of the *TRAPPC4* gene, which has been associated with splice variants linked to a neurodegenerative syndrome in infants. This finding suggests a possible involvement of TRAPPC4 in the neurodegenerative process [53].

Aminoacyl-tRNA Biosynthesis

Metabolic enrichment analysis highlighted the aminoacyl-tRNA biosynthesis pathway as being significantly perturbed. This signaling mechanism plays a crucial role in protein synthesis and in ensuring the precision of translation [54,55]. It is noteworthy that a study utilizing cerebrospinal fluid identified a strong association of aminoacyl-tRNA biosynthesis with HD [39]. Several studies have also suggested that cytosolic tRNAs play a crucial role in HD [56,57]. Elongated Gln repeat in the huntingtin protein, consisting of 40–100 repeated CAG codons, is known to cause HD. However, shorter CAG repeats can also contribute to HD, indicating the involvement of additional disease modifiers. The constant process of translation of the repeat sequence depletes charged $tRNA^{Gln}$–CUG, resulting in more frequent frameshifting translation of the huntingtin gene, which can exacerbate HD pathogenesis. The process also signifies the role of tRNA in the disease process [56–58]. Our study investigated the potential impact of epigenetic modifications, specifically DNA methylation, on the above-mentioned metabolomic changes in the striatum and frontal lobe regions of the brain in individuals with HD.

In our study, the aminoacyl-tRNA biosynthesis pathway was found to show a correlation between valine and phenylalanine with two CpGs of *SEPSECS* gene. Valine plays a vital role in protein synthesis and various cellular functions and has been found to be associated with HD. A study observed lower levels of valine in the HD patient group compared to the control group. The study also found a correlation between valine levels, weight loss, and the number of CAG repeats [59]. Also, the decreased levels of phenylalanine were identified, along with valine, in postmortem brain tissues, CSF, and serum samples of HD patients [39].

The *SEPSECS* gene codes for the enzyme Sep tRNA:Sec tRNA Synthase, which synthesizes selenocysteine. This unique amino acid has antioxidants and appears in proteins for oxidative stress defense and redox signaling [60,61]. Several studies have suggested that oxidative stress and damage contribute to the pathogenesis of HD, especially in the striatum and cerebral cortex [62–64]. In this context, it has been proposed that selenocysteine and other antioxidants might have neuroprotective effects against the oxidative stress that occurs in HD.

Moreover, recent evidence suggests that alterations in the expression of the *SEPSECS* gene and the levels of selenocysteine in the brain might be involved in the pathogenesis of HD [65–67]. Selenoproteins have been found to participate in various biological processes, such as antioxidant defense, anti-inflammation, anti-apoptosis, and immune response regulation, as well as the regulation of oxidative stress and endoplasmic reticulum stress [68]. A study by Lu et al., 2014 suggests that, when sodium selenite was given, mouse models of HD showed marked improvements in motor endurance, decreased loss of brain weight, reduced burden of mutant huntingtin aggregates, and decreased levels of brain oxidized glutathione [69]. In summary, the *SEPSECS* gene and its product selenocysteine might play an important role in the pathogenesis of HD by regulating oxidative stress and neuroprotection. Further research is needed to fully understand the mechanisms underlying the association of the metabolites noted in our study with this gene and to explore their potential therapeutic applications in HD. Notably, there were perturbations in the aminoacyl-tRNA biosynthesis pathway, which is linked to protein synthesis and translation precision [54,55]. A previous study, which used cerebrospinal fluid, also connected this pathway with HD [39]. Cytosolic tRNAs are implicated in HD, with the elongated Gln repeat in the huntingtin protein driving the disease process [56–58]. One of the genes, the *SEPSECS* gene, responsible for producing the Sep (O-Phosphoserine) tRNA:Sec (Selenocysteine) tRNA Synthase enzyme, is a focal point. This enzyme facilitates selenocysteine synthesis, a rare amino acid with antioxidant properties [60,61]. Given the oxidative stress implications in HD pathogenesis, exploring selenocysteine's neuroprotective potential is significant [62–64]. Our study sheds light on these interconnected mechanisms and their contribution to HD's complex pathology.

In conclusion, we have used an epimetabolomics approach to highlight the significance of the aminoacyl-tRNA biosynthesis pathway in the development of HD. However, this study was performed on a limited number of samples, and this is due to the difficulty of obtaining PM brain tissue. However, as a proof-of-concept study, we aimed to demonstrate the feasibility of the proposed methodology combining metabolomics and DNA methylation, which is the first of its kind in studying the pathogenesis of HD. For the analysis of methylome, we have considered known, potentially confounding, factors, such as sample age, gender, estimated proportion of neurons, and postmortem delay. The effects of nutrition or stress are assumed to be equally distributed between sample groups, and they should not have a significant impact on the outcome. We have also experimented with the use of RUV [70], the method used to estimate unknown confounding variables, but that did not impact the outcome. Further, we emphasize that the regulation of metabolites and genes is tightly intertwined and plays a critical role in maintaining cellular homeostasis in the brain. Dysregulation of these processes can lead to the accumulation of toxic metabolites, oxidative stress, and inflammation, which contribute to the development and progression of HD. Understanding the complex interplay between metabolites and genes in HD has the potential to uncover novel therapeutic targets for this devastating disease. We suggest that these biomarkers of HD should be the focus of future translational research efforts.

Supplementary Materials: The following supporting information can be downloaded at: https://www.mdpi.com/article/10.3390/genes14091752/s1. Supplementary Table S1: Striatum-Metabolites. Supplementary Table S2: Frontal-Metabolites. Supplementary Table S3: EPIC-Enrichment of genomic region. Supplementary Table S4: EPIC-DMCs. Supplementary Table S5: Epimetabolomics-Pathway-Interactions. Supplementary Table S6: Interaction of metabolites and CpGs of sig pathway. Supplementary Figure S1: Metabolomics-Striatum-Variance inflation. Supplementary Figure S2: Metabolomics-Frontal-Variance inflation. Supplementary Figure S3: Metabolomics-Striatum-Normalization. Supplementary Figure S4: Metabolomics-Frontal-Normalization. Supplementary Figure S5: Epic array variance inflation. Supplementary Figure S6: EPIC-Enrichment of genomic region. Supplementary Figure S7: Striatum-Metabolomics-Predictor. Supplementary Figure S8: Striatum-Metabolomics-Predictor. Supplementary Figure S9: Frontal lobe-Methylation-Predictor.

Author Contributions: S.V., U.R. and N.S. were involved in the methylation array process and methylation data generation; A.Y. performed sample preparation and metabolomic data collection using NMR spectroscopy; A.K. and P.R.K. worked on interpreting the results; X.P. was involved in the sample preparation; X.P. and S.M. were involved in the metabolomic data collection through mass spectrometry; A.C.R. and D.M. provided the brain specimens; J.G. and I.K. performed the data analysis; S.F.G. and B.D.G. designed the experiment; S.F.G. oversaw the experiments and provided guidance throughout the project. The manuscript was written by S.V. and S.F.G. and commented on by all authors. All authors have read and agreed to the published version of the manuscript.

Funding: This work was partly funded by the generous contribution made by the Fred A. and Barbara M. Erb Foundation. We thank Biocrates (Innsbruck, Austria) for providing the AbsoluteIDQ p180 kit for the analysis of these specimens.

Institutional Review Board Statement: Not applicable.

Informed Consent Statement: The use of post-mortem human brain samples was approved by the Manchester Brain Bank Management Committee working under conditions agreed upon with Newcastle and North Tyneside 1 Research Ethics Committee (REC Reference# 09/H0906/52 + 5). Further ethical approval was granted by the Beaumont Institutional Review Board (IRB# 2014–353).

Data Availability Statement: Data will be provided upon request to corresponding author.

Acknowledgments: This work was partly funded by the generous contribution made by the Fred A. and Barbara M. Erb Foundation. We thank Biocrates (Innsbruck, Austria) for providing the AbsoluteIDQ p180 kit for the analysis of these specimens. We also acknowledge the assistance of Muhammad Bin Nasaruddin from Queen's University, Belfast for meticulously milling the post-mortem tissue for analysis. Tissue samples were supplied by The Manchester Brain Bank, which is part of the Brains for Dementia Research program, jointly funded by Alzheimer's Research UK and Alzheimer's Society.

Conflicts of Interest: The authors declare no conflict of interest. S.F.G. receives commercial support as a consultant from Biogen, Coleman Research, and Roche, which is not related to the results of this study. No other authors have competing interests.

References

1. Stoker, T.B.; Mason, S.L.; Greenland, J.C.; Holden, S.T.; Santini, H.; Barker, R.A. Huntington's disease: Diagnosis and management. *Pract. Neurol.* **2022**, *22*, 32–41. [CrossRef]
2. Finkbeiner, S. Huntington's Disease. *Cold Spring Harb. Perspect. Biol.* **2011**, *3*, a007476. [CrossRef] [PubMed]
3. Roos, R.A. Huntington's disease: A clinical review. *Orphanet J. Rare Dis.* **2010**, *5*, 40. [CrossRef]
4. Hashimoto, M.; Watanabe, K.; Miyoshi, K.; Koyanagi, Y.; Tadano, J.; Miyawaki, I. Multiplatform metabolomic analysis of the R6/2 mouse model of Huntington's disease. *FEBS Open Bio* **2021**, *11*, 2807–2818. [CrossRef] [PubMed]
5. Medina, A.; Mahjoub, Y.; Shaver, L.; Pringsheim, T. Prevalence and Incidence of Huntington's Disease: An Updated Systematic Review and Meta-Analysis. *Mov. Disord. Off. J. Mov. Disord. Soc.* **2022**, *37*, 2327–2335. [CrossRef]
6. Scahill, R.I.; Zeun, P.; Osborne-Crowley, K.; Johnson, E.B.; Gregory, S.; Parker, C.; Lowe, J.; Nair, A.; O'Callaghan, C.; Langley, C.; et al. Biological and clinical characteristics of gene carriers far from predicted onset in the Huntington's disease Young Adult Study (HD-YAS): A cross-sectional analysis. *Lancet Neurol.* **2020**, *19*, 502–512. [CrossRef] [PubMed]
7. Blumenstock, S.; Dudanova, I. Cortical and Striatal Circuits in Huntington's Disease. *Front. Neurosci* **2020**, *14*, 82. [CrossRef]
8. Waldvogel, H.J.; Kim, E.H.; Tippett, L.J.; Vonsattel, J.P.; Faull, R.L. The Neuropathology of Huntington's Disease. *Curr. Top. Behav. Neurosci.* **2015**, *22*, 33–80. [CrossRef]
9. Gatto, E.M.; Rojas, N.G.; Persi, G.; Etcheverry, J.L.; Cesarini, M.E.; Perandones, C. Huntington disease: Advances in the understanding of its mechanisms. *Clin. Park. Relat. Disord.* **2020**, *3*, 100056. [CrossRef]
10. Chow, T.W. Personality in frontal lobe disorders. *Curr. Psychiatry Rep.* **2000**, *2*, 446–451. [CrossRef]
11. Duff, K.; Paulsen, J.S.; Beglinger, L.J.; Langbehn, D.R.; Wang, C.; Stout, J.C.; Ross, C.A.; Aylward, E.; Carlozzi, N.E.; Queller, S. "Frontal" behaviors before the diagnosis of Huntington's disease and their relationship to markers of disease progression: Evidence of early lack of awareness. *J. Neuropsychiatry Clin. Neurosci.* **2010**, *22*, 196–207. [CrossRef]
12. Vonsattel, J.P.; Myers, R.H.; Stevens, T.J.; Ferrante, R.J.; Bird, E.D.; Richardson, E.P., Jr. Neuropathological classification of Huntington's disease. *J. Neuropathol. Exp. Neurol.* **1985**, *44*, 559–577. [CrossRef]
13. Aylward, E.H.; Anderson, N.B.; Bylsma, F.W.; Wagster, M.V.; Barta, P.E.; Sherr, M.; Feeney, J.; Davis, A.; Rosenblatt, A.; Pearlson, G.D.; et al. Frontal lobe volume in patients with Huntington's disease. *Neurology* **1998**, *50*, 252–258. [CrossRef]
14. Subramanian, I.; Verma, S.; Kumar, S.; Jere, A.; Anamika, K. Multi-omics Data Integration, Interpretation, and Its Application. *Bioinform. Biol. Insights* **2020**, *14*, 1177932219899051. [CrossRef]
15. Villa, C.; Yoon, J.H. Multi-Omics for the Understanding of Brain Diseases. *Life* **2021**, *11*, 1202. [CrossRef]
16. Paananen, J.; Fortino, V. An omics perspective on drug target discovery platforms. *Brief Bioinform.* **2020**, *21*, 1937–1953. [CrossRef]
17. Lu, H.; Liu, X.; Deng, Y.; Qing, H. DNA methylation, a hand behind neurodegenerative diseases. *Front. Aging Neurosci.* **2013**, *5*, 85. [CrossRef]
18. Zsindely, N.; Siági, F.; Bodai, L. DNA Methylation in Huntington's Disease. *Int. J. Mol. Sci.* **2021**, *22*, 12736. [CrossRef]
19. Lu, A.T.; Narayan, P.; Grant, M.J.; Langfelder, P.; Wang, N.; Kwak, S.; Wilkinson, H.; Chen, R.Z.; Chen, J.; Simon Bawden, C.; et al. DNA methylation study of Huntington's disease and motor progression in patients and in animal models. *Nat. Commun.* **2020**, *11*, 4529. [CrossRef]
20. Pan, Y.; Daito, T.; Sasaki, Y.; Chung, Y.H.; Xing, X.; Pondugula, S.; Swamidass, S.J.; Wang, T.; Kim, A.H.; Yano, H. Erratum: Inhibition of DNA Methyltransferases Blocks Mutant Huntingtin-Induced Neurotoxicity. *Sci. Rep.* **2016**, *6*, 33766. [CrossRef]
21. Feehley, T.; O'Donnell, C.W.; Mendlein, J.; Karande, M.; McCauley, T. Drugging the epigenome in the age of precision medicine. *Clin. Epigenetics* **2023**, *15*, 6. [CrossRef] [PubMed]
22. Gonzalez-Covarrubias, V.; Martinez-Martinez, E.; Del Bosque-Plata, L. The Potential of Metabolomics in Biomedical Applications. *Metabolites* **2022**, *12*, 194. [CrossRef] [PubMed]
23. Skene, D.J.; Middleton, B.; Fraser, C.K.; Pennings, J.L.; Kuchel, T.R.; Rudiger, S.R.; Bawden, C.S.; Morton, A.J. Metabolic profiling of presymptomatic Huntington's disease sheep reveals novel biomarkers. *Sci. Rep.* **2017**, *7*, 43030. [CrossRef] [PubMed]
24. Zhang, A.; Sun, H.; Yan, G.; Wang, P.; Wang, X. Metabolomics for Biomarker Discovery: Moving to the Clinic. *Biomed Res. Int.* **2015**, *2015*, 354671. [CrossRef]

25. Rosas, H.D.; Doros, G.; Bhasin, S.; Thomas, B.; Gevorkian, S.; Malarick, K.; Matson, W.; Hersch, S.M. A systems-level "misunderstanding": The plasma metabolome in Huntington's disease. *Ann. Clin. Transl. Neurol.* **2015**, *2*, 756–768. [CrossRef]
26. Underwood, B.R.; Broadhurst, D.; Dunn, W.B.; Ellis, D.I.; Michell, A.W.; Vacher, C.; Mosedale, D.E.; Kell, D.B.; Barker, R.A.; Grainger, D.J.; et al. Huntington disease patients and transgenic mice have similar pro-catabolic serum metabolite profiles. *Brain* **2006**, *129*, 877–886. [CrossRef]
27. Leoni, V.; Caccia, C. The impairment of cholesterol metabolism in Huntington disease. *Biochim. Biophys. Acta* **2015**, *1851*, 1095–1105. [CrossRef]
28. Block, R.C.; Dorsey, E.R.; Beck, C.A.; Brenna, J.T.; Shoulson, I. Altered cholesterol and fatty acid metabolism in Huntington disease. *J. Clin. Lipidol.* **2010**, *4*, 17–23. [CrossRef]
29. Sbodio, J.I.; Snyder, S.H.; Paul, B.D. Transcriptional control of amino acid homeostasis is disrupted in Huntington's disease. *Proc. Natl. Acad. Sci. USA* **2016**, *113*, 8843–8848. [CrossRef]
30. Graham, S.F.; Pan, X.; Yilmaz, A.; Macias, S.; Robinson, A.; Mann, D.; Green, B.D. Targeted biochemical profiling of brain from Huntington's disease patients reveals novel metabolic pathways of interest. *Biochim. Biophys. Acta (BBA) Mol. Basis Dis.* **2018**, *1864*, 2430–2437. [CrossRef]
31. Graham, S.F.; Kumar, P.K.; Bjorndahl, T.; Han, B.; Yilmaz, A.; Sherman, E.; Bahado-Singh, R.O.; Wishart, D.; Mann, D.; Green, B.D. Metabolic signatures of Huntington's disease (HD): (1)H NMR analysis of the polar metabolome in post-mortem human brain. *Biochim. Biophys. Acta* **2016**, *1862*, 1675–1684. [CrossRef] [PubMed]
32. Mercier, P.; Lewis, M.J.; Chang, D.; Baker, D.; Wishart, D.S. Towards automatic metabolomic profiling of high-resolution one-dimensional proton NMR spectra. *J. Biomol. NMR* **2011**, *49*, 307–323. [CrossRef]
33. Urban, M.; Enot, D.P.; Dallmann, G.; Körner, L.; Forcher, V.; Enoh, P.; Koal, T.; Keller, M.; Deigner, H.-P. Complexity and pitfalls of mass spectrometry-based targeted metabolomics in brain research. *Anal. Biochem.* **2010**, *406*, 124–131. [CrossRef] [PubMed]
34. Graham, S.F.; Turkoglu, O.; Kumar, P.; Yilmaz, A.; Bjorndahl, T.C.; Han, B.; Mandal, R.; Wishart, D.S.; Bahado-Singh, R.O. Targeted Metabolic Profiling of Post-Mortem Brain from Infants Who Died from Sudden Infant Death Syndrome. *J. Proteome Res.* **2017**, *16*, 2587–2596. [CrossRef]
35. Graham, S.F.; Chevallier, O.P.; Kumar, P.; Türko Gcaron Lu, O.; Bahado-Singh, R.O. Metabolomic profiling of brain from infants who died from Sudden Infant Death Syndrome reveals novel predictive biomarkers. *J. Perinatol. Off. J. Calif. Perinat. Assoc.* **2017**, *37*, 91–97. [CrossRef] [PubMed]
36. Thomas, E.A. DNA methylation in Huntington's disease: Implications for transgenerational effects. *Neurosci. Lett.* **2016**, *625*, 34–39. [CrossRef]
37. Katada, S.; Imhof, A.; Sassone-Corsi, P. Connecting Threads: Epigenetics and Metabolism. *Cell* **2012**, *148*, 24–28. [CrossRef]
38. Kori, M.; Aydın, B.; Unal, S.; Arga, K.Y.; Kazan, D. Metabolic Biomarkers and Neurodegeneration: A Pathway Enrichment Analysis of Alzheimer's Disease, Parkinson's Disease, and Amyotrophic Lateral Sclerosis. *OMICS A J. Integr. Biol.* **2016**, *20*, 645–661. [CrossRef]
39. Herman, S.; Niemelä, V.; Emami Khoonsari, P.; Sundblom, J.; Burman, J.; Landtblom, A.-M.; Spjuth, O.; Nyholm, D.; Kultima, K. Alterations in the tyrosine and phenylalanine pathways revealed by biochemical profiling in cerebrospinal fluid of Huntington's disease subjects. *Sci. Rep.* **2019**, *9*, 4129. [CrossRef] [PubMed]
40. Young, S.N.; Shalchi, M. The effect of methionine and S-adenosylmethionine on S-adenosylmethionine levels in the rat brain. *J. Psychiatry Neurosci. JPN* **2005**, *30*, 44–48.
41. Tapia-Rojas, C.; Lindsay, C.B.; Montecinos-Oliva, C.; Arrazola, M.S.; Retamales, R.M.; Bunout, D.; Hirsch, S.; Inestrosa, N.C. Is L-methionine a trigger factor for Alzheimer's-like neurodegeneration?: Changes in Aβ oligomers, tau phosphorylation, synaptic proteins, Wnt signaling and behavioral impairment in wild-type mice. *Mol. Neurodegener.* **2015**, *10*, 62. [CrossRef] [PubMed]
42. Schöneich, C. Methionine oxidation by reactive oxygen species: Reaction mechanisms and relevance to Alzheimer's disease. *Biochim. Biophys. Acta* **2005**, *1703*, 111–119. [CrossRef]
43. Selhub, J. Homocysteine metabolism. *Annu. Rev. Nutr.* **1999**, *19*, 217–246. [CrossRef] [PubMed]
44. Andrich, J.; Saft, C.; Arz, A.; Schneider, B.; Agelink, M.W.; Kraus, P.H.; Kuhn, W.; Müller, T. Hyperhomocysteinaemia in treated patients with Huntington's disease homocysteine in HD. *Mov. Disord. Off. J. Mov. Disord. Soc.* **2004**, *19*, 226–228. [CrossRef] [PubMed]
45. Chen, Z.; Gao, M.; Su, Y.; Liu, P.; Sun, B. Running Promotes Transformation of Brain Astrocytes Into Neuroprotective Reactive Astrocytes and Synaptic Formation by Targeting Gpc6 Through the STAT3 Pathway. *Front. Physiol.* **2021**, *12*, 633618. [CrossRef] [PubMed]
46. Baier, A.; Szyszka, R. CK2 and protein kinases of the CK1 superfamily as targets for neurodegenerative disorders. *Front. Mol. Biosci.* **2022**, *9*, 916063. [CrossRef]
47. Geisinger, A.; Alsheimer, M.; Baier, A.; Benavente, R.; Wettstein, R. The mammalian gene pecanex 1 is differentially expressed during spermatogenesis. *Biochim. Biophys. Acta* **2005**, *1728*, 34–43. [CrossRef]
48. Song, W.; Li, Q.; Wang, T.; Li, Y.; Fan, T.; Zhang, J.; Wang, Q.; Pan, J.; Dong, Q.; Sun, Z.S.; et al. Putative complement control protein CSMD3 dysfunction impairs synaptogenesis and induces neurodevelopmental disorders. *Brain Behav. Immun.* **2022**, *102*, 237–250. [CrossRef]
49. Xu, A.; Wang, W.; Jiang, X. The roles of MTRR and MTHFR gene polymorphisms in congenital heart diseases: A meta-analysis. *Biosci. Rep.* **2018**, *38*, BSR20181160. [CrossRef]

50. Wang, Y.; Kerrisk Campbell, M.; Tom, I.; Foreman, O.; Hanson, J.E.; Sheng, M. PCDH7 interacts with GluN1 and regulates dendritic spine morphology and synaptic function. *Sci. Rep.* **2020**, *10*, 10951. [CrossRef]
51. Yamada, S.B.; Gendron, T.F.; Niccoli, T.; Genuth, N.R.; Grosely, R.; Shi, Y.; Glaria, I.; Kramer, N.J.; Nakayama, L.; Fang, S.; et al. RPS25 is required for efficient RAN translation of C9orf72 and other neurodegenerative disease-associated nucleotide repeats. *Nat. Neurosci.* **2019**, *22*, 1383–1388. [CrossRef] [PubMed]
52. Wild, E.J.; Tabrizi, S.J. Therapies targeting DNA and RNA in Huntington's disease. *Lancet Neurol.* **2017**, *16*, 837–847. [CrossRef] [PubMed]
53. Ghosh, S.G.; Scala, M.; Beetz, C.; Helman, G.; Stanley, V.; Yang, X.; Breuss, M.W.; Mazaheri, N.; Selim, L.; Hadipour, F.; et al. A relatively common homozygous TRAPPC4 splicing variant is associated with an early-infantile neurodegenerative syndrome. *Eur. J. Hum. Genet. EJHG* **2021**, *29*, 271–279. [CrossRef] [PubMed]
54. Rubio Gomez, M.A.; Ibba, M. Aminoacyl-tRNA synthetases. *RNA* **2020**, *26*, 910–936. [CrossRef]
55. Ibba, M.; Söll, D. Aminoacyl-tRNAs: Setting the limits of the genetic code. *Genes Dev.* **2004**, *18*, 731–738. [CrossRef]
56. Girstmair, H.; Saffert, P.; Rode, S.; Czech, A.; Holland, G.; Bannert, N.; Ignatova, Z. Depletion of cognate charged transfer RNA causes translational frameshifting within the expanded CAG stretch in huntingtin. *Cell Rep.* **2013**, *3*, 148–159. [CrossRef]
57. Lant, J.T.; Berg, M.D.; Heinemann, I.U.; Brandl, C.J.; O'Donoghue, P. Pathways to disease from natural variations in human cytoplasmic tRNAs. *J. Biol. Chem.* **2019**, *294*, 5294–5308. [CrossRef]
58. Lant, J.T.; Kiri, R.; Duennwald, M.L.; O'Donoghue, P. Formation and persistence of polyglutamine aggregates in mistranslating cells. *Nucleic Acids Res.* **2021**, *49*, 11883–11899. [CrossRef]
59. Mochel, F.; Charles, P.; Seguin, F.; Barritault, J.; Coussieu, C.; Perin, L.; Le Bouc, Y.; Gervais, C.; Carcelain, G.; Vassault, A.; et al. Early energy deficit in Huntington disease: Identification of a plasma biomarker traceable during disease progression. *PLoS ONE* **2007**, *2*, e647. [CrossRef]
60. Schmidt, R.L.; Simonović, M. Synthesis and decoding of selenocysteine and human health. *Croat. Med. J.* **2012**, *53*, 535–550. [CrossRef]
61. Zhang, Y.; Roh, Y.J.; Han, S.J.; Park, I.; Lee, H.M.; Ok, Y.S.; Lee, B.C.; Lee, S.R. Role of Selenoproteins in Redox Regulation of Signaling and the Antioxidant System: A Review. *Antioxidants* **2020**, *9*, 383. [CrossRef] [PubMed]
62. Kumar, A.; Ratan, R.R. Oxidative Stress and Huntington's Disease: The Good, The Bad, and The Ugly. *J. Huntington's Dis.* **2016**, *5*, 217–237. [CrossRef]
63. Browne, S.E.; Beal, M.F. Oxidative damage in Huntington's disease pathogenesis. *Antioxid. Redox Signal.* **2006**, *8*, 2061–2073. [CrossRef] [PubMed]
64. Machiela, E.; Jeloka, R.; Caron, N.S.; Mehta, S.; Schmidt, M.E.; Baddeley, H.J.E.; Tom, C.M.; Polturi, N.; Xie, Y.; Mattis, V.B.; et al. The Interaction of Aging and Cellular Stress Contributes to Pathogenesis in Mouse and Human Huntington Disease Neurons. *Front. Aging Neurosci.* **2020**, *12*, 524369. [CrossRef] [PubMed]
65. Pillai, R.; Uyehara-Lock, J.H.; Bellinger, F.P. Selenium and selenoprotein function in brain disorders. *IUBMB Life* **2014**, *66*, 229–239. [CrossRef]
66. Cardoso, B.R.; Roberts, B.R.; Bush, A.I.; Hare, D.J. Selenium, selenoproteins and neurodegenerative diseases. *Met. Integr. Biometal Sci.* **2015**, *7*, 1213–1228. [CrossRef]
67. Umair, M.; Alfadhel, M. Genetic Disorders Associated with Metal Metabolism. *Cells* **2019**, *8*, 1598. [CrossRef]
68. Ye, R.; Huang, J.; Wang, Z.; Chen, Y.; Dong, Y. The Role and Mechanism of Essential Selenoproteins for Homeostasis. *Antioxidants* **2022**, *11*, 973. [CrossRef]
69. Lu, Z.; Marks, E.; Chen, J.; Moline, J.; Barrows, L.; Raisbeck, M.; Volitakis, I.; Cherny, R.A.; Chopra, V.; Bush, A.I.; et al. Altered selenium status in Huntington's disease: Neuroprotection by selenite in the N171-82Q mouse model. *Neurobiol. Dis.* **2014**, *71*, 34–42. [CrossRef]
70. Risso, D.; Ngai, J.; Speed, T.P.; Dudoit, S. Normalization of RNA-seq data using factor analysis of control genes or samples. *Nat. Biotechnol.* **2014**, *32*, 896–902. [CrossRef]

Disclaimer/Publisher's Note: The statements, opinions and data contained in all publications are solely those of the individual author(s) and contributor(s) and not of MDPI and/or the editor(s). MDPI and/or the editor(s) disclaim responsibility for any injury to people or property resulting from any ideas, methods, instructions or products referred to in the content.

genes

MDPI

Article

Finding an Appropriate Mouse Model to Study the Impact of a Treatment for Friedreich Ataxia on the Behavioral Phenotype

Camille Bouchard [1,2,*,†], Catherine Gérard [1,2,†], Solange Gni-fiene Yanyabé [1,2], Nathalie Majeau [1,2], Malek Aloui [1], Gabrielle Buisson [1], Pouiré Yameogo [1,2], Vanessa Couture [1,2] and Jacques P. Tremblay [1,2,*]

1 Centre de Recherche du CHU, Québec-Université Laval, Québec, QC G1V 4G2, Canada; nathalie.majeau@crchudequebec.ulaval.ca (N.M.); malek.aloui@crchudequebec.ulaval.ca (M.A.); gabrielle.buisson@crchudequebec.ulaval.ca (G.B.)

2 Département de Médecine Moléculaire, l'Université Laval Québec, Québec, QC G1V 4G2, Canada

* Correspondence: camille.bouchard@crchudequebec.ulaval.ca (C.B.); jacques-p.tremblay@crchudequebec.ulaval.ca (J.P.T.)

† These authors contributed equally to this work.

Abstract: Friedreich ataxia (FRDA) is a progressive neurodegenerative disease caused by a GAA repeat in the intron 1 of the frataxin gene (FXN) leading to a lower expression of the frataxin protein. The YG8sR mice are Knock-Out (KO) for their murine frataxin gene but contain a human frataxin transgene derived from an FRDA patient with 300 GAA repeats. These mice are used as a FRDA model but even with a low frataxin concentration, their phenotype is mild. We aimed to find an optimized mouse model with a phenotype comparable to the human patients to study the impact of therapy on the phenotype. We compared two mouse models: the YG8sR injected with an AAV. PHP.B coding for a shRNA targeting the human frataxin gene and the YG8-800, a new mouse model with a human transgene containing 800 GAA repeats. Both mouse models were compared to Y47R mice containing nine GAA repeats that were considered healthy mice. Behavior tests (parallel rod floor apparatus, hanging test, inverted T beam, and notched beam test) were carried out from 2 to 11 months and significant differences were noticed for both YG8sR mice injected with an anti-FXN shRNA and the YG8-800 mice compared to healthy mice. In conclusion, YG8sR mice have a slight phenotype, and injecting them with an AAV-PHP.B expressing an shRNA targeting frataxin does increase their phenotype. The YG8-800 mice have a phenotype comparable to the human ataxic phenotype.

Keywords: Friedreich ataxia; mouse model; phenotype; frataxin; shRNA

Citation: Bouchard, C.; Gérard, C.; Yanyabé, S.G.-f.; Majeau, N.; Aloui, M.; Buisson, G.; Yameogo, P.; Couture, V.; Tremblay, J.P. Finding an Appropriate Mouse Model to Study the Impact of a Treatment for Friedreich Ataxia on the Behavioral Phenotype. *Genes* **2023**, *14*, 1654. https://doi.org/10.3390/genes14081654

Academic Editor: Diego Centonze

Received: 18 July 2023
Revised: 9 August 2023
Accepted: 16 August 2023
Published: 19 August 2023

Copyright: © 2023 by the authors. Licensee MDPI, Basel, Switzerland. This article is an open access article distributed under the terms and conditions of the Creative Commons Attribution (CC BY) license (https://creativecommons.org/licenses/by/4.0/).

1. Introduction

Friedreich ataxia (FRDA) is the most common inherited recessive ataxia, a progressive and degenerative neurological disease. This ataxia is due to a mutation in the frataxin gene (FXN), which is, in most cases, an expansion in a GAA repeat in the first intron [1] leading to a reduced expression of the frataxin protein. The reduction of frataxin depends on the number of GAA repeats [2,3] and the protein concentration is drastically reduced in human patients [4]. This small protein is implicated in the iron transport and metabolism in mitochondria. Frataxin allows the acceleration of a sulfur transfer step, which limits the synthesis rate of 2Fe-2S clusters in the Fe-S assembly complex. Therefore, its decrease leads first to mitochondrial dysfunction and oxidative stress and finally to the death of cells including those in the cardiac and the nervous system [5,6]. Physical symptoms mean age of onset in classical FRDA is during childhood (6–16 years old) [7,8]. Patients suffering from Friedreich ataxia have difficulties with the coordination of their movements and their balance because of the loss of neurons in the central and peripheral nervous system. The premature death of patients is due principally to the cardiac damage associated with the disease [7,9,10].

Currently, there is no effective treatment for Friedreich ataxia. There are multiple potential therapies that are being explored, focusing on frataxin directly, such as the administration of frataxin protein fused with a cell-penetrating peptide [11], the induction of frataxin expression by the TALE protein fused with VP64 [12,13], and the administration of an AAV coding for the frataxin gene [14–17]. Another gene therapy approach is the excision of the GAA repeat with zinc finger nucleases (ZFNs) or with the CRISPR/Cas9 technologies [18–20]. These gene therapies are using an AAV to deliver the CRISPR system in mice. A Phase 1/2 clinical trial using an AAVrh10 to deliver the frataxin gene (LX2006) [21] to reduce FRDA cardiac symptoms has recently been approved by the Food and Drug Administration (FDA) [17].

Salami et al. [17] used a new mouse model, called αMyhc, created to reproduce the cardiomyopathy of the FRDA patients. The mouse model is of primary importance to obtain pre-clinical results. Different mouse models have been developed to investigate potential treatments, but so far, none of these models is close to the real phenotype of Friedreich ataxia. Puccio and coworkers in 2001 [22] developed the MCK-Cre cardiac mouse model expressing the Cre recombinase under the muscle creatine kinase (MCK) promoter permitting to knock out the FNX gene leading to cardiomyopathy after 5 weeks of age [23]. That group also developed the NSE-Cre nervous system mouse model expressing the Cre recombinase under the neuron-specific promoter enolase (NSE) [24]. The NSE-Cre model showed not only neurological deficits due to frataxin absence but also cardiac deficits leading to symptoms in both systems [22]. These mouse models were previously used in our laboratory, but the symptoms were too severe compared to the disease in humans. Indeed, their death occurred between 30 to 90 days for NSE-Cre and MCK-Cre respectively [15,25]. Recently, Puccio and co-workers generated a new conditional model called Pvalbtm1(Cre)Arbr/J using the parvalbumin promotor to target proprioceptive afferent sensory neurons such as the DRG, cerebellar Purkinje cells, and interneurons in the brain, to focus on the neuropathy associated with FRDA. These mice mimic the neuropathophysiology observed in FRDA patients but with a more rapid and severe course [16]. The KIKO model (Fxn(tm1Mkn/J) has an expansion of 230 GAA in one of the mouse Fxn genes and a deletion of the exon 4 in the other Fxn gene resulting in a robust neurobehavioral phenotype [26]. In 2017, Chandran and coworkers [27] created an inducible and reversible FRDAkd mouse model. This mouse model includes a transgene containing an shRNA against mouse frataxin under an H1 promotor, which was doxycycline-inducible. This permitted a significant frataxin protein reduction in all tissues. The strong decrease in mouse frataxin protein observed under doxycycline treatment was associated with a severe phenotype. The mice showed decreased locomotor activity and less coordination and were weaker and globally a phenotype, which is more similar to the Friedreich ataxia patients. However, this mouse model is inappropriate to test certain therapies targeting the human FXN gene such as deletion of the GAA repeats by CRISPR/Cas9, but the KIKO model could be used for this application.

Each mouse model has some advantages and some disadvantages, the goal is to choose the best one for the type of treatment to be investigated. The group of Pook [28] developed two Friedreich ataxia mouse models: the YG8R and the YG8sR in which both mouse frataxin genes are knockout and a human frataxin transgene derived from a Friedreich patient is added, which is advantageous for the investigations of some potential treatments. The original YG8R contained a human FXN transgene with two consecutive GAA repeats (one containing 90 and the other 190 GAA) while the Y47R control mouse had only 9 GAA repeats [28]. The improved YG8sR contained a human FXN transgene with only one repeat containing 250–300 GAAs [29]. However, even this YG8sR model expressed only a weak phenotype.

We, therefore, assumed that the weak phenotype in YG8sR mice was due to an insufficient reduction of frataxin. We decided to accentuate the severity of the YG8sR phenotype by using a shRNA against human frataxin mRNA, an approach similar to the one used by Chandran et al., for the mouse frataxin gene [27]. We initially selected an adequate shRNA by testing the effectiveness of different shRNAs to reduce frataxin in a culture of HeLa cells. The best shRNA was inserted in an AAV-PHP.B virus, which was injected intravenously, and the mice developed a more severe phenotype. The co-injection of two AAV-PHP.B codes for an anti-FXN shRNA and the frataxin gene used as a treatment prevented the development of a more severe FRDA phenotype.

In parallel, we compared this model to the new YG8-800 developed by Jackson Laboratory from the YG8sR descendants containing 800 GAA repeats instead of 300. The hypothesis here was that since a longer GAA expansion in humans correlates with a more severe phenotype [8], the same trend would be observed in mice.

2. Materials and Methods

2.1. Plasmids Construction (Coding for FXN Gene and Anti-FXN shRNA)

Adeno-associated viruses, serotype PHP.B (AAV-PHP.B) [30], were used to deliver either a plasmid coding for anti-FXN shRNA to increase the FRDA phenotype in YG8sR mice or the frataxin gene as a treatment in YG8sR mice. This specific serotype is known to drive expression in the central nervous system with an ability to cross the blood-brain barrier (BBB).

The previously described scAAV plasmid [15] was modified to change the CAG promoter for the chicken β-actin hybrid (CBh) promoter to express the human frataxin gene ubiquitously (scAAV-CBh-FXN). To synthetize the scAAV-hSyn-FXN, a specific plasmid to drive expression in the central nervous system, the scAAV-CBh-FXN plasmid was digested with AvrII and AgeI to remove the CBh promoter. The human synapsin 1 (hSyn) promoter was excised from plasmid AAV synap FZ (Addgene #60231) by XbaI and AgeI and ligated into the plasmid to create scAAV-hSyn-FXN. Both constructions were used to compare the efficiency of both promoters to deliver the FXN gene as a potential treatment.

For the construction of anti-FXN shRNA plasmids to increase the YG8sR phenotype, scAAV-hSyn-FXN was digested with BamHI and MluI, and U6-shRE and UBC-mCherry were ligated to the plasmid using NEBuilders. The resulting plasmid was digested with AgeI and NheI, and shFXN oligonucleotides were annealed and ligated to the plasmid to create scAAV-shRNAFXN (1, 3, 4 or 6)-UBC-mCherry plasmids. The shRNA sequences were derived from predesigned Sigma sequences: TRCN0000006138 (shRNA1), TRCN0000380594 (shRNA3), TRCN0000010996 (shRNA4), TRCN0000350511 (shRNA6). The sequences and sites targeted by the various shRNAs are illustrated in Figure 1A,B. The AAV plasmid containing the shRNA sequence is illustrated in Figure 1C.

A

shRNA	sequence
shRNA1	GCTGGACTCTTTAGCAGAGTT
shRNA3	GCCATACACGTTTGAGGACTAT
shRNA4	ACGTGGCCTCAACCAGATTTG
shRNA6	GTGGACCTAAGCGTTATGACT

B

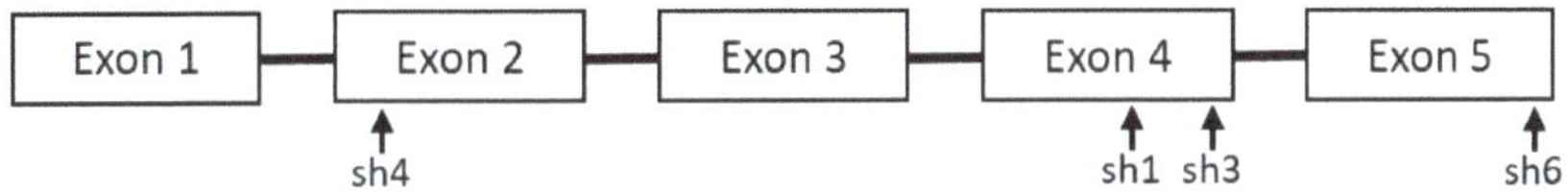

Figure 1. *Cont.*

C

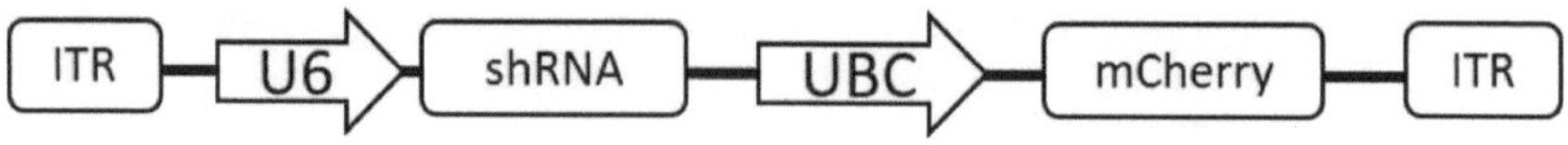

Figure 1. Specific sequences in the frataxin gene targeted by 4 shRNAs are indicated in (**A**). The sites of these 4 shRNAs are indicated in (**B**). The shRNAs were integrated into AAV vectors under the U6 promoter and these AAVs also contained a mCherry gene under the UBC promoter (**C**).

2.2. *In Vitro Testing of the shRNAs*

HeLa cells were transfected using Lipofectamine 2000 with the different plasmids coding for a shRNA and mCherry (scAAV-shFXN-mCherry) alone or in combination for a total of 2 μg of plasmids. The fluorescence of mCherry was used to assess the transfection success. The cells were collected after 72 h and the proteins were extracted. The frataxin protein concentration was measured by an ELISA test (Ab109881, Abcam, Cambridge, UK). These in vitro tests allowed us to select the best shRNA likely to reduce frataxin in vivo.

2.3. *AAV Production*

For in vivo testing, different promotors were used to control the expression of the shRNA and the FXN gene. The shRNA was placed under a U6 promotor and the mCherry gene under a UBC (Figure 1C). In the AAV coding for the frataxin gene, two promoters were tested and compared, hSyn for specific delivery to the nervous system and CBh for a ubiquitous delivery.

Adeno-associated viruses, serotype PHP.B (AAV-PHP.B) [30], were used to deliver the shRNA or the frataxin gene in vivo in YG8sR mice. The AAV-PHP.B vectors were produced by the Plateforme d'outils moléculaires, Centre de recherche CERVO (Québec, QC, Canada).

2.4. *AAV Injection*

Mice (approximately 2 months old) were injected in the tail vein (IV) with 0.9% saline solution, with an AAV coding for shRNAscr (scramble shRNA sequence, not targeting frataxin) for control mice or with an AAV coding for anti-FXN shRNA. Different doses of AAVs were used from 0.5× (i.e., 3×10^{11} viral copies) to 4× (i.e., 2.4×10^{12} viral copies) depending on the effect on the mice. To treat the 2 months old mice, an AAV coding for human frataxin (FXN) gene was administrated alone or mixed with the AAV coding for a shRNA.

2.5. *YG8sR, YG8-800 and Y47R Mice*

The YG8sR, YG8-800, and Y47R mice were obtained from Jackson Laboratories Inc and reproduced in our facilities. To summarize, YG8sR (Fxntm1Mkn Tg(FXN)YG8Pook/2J; https://www.jax.org/strain/024113) (accessed on 10 June 2023) mice, YG8-800 (Fxnem2.1Lutzy Tg(FXN)YG8Pook/800J; https://www.jax.org/strain/030395) (accessed on 10 June 2023) mice and Y47R (Fxntm1Mkn Tg(FXN)Y47Pook/J; https://www.jax.org/strain/024097) (accessed on 10 June 2023) mice [29] all have mouse FXN genes knockout but they have, respectively, a transgene containing the human FXN gene with a 250–300 GAA triplet, an 800 GAA triplet, or a 9 GAA triplet. C57Bl6J (subsequently called C57) mice were used as control mice, containing no human transgene. The C57 mice group was made of 15 mice (9 females and 6 males), the Y47R group of 21 mice (13 females and 8 males), the YG8sR group of 19 mice (11 females and 8 males), and the YG8-800 group of 15 mice (7 males and 8 females) initially at 2 months age. The housing conditions were 2–5 mice per standard ventilated cage under a 12-h light/dark cycle with access to food and water ad libitum. Neslet and Aspen's shaving are given for nidification. All experiments involving animals were approved by the Comité de Protection des Animaux de l'Université Laval (CPAUL3) (Québec, QC, Canada).

The first mice group injected with an AAV were euthanized at 3.5 months after the behavior tests, 5 weeks after the IV injection, and perfused with 0.9% saline to remove the blood from tissues. Different tissues (Tibialis anterior muscle, heart, liver, cerebrum, cerebellum, and dorsal root ganglions (DRG)) were then collected and stored at −80 °C. The subsequent group made of 15 C57 mice (9 females and 6 males), 21 Y47R mice (13 females and 8 males), 19 YG8sR mice (11 females and 8 males), and 15 YG8-800 mice (7 males and 8 females) were euthanized after 11 to 14 months with behavior tests every 3 weeks starting at 8 weeks age. The same tissues were also collected and stored at −80 °C. The euthanasia method was cardiac perfusion under isoflurane anesthesia.

2.6. Behavior Tests

The physical performances of YG8sR and YG8-800 mice were compared with those of Y47R mice and also with those of C57Bl6J (C57) mice, which are control mice with their natural mouse frataxin genes. All mice were evaluated every 3 months starting at 2 months and up to 11 months. For statistics, the females and males were pooled together because there were no significant differences in the preliminary analysis. The C57 mice group was made of 15 mice (9 females and 6 males), the Y47R group of 21 mice (13 females and 8 males), the YG8sR group of 19 mice (11 females and 8 males), and the YG8-800 group of 15 mice (7 males and 8 females).

2.6.1. Parallel Rod Floor

The mice were placed in the parallel rod floor apparatus [31] for 8 min and followed with a camera. The apparatus is a plexiglass cage with metallic parallel rods 1 cm above the metallic plate floor. The apparatus detected foot faults when the foot passed through the grid and touched the metallic plate at the bottom. Different parameters (i.e., the distance traveled, the average speed, the time spent immobile, and the foot fault number) were recorded with the ANY-maze software version 7.20 connected to the camera and floor foot detector.

2.6.2. Hanging Test

The mice were weighed before the test. This is a four-limbs hanging test, the mouse grasped a wire grid as it is inverted. The test was based on the Kondziela's inverted screen test [32,33]. The time of sustained limb tension to oppose the mouse weight was measured. The chronometer was started as soon as the grid was inverted over the cage. The time the mouse remained hanging (before the mouse fell) was noted, with a maximum of 300 s for the test. Three sets of hanging tests were conducted with a minimum of 2 min between each test. The longest time the mouse remained hanging on the screen was noted.

2.6.3. Notched Beam Test

Our test was based on the Di Bonito [34] notched beam test. We used a one-meter Plexiglass beam engraved with 18 squares of 2 cm wide/deep/long and spaced every 2 cm. The locomotion was assessed by the time required for the mouse to cross the beam over the squares. We also counted the foot faults manually with 2 observers (i.e., each time the foot touched the bottom of the square). The experiment was made in triplicates and the average number of foot faults was used.

2.6.4. Inverted T Beam/Balance Beam Test

This test used a Plexiglass balance beam, which was 1 m long and 2 cm wide with a central platform of 0.66 cm wide and 0.66 cm high based on the balance beam test of Luong and co-workers [35]. The mouse walks on the 0.66 cm top beam and steps on the bottom parts of the platform (inverted T shape) when making a foot fault. The time to cross the beam was also recorded as well as every foot fault (mouse foot slipped off the beam), which were manually counted by 2 people. The experiment was made in triplicates and the average number of foot faults was used.

2.7. Imaging of mCherry Fluorescence in Mouse Organs

The integrated imaging station IVIS Lumina LT Series III (PerkinElmer, Waltham, USA) was used to evaluate the distribution of mCherry protein in macroscopic organs using its intrinsic orange-red fluorescence. The apparatus was set in fluorescence mode with filters at 535 nm (excitation) and 620 nm (emission), small binning, and 1–30 s range of exposure time. Under the dedicated Living Image software 4.5.4, the fluorescent signal intensity was calibrated as radiant efficiency and normalized to allow intensity comparisons. No significant auto-fluorescent signal was detected. The imaging was achieved by the Plateforme Imagerie animale par bioluminescence/fluorescence, Centre de recherche CHU (Québec, QC, Canada).

2.8. DNA Extraction from Tissues

DNA was extracted from different tissues (muscle, liver, heart, brain, and dorsal root ganglions). Briefly, a part of the tissue was recovered and incubated with 50 mL proteinase K (10 mg/mL) in a lysis buffer (1.82 g TrisBase, 5 g SDS, 10 mL EDTA 0.5 M completed with distilled water to 500 mL, pH 8.0) at 56 °C until the solution became clear. Digested tissues were then mixed with a 500 µL solution of phenol/chloroform/isoamyl alcohol (25:24:1; BioShop Canada Inc., Burlington, ON, Canada) and centrifuged for 3 min at 12,000 rpm. The upper solution was recovered and mixed with the same volume of chloroform and centrifuged again. The upper solution was recovered, and 50 µL of 5 M sodium chloride was added before the addition of 1 mL 100% ethanol. After centrifugation for 8 min at 12,000 rpm, the pellets were washed in 70% alcohol before another centrifugation. Pellets were dried before DNA suspension in sterile water. The detection of the AAV content was accomplished by PCR, with a band of 410 pb and 435 pb respectively for the plasmid delivered by AAV expressing the shRNA (AAV-shRNA) or the frataxin gene (AAV-FXN) (Supplementary Figure S1). The shRNA plasmid contained a 410 pb mCherry section (Figure 1C) and the frataxin gene DNA amplified fragment was 435 pb.

2.9. Quantification of the Viral Particles by qPCR

The viral particle concentration in organs was measured by qPCR using primers that target the mCherry sequence present with the shRNA sequence in the AAV. 50 ng of the previously extracted gDNA was used per well with 6 µL Advanced Universal SYBR Green supermix, 0.5 µL 10 mM Forward mCherry primer, 0.5 µL 10 mM Reverse mCherry primer and completed to 10 µL with DNase free water. Bio-Rad CFX96 was set at SYBR only with a 3-min 98 °C polymerase activation and DNA denaturation phase, then 40 cycles of 10 s 98 °C denaturation and 20 s 60 °C annealing and extension with a melt curve analysis from 65 °C to 95 °C increasing the temperature by 0.5 °C at every 5 s. The standard curve was made with 8 samples from 101 to 108 particles. Quantified organs were TA muscle, liver, heart, cerebrum, cerebellum, and DRG. All samples were duplicated. This qPCR used primers in the mCherry and in the HPRT reference gene (Supplementary Table S1).

2.10. Quantification of mRNA Expression by qRT-PCR

Frataxin mRNA expression in tissues was compared by qRT-PCR. The RNA was first extracted by adding 1 mL of Trizol to the finely crushed tissue slice. A total of 200 µL chloroform was then added to separate the phases during 15 min centrifugation at 12,000 rpm. The supernatant was collected and precipitated in 500 µL cold (−20 °C) isopropanol centrifuged at 12,000 rpm for 10 min. The liquid was then removed from the tube to keep only the pellet to be washed twice with 1 mL 75% cold (−20 °C) ethanol and centrifuged at 10,000 rpm for 5 min. The pellet was resuspended in 20–60 µL RNAse free water. The RNA concentration was measured with a spectrophotometer and the migration on agarose gel 1% permitted to verify that the RNA was not degraded. The sample was then treated with DNAse (NEB Inc.) incubated for 10 min at 37 °C, then the DNAse was deactivated with EDTA 2.5 mM at 75 °C for 10 min. The RNA was then put in the presence of Reverse Transcriptase with dNTPs and random primers in the buffer provided by Thermo Fischer Scientific Inc. (Waltham, MA, USA) in the 'High Capacity cDNA Reverse Transcription kit'. A qPCR was then performed with the complementary DNA obtained. This qRT-PCR used primers for frataxin as well as for HPRT, a reference gene (Supplementary Table S1).

2.11. Frataxin Protein Quantification

The proteins were extracted using 300 µL of the protein extraction buffer provided by AbCam in the Human frataxin Elisa kit (ab176112, Abcam, Cambridge, MA, USA) per tissue slice (0.5 mm). The frataxin protein concentration was estimated using the PierceTM BCA Protein Assay Kit (Thermo Fischer Scientific Inc.). This kit uses a bovine serum albumin (BSA) standard of 0 mg/mL, 0.25 mg/mL, 0.625 mg/mL, 1.25 mg/mL, 1.875 mg/mL, 2.5 mg/mL and 5 mg/mL. All samples were conducted in duplicate and their optical density (OD) at 562 nm was compared to the standard curve to estimate their protein concentration.

The human frataxin protein was quantified using the Human frataxin Elisa kit (ab176112, Abcam, Cambridge, MA, USA) for in vitro and for mouse tissues. A standard curve was also made at the same time with recombinant human frataxin protein to quantify the frataxin protein in the tissues in nanogram/microgram total protein. All samples were carried out in duplicate and compared to the standard curve using their OD at 450 nm.

2.12. Statistical Analysis

The results are presented as mean ± SD. Statistical analyses were performed using GraphPad Prism7 software version 9.0 (GraphPad Inc., La Jolla, CA, USA) and detailed in each figure legend. The results were analyzed in two ways ANOVA test Tukey's multiple comparisons test (* p value < 0.05, ** p value < 0.005, *** p value < 0.0002 and **** < 0.0001) for behavior tests. For qPCR the statistics were conducted by the software for the qPCR Biorad CFX Maestro. The p values are indicated in each figure. Statistics: one-way ANOVA was used at 0.05 (95% confidence interval); * $p < 0.05$, ** $p < 0.003$, *** $p < 0.0003$, and **** $p < 0.0001$.

3. Results

3.1. In Vivo Comparison of YG8sR, YG8-800, Y47R and C57BL6J Mice

The YG8sR mouse is a model of Friedreich ataxia, having both mouse frataxin genes knockout but containing a human frataxin transgene with 250–300 GAA repeats. The YG8-800 mice are descendants from YG8sR but contain 800 GAA repeats. The control mouse for this model is the Y47R mouse, which also has both mouse frataxin genes knockout but has a human frataxin transgene with only 9 GAA repeats.

3.1.1. Behavior Tests

The physical performances of YG8sR and YG8-800 mice were compared to Y47R and C57Bl6J (C57) control mice.

The body weight (Figure 2A) was measured at each test session. The YG8sR mice had a significantly higher body weight than the Y47R mice starting at 5 months. The YG8-800 had a significantly lower weight than both control mice at 5 and 8 months. For the hanging test, where the mouse grips on an inverted grid (Figure 2B), There was a significant decrease in time hanging per body weight for YG8sR mice compared with the Y47R control mice at 5 and 8 months and also for the YG8-800 mice compared with the Y47R at 8 months.

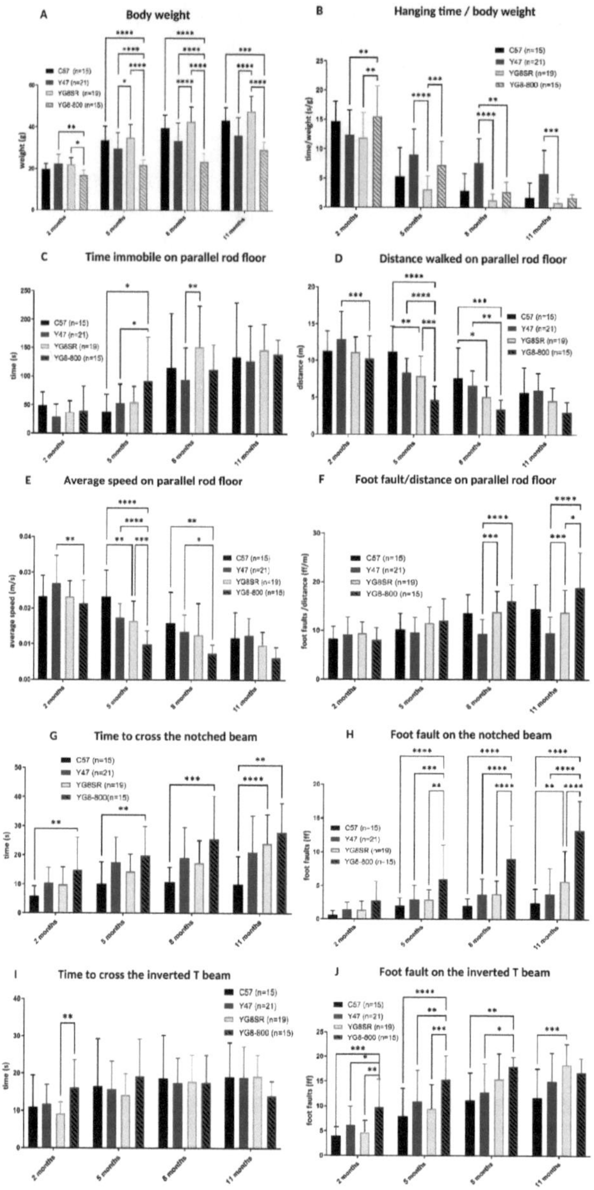

Figure 2. The C57, Y47R, YG8sR, and YG8-800 mice were followed with different behavior tests for 11 months. The body weight was determined (**A**) and served for the hanging wire test (**B**). The parallel rod floor is a cage with a wire rack on the floor and the following parameters were analyzed during 8 min: the distance traveled (**C**), the foot faults normalized with the distance traveled (**D**), the time immobile (**E**) and the average speed (**F**). The mice were also tested on two different beams. For the notched beam test, the time to cross the beam was measured (**G**) and the number of foot faults were counted (**H**). For the inversed T beam, the time to cross (**I**) and the number of foot faults (**J**) are also illustrated. * indicates a significant difference between C57 and YG8sR, † indicates a significant difference between transgenic mice (Y47R and YG8sR). The results were analyzed by a two-way ANOVA test (Tukey's multiple comparisons test). * p value < 0.05, ** p value < 0.005, *** p value < 0.0002 and **** < 0.0001. Error bars indicate standard deviation (SD).

The parallel rod floor is an apparatus made of an enclosure box with a grid on the floor and a camera above recording the distance traveled by the mouse, the average speed, the time immobile, and the foot faults during 8 min. The distance traveled by all mice decreased slowly from 2 to 11 months (Figure 2C). The distance is also significantly inferior for YG8-800 mice compared to both controls (C57 and Y47R) at 5 and 8 months and for YG8sR mice compared to the C57 model also at 5 and 8 months. The average speed (Figure 2D) was also significantly lower for the YG8-800 mice compared to both control models at 5 and 8 months. The difference was significantly inferior for the YG8sR compared to the C57 control at 5 months. The time spent immobile (Figure 2E) for YG8-800 mice was significantly higher than both controls at 5 months. At 8 months, it is the YG8sR mice that become significantly less active than the Y47R. The amount of foot faults per distance (in m) (Figure 2F) is significantly higher for both YG8sR and YG8-800 compared to the Y47R control at 8 and 11 months.

Two types of beams were also used to test the behavior: the notched beam where the mice walk from the top of a block to another block and the inverted T beam where the mice walk on a narrow surface. For the notched beam, there is a significant increase in the time needed to cross the beam for YG8-800 compared to C57 at all times and also for the YG8sR compared to C57 at 11 months (Figure 2G). The foot faults (Figure 2H) significantly increased for YG8-800 mice compared to both C57 and Y47R controls at 5, 8, and 11 months. A significant foot fault increase for YG8sR mice is also seen at 11 months compared to C57 mice. The time spent on the inverted T beam (Figure 2I) is only significantly higher for YG8-800 mice compared to YG8sR at 2 months but was not significantly different for all mice at 5, 8, and 11 months. The number of foot faults on the inverted T beam (Figure 2J) was significantly superior for YG8-800 mice compared to both controls at 2, 5, and 8 months. At 11 months, the YG8sR mice made significantly more foot faults than the C57 control mice.

To summarize, the physical performance of C57, Y47R, and YG8sR mice from 2 to 11 months old were similar for many tests. The YG8sR mice showed a mild movement coordination impairment phenotype and the YG8-800 mice, a more severe one.

3.1.2. Frataxin Concentration in Mouse Organs

The human frataxin protein concentration was measured in several tissues, i.e., the Tibialis anterior muscle, the liver, the heart, the cerebellum, the cerebrum, and dorsal root ganglions (Figure 3A). Even if there were great variations between mice, the frataxin concentration was significantly lower in the YG8sR mice compared with the Y47R mice. Indeed, the frataxin concentration in the YG8sR mice was only 13% to 20% of the frataxin concentration in the Y47R mice, depending on the tissues (Figure 3B).

Our aim was to identify a mouse model with clear ataxic phenotypes, which was not the case for the YG8sR mice. Even if the frataxin concentration in tissues is low, we thus decided to further decrease the frataxin concentration with an shRNA against human frataxin (the YG8sR mice have a human frataxin transgene) to trigger a phenotype earlier, which might be closer to the human pathology.

We also added the YG8-800 mouse model to our study to evaluate its phenotype, since it had not been characterized yet. The decrease in human frataxin is more severe in YG8-800 mice than in the YG8sR mice. When compared to Y47R mice, YG8-800 mice have between 0.9% and 16.3% of the Y47R human frataxin concentration in tissues (Figure 3B). This is therefore the aimed concentration for YG8sR mice injected with a shRNA targeting the human frataxin gene since this frataxin level induces an ataxic phenotype which is comparable to the human patients.

A

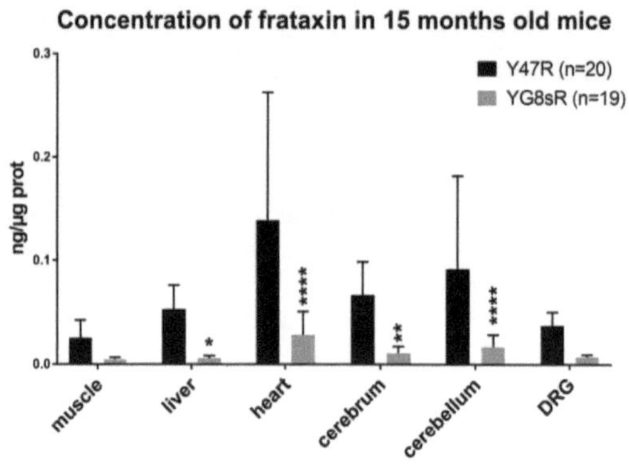

B **Percentage of frataxin compared to the Y47R mice**

	Muscle	Heart	Liver	Cerebrum	Cerebellum	DRG
YG8sR/Y47	18,4%	23,6%	13,7%	16,6%	19,2%	20,2%
YG8-800/Y47	0,9%	2,9%	5,3%	10,8%	16,3%	6,9%

Figure 3. Human frataxin concentration was measured with an ELISA kit in different organs of YG8sR, YG8-800, and Y47R mice. YG8sR compared to Y47R mice (**A**). The frataxin concentrations in YG8sR mice were between 13.7 and 23.6% of those of Y47R mice and were significantly different in the liver, the heart, the cerebrum and the cerebellum. The frataxin concentrations in YG8-800 mice were between 0.9% and 16.3% of those of Y47R mice and were significantly different in the liver, the heart, the cerebrum, the cerebellum and dorsal root ganglions (**B**). The results were analyzed by a two-way ANOVA test (Tukey's multiple comparisons test). * p value < 0.05, ** p value < 0.005, and **** < 0.0001. Error bars indicate standard deviation (SD).

3.2. In Vitro and in Vivo Testing of Different shRNAs against Frataxin

3.2.1. In Vitro Testing of 4 shRNAs against Frataxin

The sequence of one of four shRNAs (Figure 1A) against the human frataxin mRNA or a control scramble shRNA (shRNAscr) was inserted under the control of a U6 promoter in an AAV-PHP.B vector. The plasmids coding for these shRNAs also contained a mCherry gene under a UBC promoter to monitor the transfection efficiency by fluorescence (Figure 1C). The different plasmids containing the shRNA were transfected in HeLa cells. After 72 h of incubation, the cells were recovered, and proteins were extracted to quantify the frataxin concentration. The transfection of the control shRNAscr did not affect the frataxin concentration in the HeLa cells compared to the untransfected cells, which confirmed that the transfection of the plasmids by themselves did not affect human frataxin expression (Figure 4). In contrast, each shRNA targeting the frataxin mRNA led to a significant decrease in frataxin concentration in the transfected cells. The shRNA6 decreased the frataxin concentration close to 50% of controls and the shRNA4 to about 40%. The shRNA1 and the shRNA3 reduced frataxin concentration the most, to a concentration of 28.8% and 28.3% respectively of the control cells frataxin. Combinations of two shRNAs were also tested in HeLa cells to evaluate if their association could decrease even more the frataxin concentration. All combinations also significantly reduced the frataxin protein concentration compared to the control but none of them decreased the frataxin significantly more than one shRNA alone.

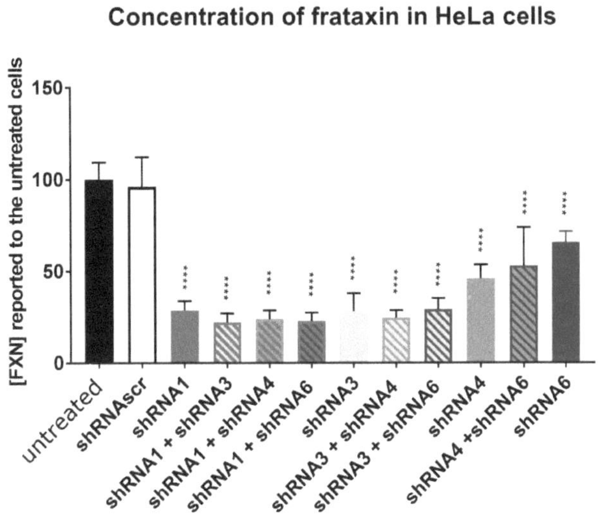

Figure 4. HeLa cells were transfected with plasmids coding for various shRNAs and the frataxin concentration was determined 72 h after transfection. The results were analyzed by a two-way ANOVA test (Tukey's multiple comparisons test). Control cells were untreated. **** $p < 0.0001$. Error bars indicate standard deviation (SD).

Following these results, we selected the three shRNAs, two that were more effective for decreasing the frataxin concentration, i.e., shRNA1 and shRNA3, and also shRNA6 although it was less effective. We also kept the shRNAscr as a negative control. These shRNAs were also integrated into an AAV-PHP.B virus to be tested in vivo. Error bars indicate standard deviation (SD).

3.2.2. In Vivo Tests of the shRNAs against Frataxin

The Impact of Anti-FXN shRNAs on the Coordination of YG8sR Mice

To increase the severity of the impaired coordination phenotype, we administrated intravenously in YG8sR mice an AAV-PHP.B coding for one anti-FXN shRNA previously validated in vitro or for the negative control shRNAscr. Different concentrations of AAVs coding for an shRNA were tested in vivo to obtain a progressive worsening phenotype, which would be severe enough to observe a loss of coordination 5 weeks after injection. The AAV concentrations chosen were 2× and 4× (respectively 12 and 24 $\times$ 10^{11} viral particles) for shRNA1 and shRNA6 and 0.5× and 1× (respectively 3 and 6 $\times$ 10^{11} viral particles) for shRNA3 (Supplementary Figure S2). The concentration of shRNA3 injected was lower than for shRNA1 because, in preliminary experiments, the mice injected with AAV shRNA3 had to be sacrificed earlier, around 15 days after the AAV injection at 2× due to the severity of the mouse illness. For the shRNAscr, three concentrations were tested 6, 12, and 24 $\times$ 10^{11} viral particles.

Different behavior tests were made before the AAV injection as training for the mice at 5 weeks after injection. At 5 weeks after injection, no significant differences were observed in the parallel rod floor (results not shown). Two beam tests were used, the notched beam test (Supplementary Figure S2A,B) where mice crossed the beam on the top of blocks, and the inverted T beam (walk beam) test (Supplementary Figure S2C,D), where the mice tried to cross the beam on a thin top. The mice were also tested for their muscle strength with the hanging test. Based on all these tests, the shRNA3 was determined as the best shRNA to induce an ataxic phenotype (Supplementary Figure S2). We, therefore, focused on the shRNA3 for the following studies.

The results obtained with the mice injected with the AAV-PHP.B-shRNAscr are similar to those obtained with the untreated mice, i.e., saline-injected mice (Figure 5). The two doses of AAV shRNA3 led to a more severe disease phenotype by increasing the time to cross the two beams and increasing the number of foot faults (Figure 5A–D) but also by decreasing the time the mice stayed hanging on the grid (Figure 5E).

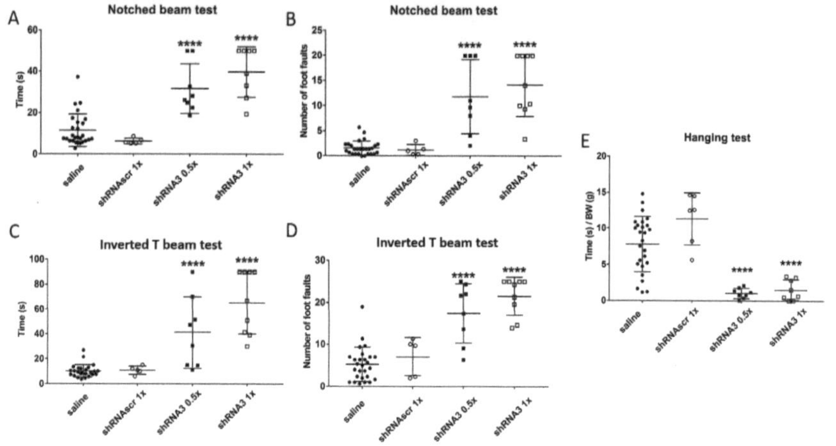

Figure 5. The mice injected with an AAV containing the different shRNAs were monitored with various behavior tests. The time taken by the YG8sR mice to cross the notched beam (**A**) and the inversed T beam (**C**) is illustrated 5 weeks after the injection. The number of foot faults made during the notched beam crossing (**B**) and the inversed T beam crossing (**D**) is also illustrated. In (**E**), the hanging test is quantified as the ratio of hanging time/body weight. The results were analyzed by a two-way ANOVA test (Sidak's multiple comparisons test). **** p value < 0.0001 compared to the saline controls. Error bars indicate standard deviation (SD). The number of animals per group is: saline n = 28, shRNAscr 1 × n = 5, shRNA3 0.5 × n = 8 and shRNA3 1 × n = 9.

The best shRNA was the shRNA3 to induce not only a loss of coordination observed on the beams but also a loss of muscle strength determined with the hanging wire test (Figure 5E) and that, at a lower concentration of virus than the shRNA1 administration.

Virus Detection

At the time of sacrifice, different organs were removed: the Tibialis anterior muscle (M), the cerebrum (Ce), the cerebellum (Cr), the liver (L), the heart (H), and the spine containing the dorsal root ganglions (DRG). All the organs were placed into the IVIS Lumina LT Series III and excited to detect the fluorescence of the mCherry protein (Supplementary Figure S3A). Fluorescence was observed in the tissues of the mice treated with all the shRNAs against frataxin or with the shRNAscr but not in the tissues of the control mice injected with saline (Supplementary Figure S3A right side). The fluorescence was most intense in the brain followed by the liver, the heart, and the spine. However, the low fluorescence of the spine may be due to the fact that the bones and tissues surrounding it were not removed. No fluorescence was detected in the muscles. Since these data provide indirect evidence of shRNA concentration, they are available in Supplementary Figure S3.

The genomic DNA was extracted from each tissue and a PCR was made to detect the expression of the mCherry gene contained in the AAV. The AAV-PHP.B serotype was used, which is known to cross the blood-brain barrier and concentrate in the nervous system. A positive control was made with the direct plasmid to confirm the band of mCherry at 410 pb. A band was observed in each tissue of mice injected with each shRNA and the shRNAscr (Supplementary Figure S3B). The band was more intense in the nervous tissues (cerebrum, cerebellum, and DRG) and the band was the lightest in the muscles of the injected mice. No band was observed in the control mice injected with saline (Supplementary Figure S1).

A q-PCR was also conducted to estimate the number of viral particles in each tissue, also by detecting the mCherry expression (Supplementary Figure S3C). The number of viral particles was the highest in the DRG, while the cerebrum and cerebellum had similar but slightly lower concentrations of AAV. Lower concentrations of AAVs were progressively detected in the liver, the heart, and muscle of the injected mice.

3.2.3. Quantification of the Frataxin Expression and Concentration

The frataxin mRNA concentration was evaluated in three organs, i.e., the liver, as an organ of detoxification, the cerebrum, and the cerebellum for the nervous system. RNA was extracted from these tissues and a qRT-PCR was made to quantify the expression of human frataxin in these tissues. The proteins of the tissues were also extracted and the human frataxin concentration was quantified and expressed as a percentage reported to the saline-treated mice. The concentrations of the frataxin mRNA and the frataxin protein were similar for the mice injected with saline or with the scramble shRNA, except for the liver (Figure 6A,B). In the liver, both frataxin mRNA and frataxin protein concentrations were significantly reduced by 37.7% and 50% respectively in the mice injected with the shRNA3 (Figure 6A,B). The shRNAscr induced a significant increase of the frataxin mRNA compared to the saline control in the liver, thus if compared to the shRNAscr and the shRNA3, a significant reduction of close to 60% was measured with the shRNA3. The frataxin mRNA expression was also significantly reduced by 29.3% in the cerebellum but no difference was observed for the protein with the control mice (Figure 6C,D). No significant differences were observed in the cerebrum neither for the mRNA expression nor for the frataxin concentration (Figure 6E,F).

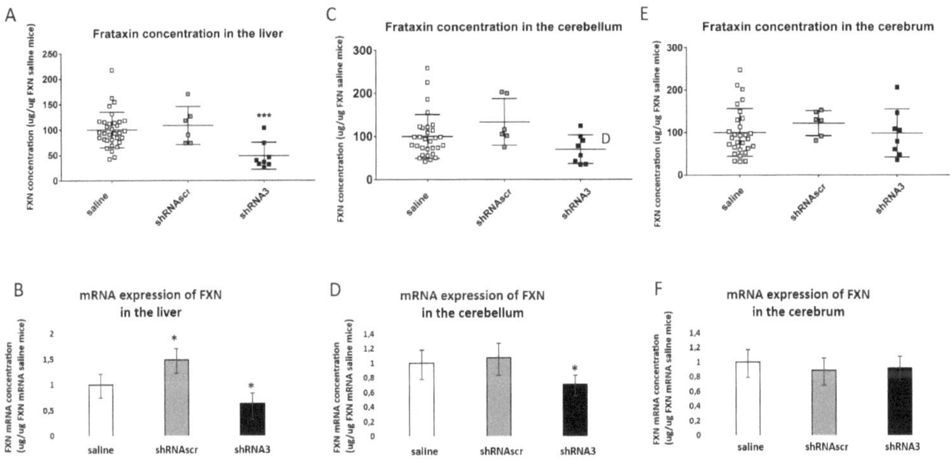

Figure 6. The concentration of frataxin protein was quantified by ELISA (**A**,**C**,**E**) saline n = 35, shRNAscr n = 6, shRNA3 n = 8, and the frataxin mRNA by qRT-PCR (**B**,**D**,**F**) in the liver, the cerebellum, and the cerebrum of mice saline n = 7, shRNAscr n = 6, shRNA3 n = 7. The results of protein concentration were analyzed by a two-way ANOVA test (Tukey's multiple comparisons test). * p value < 0.05, and *** p value < 0.0002. Error bars indicate standard deviation (SD). The results from the frataxin mRNA expression were analyzed using the Biorad CFX Maestro software version 2.2.

3.2.4. Comparison between Two Promoters for the Treatment with an AAV Coding for Human Frataxin

With the intent to develop a treatment by delivering the human frataxin gene systemically in mice, a part of this project was to use AAV.PHP.B with neuro-specific and non-neuro-specific promoters to deliver the human frataxin gene and study the impact on the movement coordination. The AAV.PHP.B allows to cross the BBB but is not neuro-specific. Therefore, AAV.PHP.B expressing human frataxin under two different promoters were tested, i.e., CBh-FXN for a more general expression and hSyn-FXN for an expression focused on the nervous system. The recombinant FXN gene present in the AAV has an almost identical nucleotide sequence to the human FXN gene expressed in the mouse but with some differences in the nucleotide sequence so that the shRNA3 cannot reduce the human FXN expression induced by the virus. These two AAVs coding for the human FXN were tested alone or in combination with AAVs coding for the shRNAscr or the shRNA3. Because all AAVs were of the same serotype (i.e., AAV-PHP.B), the treatment (CBh-FXN or hSyn-FXN) and the shRNA were injected at the same time, and mouse behavior tests were made 5 weeks after the AAV injection (Figure 7). In these behavior tests, the treatment with an AAV coding for frataxin under the CBh or the hSyn promoter alone or in association with the shRNAscr did not affect the behavior of the YG8sR mice for both beam tests (Figure 7). In the notched beam, both treatments (i.e., FNX under the CBh or the hSyn promoter) co-injected with the shRNA3 counterbalanced the shRNA inhibitory effect. Indeed, the time and the number of foot faults made by crossing the notched beam of the co-injected mice were similar to the performance of the saline-treated mice (Figure 7A,B). In the inverted T beam, although the treated mice crossed the beam as rapidly as the controls, the number of foot faults stayed similar to the saline control mice only with the injection of CBh-FXN (Figure 7C,D). Figure 7E represents the results of the hanging wire test. Both treatments, CBh-FXN and hSyn-FXN, restored the ability of the mice to remain gripped for an extended period of time. However, when examining the individual mouse performance for the hSyn-FXN treatment, 3 out of 4 mice fell very fast, and only one stayed longer. On the contrary, with the CBh-FXN treatment, all mice stayed on the grid much longer. The treatment with the CBh-FXN alone or with shRNA co-injection even increased the time that the mice stayed hanging compared to the saline control mice.

Before analysis of the tissues, genomic DNA was extracted, and PCRs were made to detect the mCherry gene present in the shRNA vector and the human FXN gene in mice that received the AAV coding for that gene to verify that each mouse was well injected with the right treatment (Supplementary Figure S1).

The expression of human frataxin mRNA was also quantified 5 weeks after the IV injection of the AAV and reported as fold increases compared to saline-treated mice (Table 1). In the cerebrum and cerebellum, the frataxin increase was comparable between CBh-FXN and hSyn-FXN alone or in co-injection with the shRNA3. On the contrary, in the liver, the concentration of frataxin was greatly increased with the CBh promoter comparatively to the hSyn promoter which does not significantly affect the expression and concentration of frataxin.

Table 1. The concentration of human frataxin was determined and expressed as fold modification relative to the untreated mice for the liver, the cerebellum, and the cerebrum of mice injected with the different AAVs. The control shRNAscr did not significantly modify the frataxin concentration in the 3 tissues. An AAV shRNA3 reduced the expression of frataxin in the liver and cerebellum. The treatment with an AAV coding for frataxin under the CBh promoter increased the expression of frataxin in all three tissues but when the promoter was hSyn the expression was increased in the cerebrum and the cerebellum and only slightly in the liver. The co-injection of AAV-PHP.B coding for the shRNA3 and FXN reduced the expression of FXN compared with AAV-PHP.B coding only for FXN under the same promoter (either CBh or hSyn).

Frataxin Concentration Normalized with Saline Treated Mice							
FXN/FXN saline	shRNAscr	sshRNA3	CBh-FXN	hSyn-FXN	shRNA3 + CBh-FXN	shRNA3 + hSyn-FXN
Liver	1.1 ± 0.4	0.5 ± 0.3	353 ± 245	1.5 ± 0.4	330 ± 238	0.7 ± 0.15
Cerebrum	1.3 ± 0.3	1.0 ± 0.6	134 ± 75	284 ± 197	45 ± 28	164 ± 105
cerebellum	1.4 ± 0.6	0.7 ± 0.3	264 ± 170	113 ± 78	103 ± 56	76 ± 84

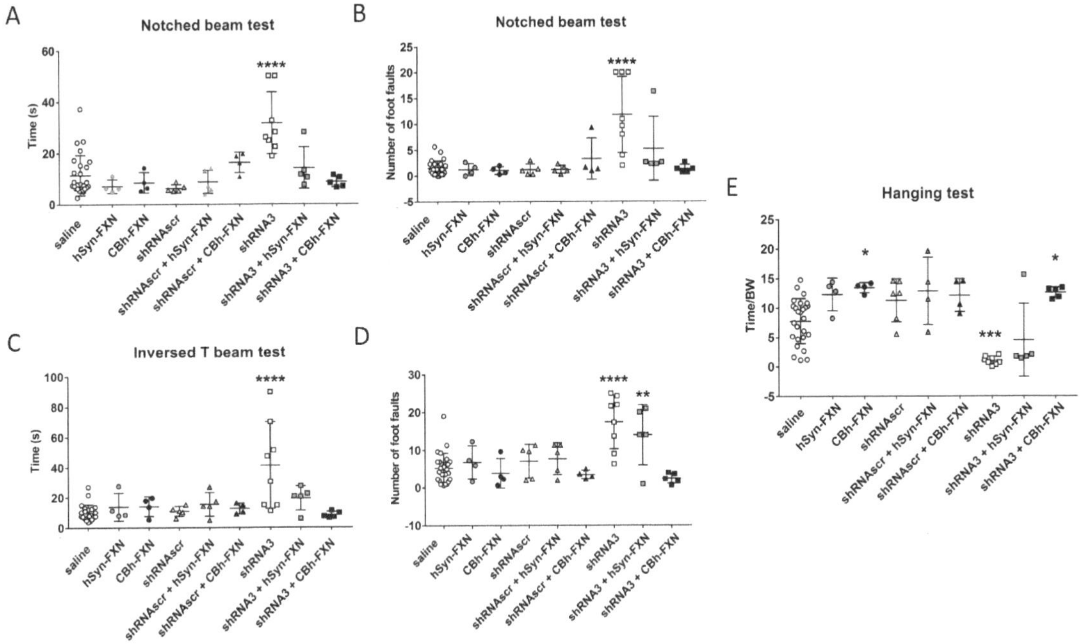

Figure 7. The mice injected with an AAV containing either an shRNA or the frataxin gene or injected with both AAVs were followed with behavior tests. The time taken by the YG8sR mice to cross the notched beam (**A**) and the inversed T beam (**C**) is illustrated 45 days after the injection of AAVs. The number of foot faults made during the notched beam crossing (**B**) and the inversed T beam crossing (**D**) is also illustrated. In (**E**), the results of the hanging test are presented as the ratio of hanging time/body weight. The number of animals per group is saline n = 28, hSyn-FXN n = 4, CBh-FXN n = 4, shRNAscr n = 5, shRNAscr + hSyn-FXN n = 4, shRNAscr + CBh-FXN n = 4, shRNA3 n = 8, shRNA3 + hSyn-FXN n = 5, shRNA3 + CBh-FXN n = 5. The results were analyzed by a two-way ANOVA (Tukey's multiple comparisons test). * p value < 0.05, ** p value < 0.005, and **** < 0.0001. Error bars indicate standard deviation (SD).

4. Discussion

Initially, Virmouni and co-workers developed a new mouse model containing a human FNX transgene with about 200 GAA repeats, here called YG8sR(200) for an easier discussion [29]. Even if the behavior of these YG8sR(200) mice on the rotarod was significantly different from C57 and Y47R mice, by pooling males and females together, the differences were not as clear as the results obtained when the sexes were analyzed separately. Moreover, the YG8sR(200) mice showed reduced locomotion activity over time but the most significant differences were observed with more specific tests such as the beam tests, the hang-wire test, and the grip test at 12 months. Indeed, at that age, the YG8sR(200) mice took more time to cross the two beams, fell faster on the hang-wire test, and showed less grip strength compared to the Y47R or C57 mice. The human frataxin mRNA, as well as the human frataxin protein concentrations in YG8sR(200) mice, were also decreased in the brain, the cerebellum, and the liver at 12 months compared to Y47R and C57 mice [29]. We have purchased the YG8sR mice from Jackson Laboratories Inc. and those that we have used contained between 250 and 300 GAA repeats (https://www.jax.org/strain/024113) (accessed on 10 June 2023). In our studies, the YG8sR females performed better than the YG8sR males in the behavior tests. Nevertheless, no great difference was observed compared to Y47R mice in these tests. The concentration of human frataxin protein was reduced in YG8sR in all tissues tested compared to Y47R and even more than in the YG8sR(200) for the cerebellum (i.e., 70% in the YG8sR(200) and 19% in YG8sR). Considering that YG8sR mice have a human frataxin transgene containing 250–300 GAA repeats responsible for low human frataxin expression, these mice remain a good model to study the effects of frataxin reduction [36], or to try to increase the frataxin expression by some treatments [37,38] or even through genetic modifications using the CRISPR-Cas9 system to delete the GAA repeat [19,20]. All these studies are essential to understanding the mechanism and interaction between the frataxin protein and other proteins or enzymes but at some point, it becomes essential to have a good mouse model that recreates the symptoms of Friedreich ataxia and allows for the detection of functional improvements with treatment.

The use of shRNAs is a method used to decrease the expression of a protein and can be tissue-specific by using a specific promoter. For example, by using a member of the cadherin family that is exclusively expressed in the renal tubular epithelial cells, i.e., Ksp-cadherin, Xu et al. [39] observed 2 weeks after IV injection of an AAV9-shRNA-ALDH2, a decrease of the aldehyde dehydrogenase 2 protein (ALDH2) only in the kidney although this protein is also expressed in the liver and the heart [39]. Chandran and co-workers in 2017 [27] used shRNAs to decrease the mouse frataxin and create a mouse model for FRDA called FRDAkd. In their study, different shRNAs against frataxin were tested and they decided to use one, which induced the strongest reduction of frataxin in vitro. Among the different shRNAs that they tested, the one decreasing the most the frataxin expression was also targeting exon 4, as in our case for our sh1 and sh3 [27].

Piras et al., in 2013 suggested that gene knockdown is generally less predictable than gene over-expression [40]. In our study, the concentration of viral particles in the brain (cerebrum and cerebellum) was as high as in the liver but the frataxin concentration was significantly reduced only in the liver. In our case, the AAV serotype PHP.B was used. In the first study on AAV-PHP.B, the bio-distribution after IV injection in mice was compared to AAV9 and was similar between the two AAVs in the liver, the heart, and the skeletal muscle but the expression of GFP was lower with the AAV-PHP.B than with AAV9, even in the liver, where viral particles were as high as in the nervous system [30].

Chandran and co-workers also created a new mouse model for Friedreich Ataxia. They incorporated in a defined genomic locus, a single copy of a doxycycline-inducible anti-FXN shRNA to silence frataxin. Different behavior tests were carried out at 12 and 24 weeks of age with doxycycline treatment and they observed neurological deficits. After 20 weeks of age, the mice showed a drastic reduction of frataxin protein in the eight organs tested (liver, heart, kidney, brain, muscle, pancreas, lung and spinal cord) [27]. In our experiments, the effect of shRNA was faster and the frataxin mRNAs and protein concentrations were estimated only 5 weeks after the AAV-shRNA injection. However, after 5 weeks, a reduction of frataxin protein was observed in the liver but not in the other organs. Nevertheless, the frataxin mRNA was significantly reduced in the liver but also in the cerebellum. It is probable that 5 weeks is too short to observe an effect on the frataxin concentration on other organs than the liver because in the Chandran et al. study, the investigators showed a decrease of the frataxin protein concentration only for the liver in a time series starting at 3 weeks of age, but in other organs after 20 weeks [27].

The important question is whether we can reverse a phenotype even in patients who have lived several years with the disease by increasing frataxin expression. Many researchers tried to answer this question. Chandran et al. [27] with their inducible mouse model FRDAkd were able to stop the treatment with the doxycycline to increase again the expression of frataxin. After 13 weeks of doxycycline treatment followed by 12 weeks without this treatment, the mice recovered a healthy phenotype close to control mice [27]. Therefore, it appears that the coordination impairment is reversible when frataxin is increased. Another method to increase the concentration of frataxin in mice is to inject an AAV coding for the frataxin gene. We have worked on the delivery of the frataxin gene by AAVs since 2014 [15]. We initially used an AAV9 for the delivery in NSE-cre and MCK-cre mice, i.e., the mouse models created by Puccio H. in 2001 [22]. We found a great improvement in heart function and behavior tests by intraperitoneal administration of AAV9-FXN at an early age, between 5 and 9 days of age [15]. In the present study, our goal was to focus on the neuronal dysfunction, therefore, the AAV-PHP.B was chosen to target the nervous system with the shRNA. A co-injection (AAV-shRNA and AAV-FXN) was necessary to avoid a hyperacute immune response that could be provoked by a second AAV injection [41]. In our study, the co-injection of AAV-PHP.B-CBh-FXN with an AAV-PHP.B-antiFXNshRNA improved the behavior of mice compared to those treated with the AAV-PHP.B shRNA against frataxin alone. Puccio and co-workers created in 2018 a mouse model to specifically target cells expressing parvalbumin, called parvalbumin conditional knockout mouse model (Pvalb cKO mouse) [16]. The parvalbumin promoter is specific to proprioceptive neurons. These mice reflected the FRDA neuropathophysiology. They injected intravenously into 3.5 weeks old mice an AAV9-CAG-FXN and obtained a significant coordination improvement in the notched beam but not reaching the phenotype of their control mice (Fxn+/L3). Moreover, in the wire-hanging test, the treated Pvalb cKO mice were similar to the WT. Our aim was to have a comparable coordination impairment to this model in a mouse model containing the human gene with an expanded GAA to evaluate the impact of a genetic treatment on the coordination of mice. We used AAV. PHP.B to cross the BBB, but tissue analysis revealed not only the efficiency of the AAV9 on the sensory neuropathy but also a non-homogenous distribution of the AAVrh10 in the cerebellum leading to a partial rescue of Purkinje cells [16]. AAV8 was also tested in the FRDAkd mouse model (described above) in 2022 [42]. Ten weeks after dox treatment the mice received an intravenous injection of AAV8 coding for frataxin. Unfortunately, restoration of frataxin with this AAV failed to improve the behavior, metabolic and histological features of this mouse model [42].

The different articles using an AAV coding for frataxin were often using a strong promoter such as CMV or CBh which led to overexpression of frataxin in different organs. A recent review by Sivakumar and Cherqui in 2022 highlighted the controversial effects of frataxin overexpression [43]. In the past, in our studies, even if we obtained overexpression of frataxin in the tissues of the mouse models (and especially in the liver), we never found clear evidence that overexpression was harmful to the mice. This may depend on the frataxin concentration which was obtained. Indeed, Puccio's group published in 2020 [44] a study showing that a dose-dependent harmful effect was observed in mouse heart function started after a 20-fold increase in frataxin concentration. However, the harmful effect was tissue-specific since a 90-fold increase of frataxin in the liver of mice, did not induce toxicity. In the present study, we tested two promoters, a general and strong (CBh) one and a weak one more specific for the nervous system (hSyn). The frataxin expression under a CBh promoter showed at 5 weeks post injection a significant improvement in the behavior tests compared to the hSyn promoter. Nevertheless, the period of testing was just 5 weeks post-injection and it is possible that this follow-up period was too short to observe an improvement in the behavior with the hSyn promoter. Moreover, the hSyn promoter, due to its specificity, did not induce any increase of frataxin in the liver even if the number of viral particles was high.

It is important to find an appropriate mouse model of FRDA for what is being tested, e.g., to measure the impact of treatment on the behavioral phenotype. Mercado-Ayón et al., (2022) [42], demonstrated that the symptoms obtained in their FRDAkd mouse model are not only the result of the reduction of frataxin expression but that the doxorubicin treatment has also an impact on the symptoms. In our study, the shRNA was provided by an AAV and no additional treatment was necessary. The only disadvantage of our approach is that we had to co-inject the treatment and the shRNA because the same AAV serotype was used for both and thus a delayed administration of the AAV was not possible because of the immune response induced by the first administration.

The aim of this study was to decrease the frataxin in YG8sR mice to increase the severity of their phenotype. Therefore, both YG8sR injected with an shRNA and YG8-800 mice phenotypes were compared. The results have shown that shRNA targeting frataxin does decrease its expression and increases the ataxic phenotype of YG8sR mice. The question was if this method is now the optimal one knowing that the YG8-800 mouse model is available. The complete characterization of this model has been carried out in our laboratory [45] showing that the phenotype of the YG8-800 mice is comparable to the human patients since the movement coordination progressively decreases and a cardiac hypertrophy is noticed. When compared, the YG8sR mice injected with the shRNA3 (Figure 5) have a phenotype comparable to the YG8-800 mice (Figure 2G–J). In fact, at 3–4 months, the YG8sR +shRNA3 need an average of 35 s to cross the notched beam and make 12-foot faults, which is slightly more severe than the YG8-800 mice 20 s and 7-foot faults at 5 months (saline control needing in average 10 s and making around 2.5-foot faults). The same trend is noticed for the inverted T beam, on which the shRNA-treated mice need on average 57 s and make 19-foot faults, whereas YG8-800 mice need 18 s and make 14-foot faults (saline needs 9 s and make 6-foot faults). In conclusion, both YG8sR injected with shRNA3 and YG8-800 offer a phenotype that can be compared to human patients. The YG8-800 mouse model is recent and has only a few articles about it. In September 2022, our team published the characterization of this mouse model including the cardiac hypertrophy which is also part of the human phenotype [45]. In January 2023, Kalef-Ezra et al. confirmed the impaired coordination and reduced frataxin levels of the mouse model [46].

We also studied the possibility of using AAV.PHP.B to deliver the frataxin gene. Our results on YG8sR mice were promising since they reestablished the coordination and more tests will be conducted in YG8-800 mice without needing to co-inject with an anti-FXN shRNA.

5. Conclusions

In our study, we were able to rapidly increase the severity of the YG8sR phenotype with an AAV coding for a sh-RNA against frataxin. Indeed, only 5 weeks after this AAV injection, the mice demonstrated significantly more severe locomotion and balance symptoms. A co-injection of an AAV coding for the same shRNA and of an AAV (same serotype) coding for frataxin prevented the development of disease symptoms. AAV-based gene therapy remains for the moment the best method to deliver a DNA sequence and the development of new serotypes opens the possibility of developing more tissue-specific delivery.

Supplementary Materials: The following supporting information can be downloaded at: https://www.mdpi.com/article/10.3390/genes14081654/s1. Figure S1: For each mouse, two PCRs were made to confirm the treatment received by the presence of mCherry (shRNA) on the top and/or AAV-FXN on the bottom. Figure S2: The mice injected with an AAV containing different shRNAs were monitored with various behavior tests. Figure S3: mCherry and AAV repartition. Table S1: Sequences of primers.

Author Contributions: C.G. and J.P.T. conceived, designed, and conducted experiments; N.M. constructed AAV vectors; C.G., C.B., N.M., P.Y. and V.C. sacrificed the mice; N.M. and P.Y. analyzed in vitro studies and C.G. summarized and analyzed experimental data from in vivo studies with the help during the summer 2019 from G.B for the quantification of total and frataxin proteins in some tissues, C.G. was in charge of the behavior; and C.B., C.G. and J.P.T. wrote the manuscript. S.G.-f.Y., M.A. and G.B. helped with experiments such as mouse euthanasia and tissue collection. All authors have read and agreed to the published version of the manuscript.

Funding: This research was funded by Ataxia Canada (grant 130852), the Canadian Institute of Health Research (grant 418059) and the ThéCell FRQS network (grant 91977).

Institutional Review Board Statement: All experiences involving animals were approved by the Comité de Protection des Ani-maux de l'Université Laval (CPAUL3) (Québec, QC, Canada).

Informed Consent Statement: Not applicable.

Data Availability Statement: Data is contained within the article or supplementary material.

Acknowledgments: This work was supported by grants from Ataxia Canada, the Canadian Institute of Health Research and the ThéCell FRQS network. We would like to thank Nizar Chetoui and Marc Bazin for their help. A big thank to the personal at the animal care and especially to Caroline Genest, Jessina Chrétien, Stephanie Bernard and Martin Saint-Pierre for their availability and flexibility.

Conflicts of Interest: The authors declare no conflict of interest.

References

1. Li, L.; Shen, X.; Liu, Z.; Norrbom, M.; Prakash, T.P.; O'Reilly, D.; Sharma, V.K.; Damha, M.J.; Watts, J.K.; Rigo, F.; et al. Activation of frataxin protein expression by antisense oligonucleotides targeting the mutant expanded repeat. *Nucleic Acid Ther.* **2018**, *28*, 23–33. [CrossRef] [PubMed]
2. Filla, A.; De Michele, G.; Cavalcanti, F.; Pianese, L.; Monticelli, A.; Campanella, G.; Cocozza, S. The relationship between trinucleotide (GAA) repeat length and clinical features in Friedreich ataxia. *Am. J. Hum. Genet.* **1996**, *59*, 554. [PubMed]
3. Delatycki, M.B.; Paris, D.B.; Gardner, R.M.; Nicholson, G.A.; Nassif, N.; Storey, E.; MacMillan, J.C.; Collins, V.; Williamson, R.; Forrest, S.M. Clinical and genetic study of Friedreich ataxia in an Australian population. *Am. J. Med. Genet.* **1999**, *87*, 168–174. [CrossRef]
4. Gottesfeld, J.M.; Rusche, J.R.; Pandolfo, M. Increasing frataxin gene expression with histone deacetylase inhibitors as a therapeutic approach for Friedreich's ataxia. *J. Neurochem.* **2013**, *126*, 147–154. [CrossRef]
5. Fox, N.G.; Das, D. Frataxin accelerates [2Fe-2S] cluster formation on the human Fe–S assembly complex. *Biochemistry* **2015**, *54*, 3880–3889. [CrossRef]
6. Castro, I.H.; Pignataro, M.F.; Sewell, K.E.; Espeche, L.D.; Herrera, M.G.; Noguera, M.E.; Dain, L.; Nadra, A.D.; Aran, M.; Smal, C.; et al. Frataxin structure and function. In *Macromolecular Protein Complexes II: Structure and Function*; Springer: Berlin, Germany, 2019; pp. 393–438.
7. Cook, A.; Giunti, P. Friedreich's ataxia: Clinical features, pathogenesis and management. *Br. Med. Bull.* **2017**, *124*, 19–30. [CrossRef]
8. Bürk, K. Friedreich Ataxia: Current status and future prospects. *Cerebellum Ataxias* **2017**, *4*, 1–9. [CrossRef]
9. Lynch, D.R.; Farmer, J.M.; Tsou, A.Y.; Perlman, S.; Subramony, S.H.; Gomez, C.M.; Ashizawa, T.; Wilmot, G.R.; Wilson, R.B.; Balcer, L.J. Measuring Friedreich ataxia: Complementary features of examination and performance measures. *Neurology* **2006**, *66*, 1711–1716.
10. Weidemann, F.; Störk, S.; Liu, D.; Hu, K.; Herrmann, S.; Ertl, G.; Niemann, M. Cardiomyopathy of Friedreich ataxia. *J. Neurochem.* **2013**, *126* (Suppl. S1), 88–93. [CrossRef]
11. Chikh, A. *Développement d'une Thérapie pour l'Ataxie de Friedreich Basée sur L'administration des Protéines Tat-Frataxine et Pep-1-Frataxine*; Université Laval: Québec, QC, Canada, 2015.
12. Chapdelaine, P.; Gerard, C.; Sanchez, N.; Cherif, K.; Rousseau, J.; Ouellet, D.L.; Jauvin, D.; Tremblay, J.P. Development of an AAV9 coding for a 3XFLAG-TALEfrat# 8-VP64 able to increase in vivo the human frataxin in YG8R mice. *Gene Ther.* **2016**, *23*, 606–614. [PubMed]
13. Cherif, K.; Gérard, C.; Rousseau, J.; Ouellet, D.L.; Chapdelaine, P.; Tremblay, J.P. Increased frataxin expression induced in Friedreich ataxia cells by platinum TALE-VP64s or platinum TALE-SunTag. *Mol. Ther.-Nucl. Acids* **2018**, *12*, 19–32. [CrossRef] [PubMed]
14. Perdomini, M.; Belbellaa, B.; Monassier, L.; Reutenauer, L.; Messaddeq, N.; Cartier, N.; Crystal, R.G.; Aubourg, P.; Puccio, H. Prevention and reversal of severe mitochondrial cardiomyopathy by gene therapy in a mouse model of Friedreich's ataxia. *Nat. Med.* **2014**, *20*, 542–547. [CrossRef] [PubMed]
15. GGérard, C.; Xiao, X.; Filali, M.; Coulombe, Z.; Arsenault, M.; Couet, J.; Li, J.; Drolet, M.C.; Chapdelaine, P.; Chikh, A.; et al. An AAV9 coding for frataxin clearly improved the symptoms and prolonged the life of Friedreich ataxia mouse models. *Mol. Ther.-Methods Clin. Dev.* **2014**, *1*, 14044. [CrossRef]
16. Piguet, F.; de Montigny, C.; Vaucamps, N.; Reutenauer, L.; Eisenmann, A.; Puccio, H. Rapid and complete reversal of sensory ataxia by gene therapy in a novel model of Friedreich ataxia. *Mol. Ther.* **2018**, *26*, 1940–1952. [CrossRef] [PubMed]
17. SSalami, C.O.; Jackson, K.; Jose, C.; Alyass, L.; Cisse, G.I.; De, B.P.; Stiles, K.M.; Chiuchiolo, M.J.; Sondhi, D.; Crystal, R.G.; et al. Stress-induced mouse model of the cardiac manifestations of Friedreich's ataxia corrected by AAV-mediated gene therapy. *Hum. Gene Ther.* **2020**, *31*, 819–827. [CrossRef] [PubMed]
18. Li, Y.; Polak, U.; Bhalla, A.D.; Rozwadowska, N.; Butler, J.S.; Lynch, D.R.; Dent, S.Y.; Napierala, M. Excision of expanded GAA repeats alleviates the molecular phenotype of Friedreich's ataxia. *Mol. Ther.* **2015**, *23*, 1055–1065. [CrossRef]
19. Ouellet, D.L.; Cherif, K.; Rousseau, J.; Tremblay, J.P. Deletion of the GAA repeats from the human frataxin gene using the CRISPR-Cas9 system in YG8R-derived cells and mouse models of Friedreich ataxia. *Gene Ther.* **2017**, *24*, 265–274. [CrossRef]
20. Yaméogo, P.; Gérard, C.; Majeau, N.; Tremblay, J.P. Removal of the GAA repeat in the heart of a Friedreich's ataxia mouse model using CjCas9. *Gene Ther.* **2023**, 1–8. [CrossRef]
21. ClinicalTrials.gov. Gene Therapy for Cardiomyopathy Associated with Friedreich's Ataxia. 2022. Available online: https://clinicaltrials.gov/ct2/show/study/NCT05445323 (accessed on 5 May 2023).
22. Puccio, H.; Simon, D.; Cossée, M.; Criqui-Filipe, P.; Tiziano, F.; Melki, J.; Hindelang, C.; Matyas, R.; Rustin, P.; Koenig, M. Mouse models for Friedreich ataxia exhibit cardiomyopathy, sensory nerve defect and Fe-S enzyme deficiency followed by intramitochondrial iron deposits. *Nat. Genet.* **2001**, *27*, 181–186. [CrossRef]
23. Seznec, H.; Simon, D.; Monassier, L.; Criqui-Filipe, P.; Gansmuller, A.; Rustin, P.; Koenig, M.; Puccio, H. Idebenone delays the onset of cardiac functional alteration without correction of Fe-S enzymes deficit in a mouse model for Friedreich ataxia. *Hum. Mol. Genet.* **2004**, *13*, 1017–1024. [CrossRef]

24. Frugier, T.; Tiziano, F.D.; Cifuentes-Diaz, C.; Miniou, P.; Roblot, N.; Dierich, A.; Le Meur, M.; Melki, J. Nuclear targeting defect of SMN lacking the C-terminus in a mouse model of spinal muscular atrophy. *Hum. Mol. Genet.* **2000**, *9*, 849–858. [CrossRef]
25. Filali, M.; Lalonde, R.; Gérard, C.; Coulombe, Z.; Tremblay, J.P. Sensorimotor skills in Fxn KO/Mck mutants deficient for frataxin in muscle. *Brain Res.* **2015**, *1608*, 91–96. [CrossRef] [PubMed]
26. McMackin, M.Z.; Henderson, C.K.; Cortopassi, G.A.; Henderson, G.A. Cortopassi, Neurobehavioral deficits in the KIKO mouse model of Friedreich's ataxia. *Behav. Brain Res.* **2017**, *316*, 183–188. [CrossRef]
27. Chandran, V.; Gao, K.; Swarup, V.; Versano, R.; Dong, H.; Jordan, M.C.; Geschwind, D.H. Inducible and reversible phenotypes in a novel mouse model of Friedreich's Ataxia. *eLife* **2017**, *6*, e30054. [CrossRef] [PubMed]
28. Anjomani Virmouni, S.; Sandi, C.; Al-Mahdawi, S.; Pook, M.A. Cellular, molecular and functional characterisation of YAC transgenic mouse models of Friedreich ataxia. *PLoS ONE* **2014**, *9*, e107416. [CrossRef] [PubMed]
29. Anjomani Virmouni, S.; Ezzatizadeh, V.; Sandi, C.; Sandi, M.; Al-Mahdawi, S.; Chutake, Y.; Pook, M.A. A novel GAA-repeat-expansion-based mouse model of Friedreich's ataxia. *Dis. Model. Mech.* **2015**, *8*, 225–235. [PubMed]
30. Deverman, B.E.; Pravdo, P.L.; Simpson, B.P.; Kumar, S.R.; Chan, K.Y.; Banerjee, A.; Wu, W.L.; Yang, B.; Huber, N.; Pasca, S.P.; et al. Cre-dependent selection yields AAV variants for widespread gene transfer to the adult brain. *Nat. Biotechnol.* **2016**, *34*, 204–209. [CrossRef] [PubMed]
31. Kamens, H.M.; Crabbe, J.C. The parallel rod floor test: A measure of ataxia in mice. *Nat. Protoc.* **2007**, *2*, 277–281. [CrossRef]
32. Kondziella, W. A new method for the measurement of muscle relaxation in white mice. *Arch. Int. Pharmacodyn. Ther.* **1964**, *152*, 277–284.
33. Deacon, R.M. Measuring the strength of mice. *JoVE* **2013**, *76*, e2610.
34. Di Bonito, M.; Boulland, J.L.; Krezel, W.; Setti, E.; Studer, M.; Glover, J.C. Loss of projections, functional compensation, and residual deficits in the mammalian vestibulospinal system of Hoxb1-deficient mice. *ENeuro* **2015**, *2*. [CrossRef] [PubMed]
35. Luong, T.N.; Carlisle, H.J.; Southwell, A.; Patterson, P.H. Assessment of motor balance and coordination in mice using the balance beam. *JoVE 2011*, e2376.
36. Muñoz-Lasso, D.C.; Mollá, B.; Sáenz-Gamboa, J.J.; Insuasty, E.; de la Iglesia-Vaya, M.; Pook, M.A.; Pallardó, F.V.; Palau, F.; Gonzalez-Cabo, P. Frataxin Deficit Leads to Reduced Dynamics of Growth Cones in Dorsal Root Ganglia Neurons of Friedreich's Ataxia YG8sR Model: A Multilinear Algebra Approach. *Front. Mol. Neurosci.* **2022**, *15*, 912780. [CrossRef] [PubMed]
37. Santoro, A.; Anjomani Virmouni, S.; Paradies, E.; Villalobos Coa, V.L.; Al-Mahdawi, S.; Khoo, M.; Porcelli, V.; Vozza, A.; Perrone, M.; Denora, N.; et al. Effect of diazoxide on Friedreich ataxia models. *Hum. Mol. Genet.* **2018**, *27*, 992–1001. [CrossRef] [PubMed]
38. Rodríguez-Pascau, L.; Britti, E.; Calap-Quintana, P.; Dong, Y.N.; Vergara, C.; Delaspre, F.; Medina-Carbonero, M.; Tamarit, J.; Pallardó, F.V.; Gonzalez-Cabo, P.; et al. PPAR gamma agonist leriglitazone improves frataxin-loss impairments in cellular and animal models of Friedreich Ataxia. *Neurobiol. Dis.* **2021**, *148*, 105162. [CrossRef] [PubMed]
39. Xu, T.; Guo, J.; Wei, M.; Wang, J.; Yang, K.; Pan, C.; Pang, J.; Xue, L.; Yuan, Q.; Xue, M.; et al. Aldehyde dehydrogenase 2 protects against acute kidney injury by regulating autophagy via the Beclin-1 pathway. *JCI Insight* **2021**, *6*, e138183.
40. Piras, B.A.; O'Connor, D.M.; French, B.A. Systemic delivery of shRNA by AAV9 provides highly efficient knockdown of ubiquitously expressed GFP in mouse heart, but not liver. *PLoS ONE* **2013**, *8*, e75894. [CrossRef]
41. Earley, J.; Piletska, E.; Ronzitti, G.; Piletsky, S. Evading and overcoming AAV neutralization in gene therapy. *Trends in Biotechnol.* **2022**, *16*, 819569. [CrossRef]
42. Mercado-Ayón, E.; Warren, N.; Halawani, S.; Rodden, L.N.; Ngaba, L.; Dong, Y.N.; Chang, J.C.; Fonck, C.; Mavilio, F.; Lynch, D.R.; et al. Cerebellar Pathology in an Inducible Mouse Model of Friedreich Ataxia. *Front. Neurosci.* **2022**, *16*, 373. [CrossRef]
43. Sivakumar, A.; Cherqui, S. Advantages and Limitations of Gene Therapy and Gene Editing for Friedreich's Ataxia. *Front. Genome Edit.* **2022**, *4*, 903139. [CrossRef]
44. Belbellaa, B.; Reutenauer, L.; Messaddeq, N.; Monassier, L.; Puccio, H. High levels of frataxin overexpression lead to mitochondrial and cardiac toxicity in mouse models. *Mol. Ther.-Methods Clin. Dev.* **2020**, *19*, 120–138. [CrossRef] [PubMed]
45. Gérard, C.; Archambault, A.F.; Bouchard, C.; Tremblay, J.P. A promising mouse model for Friedreich Ataxia progressing like human patients. *Behav. Brain Res.* **2023**, *436*, 114107. [CrossRef] [PubMed]
46. Kalef-Ezra, E.; Edzeamey, F.J.; Valle, A.; Khonsari, H.; Kleine, P.; Oggianu, C.; Al-Mahdawi, S.; Pook, M.A.; Anjomani Virmouni, S. A new FRDA mouse model [Fxn null: YG8s(GAA) > 800] with more than 800 GAA repeats. *Front. Neurosci.* **2023**, *17*, 930422. [CrossRef] [PubMed]

Disclaimer/Publisher's Note: The statements, opinions and data contained in all publications are solely those of the individual author(s) and contributor(s) and not of MDPI and/or the editor(s). MDPI and/or the editor(s) disclaim responsibility for any injury to people or property resulting from any ideas, methods, instructions or products referred to in the content.

genes

MDPI

Article

Genome-Wide Epistasis Study of Cerebrospinal Fluid Hyperphosphorylated Tau in ADNI Cohort

Dandan Chen [1,2,†], Jin Li [1,†], Hongwei Liu [3], Xiaolong Liu [1], Chenghao Zhang [1], Haoran Luo [1], Yiming Wei [1], Yang Xi [3], Hong Liang [1,*] and Qiushi Zhang [3,*]

1 College of Intelligent Systems Science and Engineering, Harbin Engineering University, Harbin 150001, China; denise0620@163.com (D.C.); lijin@hrbeu.edu.cn (J.L.)
2 School of Automation Engineering, Northeast Electric Power University, Jilin 132012, China
3 School of Computer Science, Northeast Electric Power University, Jilin 132012, China
* Correspondence: lh@hrbeu.edu.cn (H.L.); zhangqiushi@neepu.edu.cn (Q.Z.)
† These authors contributed equally to this work.

Abstract: Alzheimer's disease (AD) is the main cause of dementia worldwide, and the genetic mechanism of which is not yet fully understood. Much evidence has accumulated over the past decade to suggest that after the first large-scale genome-wide association studies (GWAS) were conducted, the problem of "missing heritability" in AD is still a great challenge. Epistasis has been considered as one of the main causes of "missing heritability" in AD, which has been largely ignored in human genetics. The focus of current genome-wide epistasis studies is usually on single nucleotide polymorphisms (SNPs) that have significant individual effects, and the amount of heritability explained by which was very low. Moreover, AD is characterized by progressive cognitive decline and neuronal damage, and some studies have suggested that hyperphosphorylated tau (P-tau) mediates neuronal death by inducing necroptosis and inflammation in AD. Therefore, this study focused on identifying epistasis between two-marker interactions at marginal main effects across the whole genome using cerebrospinal fluid (CSF) P-tau as quantitative trait (QT). We sought to detect interactions between SNPs in a multi-GPU based linear regression method by using age, gender, and clinical diagnostic status (cds) as covariates. We then used the STRING online tool to perform the PPI network and identify two-marker epistasis at the level of gene–gene interaction. A total of 758 SNP pairs were found to be statistically significant. Particularly, between the marginal main effect SNP pairs, highly significant SNP–SNP interactions were identified, which explained a relatively high variance at the P-tau level. In addition, 331 AD-related genes were identified, 10 gene–gene interaction pairs were replicated in the PPI network. The identified gene-gene interactions and genes showed associations with AD in terms of neuroinflammation and neurodegeneration, neuronal cells activation and brain development, thereby leading to cognitive decline in AD, which is indirectly associated with the P-tau pathological feature of AD and in turn supports the results of this study. Thus, the results of our study might be beneficial for explaining part of the "missing heritability" of AD.

Keywords: Alzheimer's disease; epistasis; hyperphosphorylated tau (P-tau); PPI; ADNI

Citation: Chen, D.; Li, J.; Liu, H.; Liu, X.; Zhang, C.; Luo, H.; Wei, Y.; Xi, Y.; Liang, H.; Zhang, Q. Genome-Wide Epistasis Study of Cerebrospinal Fluid Hyperphosphorylated Tau in ADNI Cohort. *Genes* **2023**, *14*, 1322. https://doi.org/10.3390/genes14071322

Academic Editors: Diego Centonze and Claudia Ricci

Received: 1 June 2023
Revised: 19 June 2023
Accepted: 20 June 2023
Published: 23 June 2023

Copyright: © 2023 by the authors. Licensee MDPI, Basel, Switzerland. This article is an open access article distributed under the terms and conditions of the Creative Commons Attribution (CC BY) license (https://creativecommons.org/licenses/by/4.0/).

1. Introduction

Alzheimer's disease (AD) is an insidious neurodegenerative disorder. The currently available therapies do not slow disease progression, provides short-term symptomatic relief only [1,2]. Genome-wide association studies (GWAS) and related techniques are gradually discovering variants of causal gene variants that contribute to complex human diseases. However, after years of GWAS efforts by countless researchers, these findings can explain only a small fraction of the heritability and the genetic factors of many human diseases and traits failed to be discovered, the so-called "missing heritability" [3,4]. Many research suggest that "missing heritability" in AD remains extensive with an estimated

25% of phenotypic variance unexplained by known variants, which may be explained by epistasis [5–7]. A single nucleotide polymorphism (SNP) is defined as single nucleotide alteration in a DNA sequence among individuals [8,9]. As a major drawback, in common GWAS, the focus is usually on SNPs that have significant individual effects [10,11]. Epistasis is the phenomenon about the interaction alleles of different loci when expressing a certain phenotype, and it cannot be attributed to the additive combination of effects corresponding to the individual loci. If the effect of one variant affecting a complex trait depends on the genotype of a second variant affecting that trait, epistasis will occurs [12]. To be specfic, epistasis leads to complex phenotypic effects, in which the effect of one locus is masked by the effects on another locus or the joint effects of two SNPs may be significant whereas they are ineffective separately [13,14]. Therefore, epistasis detection is expected to explain the "missing heritability" of many complex diseases such as AD, diabetes, and hypertension [15–17].

As the number of interactions grows exponentially with the number of variants, computational limitation is a bottleneck [18]. Most methods of epistasis detection choose to refrain from the brute force search in the SNP–SNP interaction space and try to reduce computational burden using dimensionality reduction screening and priori knowledge [19,20]. However, using more subjective priori knowledge or random factors for dimension reduction search will lead to signal loss because the risk of epistatic interaction is unknown [21]. Moreover, most methods of epistasis detection have been designed for case-control tasks over the past decade, with few on quantitative traits (QT) [19]. Compared to case–control status, QT has increased statistical significance and could better track AD progression [22]. Therefore, mining more potential loci by QT, which are implicated in AD without dropping signals across the whole genome, is urgently needed.

Emerging data have suggested that the prevailing amyloid cascade hypothesis is insufficient to explain many aspects of AD pathogenesis, neuroinflammation also plays an important role in the pathogenesis of AD [2,23]. AD is characterized by extracellular amyloid-β (Aβ) peptides in cortical Aβ plaques, intracellular phosphorylated tau protein as neurofibrillary tangles, and neuronal as well as axonal degeneration. These key hallmarks of AD can be measured in vivo with positron emission tomography (PET) imaging and biofluid markers including plasma and CSF assays [24]. CSF tau, P-tau, and Aβ_{42} are established biomarkers for AD and have been widely used as QT for genetic analyse [25]. Furthermore, accumulated P-tau may be the primary contributor to neurodegeneration during AD, and neuroinflammation is a central mechanism involved in neurodegeneration as observed in AD and might play a critical role in inducing neurodegeneration [26–28]. Moreover, P-tau, one of the three candidate CSF QT, has been widely studied in the Alzheimer's Disease Neuroimaging Initiative (ADNI) cohort. For example, heterogeneity in p-tau species carries predictive power in the identification of disease severity in incipient AD [29]. P-tau might start neurodegenerative processes and are necessary for cognitive decline [30]. Tau pathology is an initiating factor in sporadic AD [31]. Increase of p181tau levels might predict preclinical AD in cognitively normal elderly [32].

Therefore, CSF P-tau was used as a QT to advance statistical power and biological interpretation in this study. Then, we performed genome-wide epistasis detection in ADNI cohort based on a multi-GPU method, which has a better detection power outperform other competitive approaches.

2. Materials and Methods

2.1. Genotyping Data and Subjects Processing

Data used in this study were obtained from the ADNI database. The ADNI is a longitudinal multi-center study designed to be used for the early detection and tracking of AD, which was founded in 2004 under the leadership of Dr. Michael W. Weine and supported by the Foundation for the National Institutes of Health ($27 million) and the National Institute on Aging ($40 million). The primary goal of ADNI was developing biomarkers as outcome measures for clinical trials, examining biomarkers in earlier stages of the disease, and developing biomarkers as predictors of cognitive decline, etc. The SNP data were collected from the Illumina 2.5M array and the Illumina OmniQuad array including ADNI-1, ADNI-GO, and ADNI-2 cohorts, which can be downloaded from the LONI website (https://adni.loni.usc.edu, accessed on 19 June 2023). A total of 687,414 SNPs were involved in the study. To get pure SNP data, genetic analysis tool PLINK v1.90 was used to filter the SNPs according to the following quality control (QC) criteria: (1) SNPs on chromosome 1–22; (2) minimum call rate for SNPs and subjects $\geq 95\%$; (3) minimum allele frequencies (MAF) $\geq 5\%$; (4) Hardy-Weinberg equilibrium (HWE) test $p \geq 10^{-6}$. After the QC, a total of 563,980 SNPs participated in this study.

Subjects were checked by the following QC flow: (1) call rate per subjects $\geq 90\%$; (2) gender check; (3) identity check. Then, EIGENSTRAT was used to perform the population stratification analysis [33]. The population stratification analysis yielded 89 subjects who were non-Hispanic Caucasians. These 89 participants were excluded from the analysis. Finally, 1079 subjects passed the QC.

CSF P-tau phenotype was used as QT in this study. The QC criteria of phenotype was based on two principles: baseline consistency principle and normal distribution principle. Out of the 1079 subjects retained after the QC, 860 subjects had both genotype data and phenotype (CSF P-tau). These subjects (N = 860) including 201 cognitive normal cognition (CN), 84 significant memory impairment (SMC), 251 early mild cognitive impairment (EMCI), 209 late mild cognitive impairment (LMCI), and 115 AD subjects.

Overall, 860 valid P-tau of CSF subjects and 563,980 remained for the subsequent genome-wide SNP-SNP interaction analysis.

2.2. Genome-Wide SNP-SNP Interaction Analysis

In this study, including as covariates in the liner regression analysis such as age, gender and clinical diagnostic status (cds), we consider the linear regression model of additive main effect of two SNPs,

$$L_{1,2} = \alpha_0 + \alpha_1 \times SNP_1 + \alpha_2 \times SNP_2 + age + gender + cds + \varepsilon_i \quad (1)$$

where α_0, α_1 and α_2 are regression coefficients; ε_i is a residual that follows a normal distribution with mean zero and variance σ_2. Therefore, the sum of both additive main effect and SNP_1-SNP_2 interaction is then given by

$$L_{1,2}{}^S = \alpha_0 + \alpha_1 \times SNP_1 + \alpha_2 \times SNP_2 + \alpha_{1,2} \times SNP_1 \times SNP_2 + age + gender + cds + \varepsilon_i \quad (2)$$

where α_0, α_1, α_2 and $\alpha_{1,2}$ are regression coefficients. We set Y = Y_1 Y_2 ... Y_n, where n is the number of subjects in the sample, then the signal of SNPs is set in the form of S_i = S_{1j} S_{2j} ... S_{nj} , where j = 1 or 2; S_{ij} is the genotype of the allele on the SNP_j of the ith subject; S_{ij} = 0, 1 or 2. For each SNP-SNP interaction pair, the interaction effect was evaluated by two linear regression models according to the CSF P-tau quantitative trait. In practice, we test the significance of the interaction terms using an F test, then the p-value would be calculated.

2.3. Bioinformatics Analyses

To further explain the biological functions of SNP-SNP interaction pairs with significant interactions, all SNPs were mapped to the corresponding genes according to position based on the Homo sapiens genome assembly GRCh37 (hg19), and SNPs not located within the gene region were mapped to nearby genes by position offset of 100 kb. Moreover, the gene databases National Center for Biotechnology Information Phenotype-Genotype Integrator (NCBI PheGenI) was used to analyze the association of the candidate genes with phenotype trait, and Reactome 2022 was used to conduct pathway analysis for discover associated biological processes. For the gene-gene interaction pairs after mapping from the SNP level, functional enrichment processes were performed by PPI network enrichment analysis through the STRING database.

3. Results

3.1. SNP-SNP Interaction Results

In this study, a genome-wide SNP-SNP interaction detection using CSF P-tau as the intermediate quantitative phenotype was implemented. With the assistance of the GEEpiQt tool [34], we completely detected all SNP-SNP interaction pairs with significant interaction across the whole genome. According to the set *p*-value criteria, 758 SNP interaction pairs passed the significance requirement. After the GWAS analysis was conducted on the SNPs of 758 interaction pairs, all the results indicated that the statistical significance of the interaction effect was much higher than the main effect. The interaction effects and main effects of 758 significant SNP-SNP interaction pairs are shown in Figure 1.

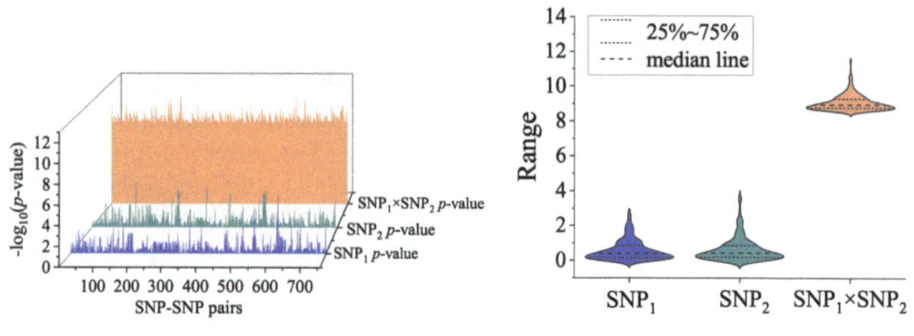

(a) (b)

Figure 1. (**a**) The 3D waterfall plot reveals interaction effects and main effects of each SNP with $-\log_{10}$ (*p*-values). The number of significant SNP-SNP pairs was 758. For each interaction pair, the blue waterfall represents main effects of SNP_1; the green waterfall represents main effects of SNP_2; and the orange waterfall represents the interaction effects of SNP_1–SNP_2 pairs. (**b**) The violin plot shows the distribution state and probability density of SNP-SNP interaction effect, main effect of SNP_1 and SNP_2.

To further confirm the association of SNP interactions with quantitative phenotypes, the explained variance of genetic epistasis of CSF P-tau was calculated by IBM SPSS 24.0. Two general linear models were used, with age, sex and disease diagnosis status added as covariates, and interaction terms were added to one model for calculating the main effect and interaction effect on phenotypic explanatory rate separately. The R square of the interaction terms and the additive terms are shown in Figure 2. The top 10 R-square of SNP-SNP interaction pairs and results of post hoc analysis on P-tau level are seen in Table 1. Age, gender, and cds accounted for 9.3% of variance on the P-tau level. Moreover, Table 1 gives the proportion of additional variance in P-tau level explained by the combined main effect and interaction effect of SNP_1 and SNP_2 after accounting for age, gender, cds, SNP_1 and SNP_2. The percentages of each interaction pair are as follows. For rs2291948 (*APOOP5*)—rs2619171, the interaction term accounted for 5.6% of variance, and the main effects accounted for 0.1% of variance (5.7% combined). For rs17069204 (*SEC63*)—rs4983187 (*LINC02588*), the interaction term accounted for 5.5% of variance, and the main effects accounted for 0.8% of variance (6.3% combined). For rs6882813—rs17416058, the interaction term accounted for 5.5% of variance, and the main effects accounted for 0.7% of variance (6.2% combined). For rs129600 (*PPARA*)—rs6602151 (*RSU1*), the interaction term accounted for 5.4% of variance, and the main effects accounted for 0.5% of variance (5.9% combined). For rs6796502 (*PRSS42P*)—rs6999890 (*SLC45A4*), the interaction term accounted for 5.3% of variance, and the main effects accounted for 0.4% of variance (5.7% combined). For rs1412839 (*PDPN*)—rs2397718, the interaction term accounted for 5.3% of variance, and the main effects accounted for 0.5% of variance (5.8% combined). For rs2219872 (*GRIP1*)—rs2647911 (*C12orf66*), the interaction term accounted for 5.2% of variance, and the main effects accounted for 1.3% of variance (6.5% combined). For rs9320250 (*OSTM1*)—rs4983187 (*LINC02588*), the interaction term accounted for 5.2% of variance, and the main effects accounted for 1.1% of variance (6.3% combined). For rs10802434 (*SCCPDH*)—rs12470444 (*NRP2*), the interaction term accounted for 5.2% of variance, and the main effects accounted for 0.6% of variance (5.8% combined). For rs2487643 (*PDPN*)—rs2397718, the interaction term accounted for 5.2% of variance, and the main effects accounted for 0.6% of variance (5.8% combined).

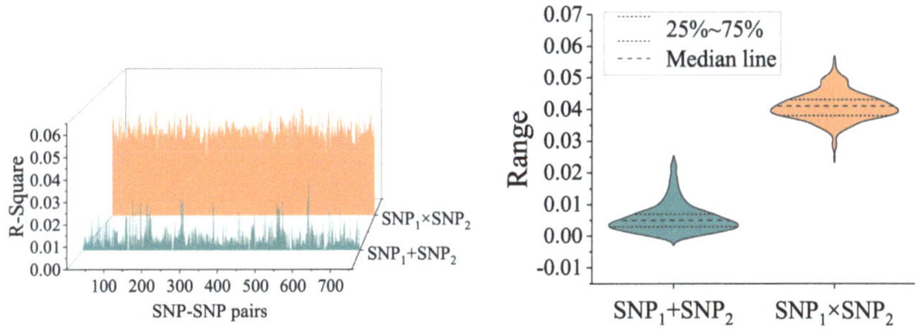

(**a**) (**b**)

Figure 2. (**a**) The 3D waterfall plot reveals interaction and additive terms with R square in the linear regression model. The orange area represents the variance explained by interaction term on P-tau. The green area represents the variance explained by additive term on P-tau. (**b**) The violin plot shows the distribution state and probability density of interaction R square, additive terms of R square the two SNPs.

Table 1. Top10 R Square Of Snp-Snp Interaction Pairs.

NO	$SNP_1 \times SNP_2$	GENE	CHR	*p*-Value		Explained Variance (R Square)		
				GWAS	Interaction	Age + Gender + cdsr [1]	$SNP_1 + SNP_2$ [2]	$SNP_1 \times SNP_2$ [3]
1	rs2291948 rs2619171	*APOOP5* -	16 15	0.963536 0.911948	1.70×10^{-9}	0.093	0.001	0.056
2	rs17069204 rs4983187	***SEC63*** *LINC02588*	6 14	0.0222884 0.592583	6.73×10^{-11}	0.093	0.008	0.055
3	rs6882813 rs17416058	- -	5 11	0.420613 0.212032	2.86×10^{-10}	0.093	0.007	0.055
4	rs129600 rs6602151	***PPARA*** ***RSU1***	22 10	0.635138 0.870792	4.92×10^{-10}	0.093	0.005	0.054
5	rs6796502 rs6999890	*PRSS42P* *SLC45A4*	3 8	0.676528 0.454329	4.80×10^{-11}	0.093	0.004	0.053
6	rs1412839 rs2397718	***PDPN*** -	1 5	0.251693 0.785923	6.00×10^{-10}	0.093	0.005	0.053
7	rs2219872 rs2647911	*GRIP1* *C12orf66*	12 12	0.016483 0.292968	3.52×10^{-12}	0.093	0.013	0.052
8	rs9320250 rs4983187	*OSTM1* *LINC02588*	6 14	0.0269661 0.592583	4.37×10^{-10}	0.093	0.011	0.052
9	rs10802434 rs12470444	***SCCPDH*** ***NRP2***	1 2	0.873933 0.291239	8.49×10^{-10}	0.093	0.006	0.052
10	rs2487643 rs2397718	***PDPN*** -	1 5	0.231468 0.785923	1.31×10^{-9}	0.093	0.006	0.052

[1] Age + gender + cds: percent of variance in P-tau level explained by age, gender and cds. [2] $SNP_1 + SNP_2$: percent of additional variance in P-tau level explained by the combined main effect of SNP_1 and SNP_2 after accounting for age, gender and diagnosis. [3] $SNP_1 \times SNP_2$: percent of additional variance in P-tau level explained by the interaction effect of SNP_1 and SNP_2 after accounting for age, gender, diagnosis, SNP_1 and SNP_2. The genes with bold italics in the table are AD-related genes.

3.2. Functional Annotations for Significant Interaction Pairs

A total of 1161 SNPs were mapped onto 578 genes. Then, gene set enrichment analysis was performed based on HDSigDB Human 2021 in Enrichr. The results of enrichment analysis indicate that 331 genes have been shown associated with AD, and the number of unconfirmed AD-related genes is 247. In order to make more sensible explanations at the gene level in subsequent study, these gene pairs were categorized according to the relationship of genes with AD on both sides. The interactions were classified into three categories: both genes in each pair are AD-related, only one gene in each pair is AD-related, and none gene in each pair is associated with AD, as shown in Figure 3.

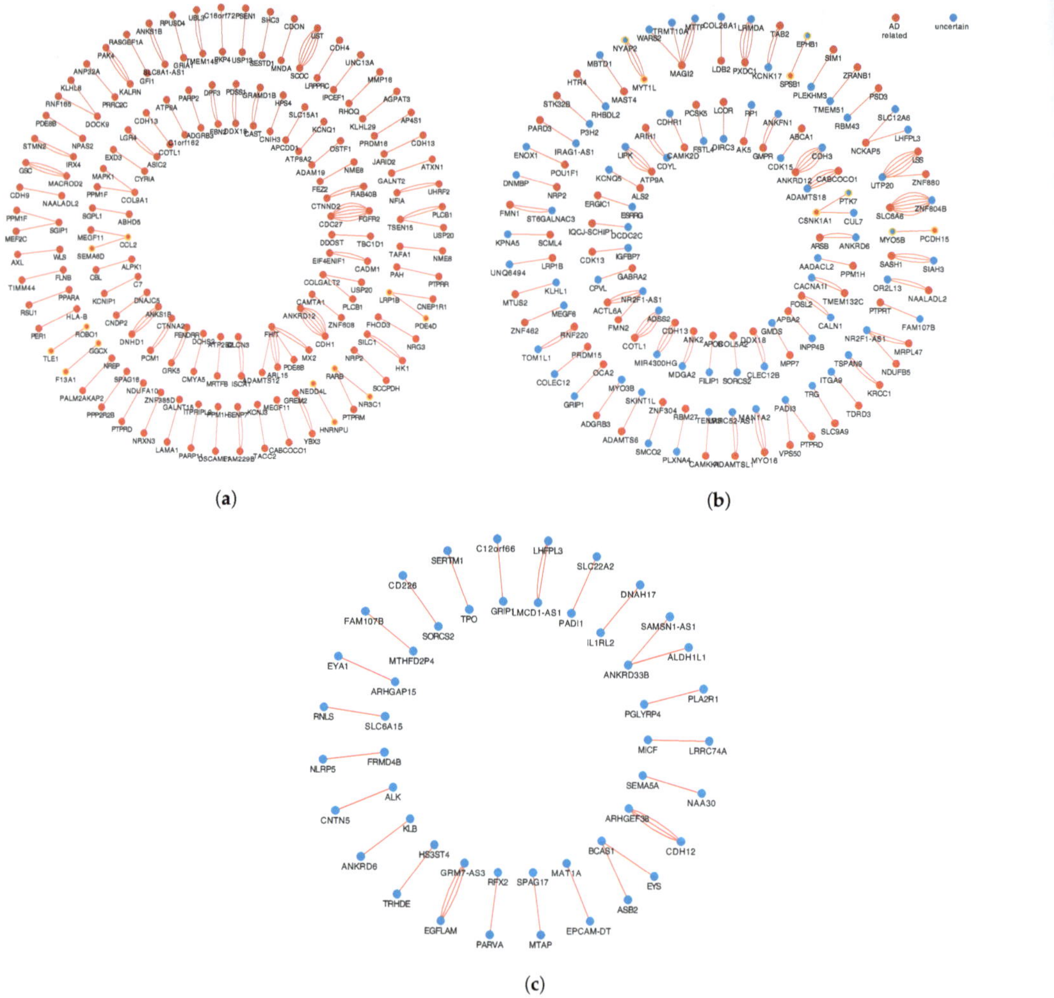

Figure 3. Three categories of different relationships with AD correlation in a gene-gene interaction pair. (**a**) The 1st relationship, include 95 gene pairs, both genes in each pair are AD-related. (**b**) The 2nd relationship, includes 79 gene pairs, only one gene in each pair is AD-related. (**c**) The 3rd relationship, includes 25 gene pairs, none gene in each pair is associated with AD. The red spots represent AD-related genes. The blue spots represent AD-unrelated genes. Multiple lines between two points represent multiple pairs of SNP mapped ont to the same pair of genes.

Moreover, the gene databases National Center for Biotechnology Information Phenotype-Genotype Integrator (NCBI PhenGenI) and Reactome 2022 were used to analyze the association of the detected genes with CSF P-tau quantitative traits. Pathway enrichment analysis of genes mapped by the identified SNPs based on their enrichment adjusted *p*-values is presented in Figure 4. The Reactome enrichment analysis showed that three of the top ten pathways were significant, Cell-Cell Communication, Cell Junction Organization and Neuronal System pathways. Among them, there is evidence that "Cell-Cell Communication" is associated with the abnormal accumulation of phosphorylated tau protein in Alzheimer's disease. In the Neuronal System pathway, tau protein is normally involved in stabilizing microtubules in neurons, but in Alzheimer's disease, it can become hyperphosphorylated and form aggregates called neurofibrillary tangles. These tangles can disrupt the transport of nutrients and other substances within neurons, which can further damage the pathways of the neuronal system. The results of the PhenGenI Association enrichment analysis were found significant in several diseases, such as Platelet Function Tests, Alzheimer Disease, and Stroke.

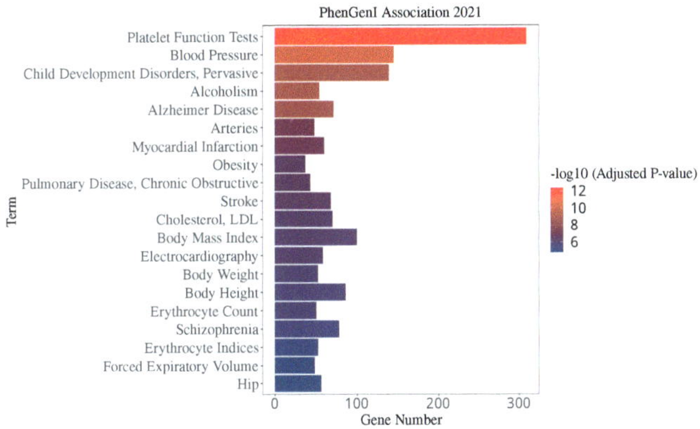

(a)

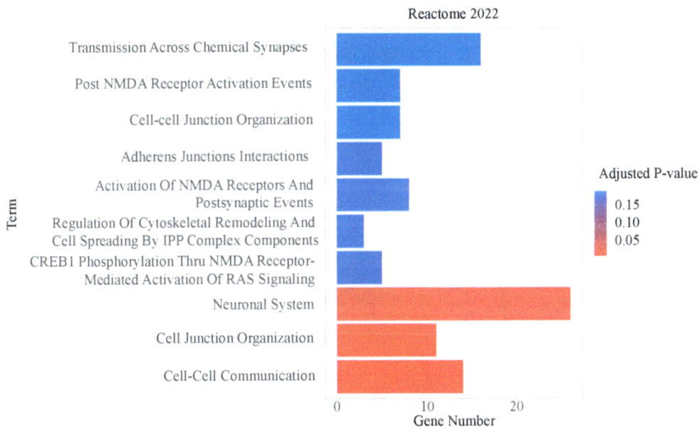

(b)

Figure 4. Significantly enriched pathways of studied genes. x-axis indicates the number of overlapped genes of related pathway, y-axis indicates significant pathways. The gradient of the color represents the level of significance. The red bars represent high significance. (**a**) Top 20 pathways identified by PhenGenI Association 2021 enrichment, −log10 adjusted p-value. (**b**) Top 10 pathways identified by Reactome 2022, adjusted p-value.

3.3. Potential Interactions via Protein-Protein Interaction Analysis

To investigate and validate the potential interactions additionally, we submitted 174 gene-gene interaction pairs to the STRING database for PPI enrichment analysis. All the 174 gene-gene interaction pairs were selected from the 1st relationship shows in Figure 3a and the 2nd relationship as shows in Figure 3b. Notably, 10 of the gene interaction pairs overlapped with PPI networks in the database, as shown in Figure 5, including: *SPSB1-EPHB1, HNRNPU-NEDD4L, MYT1L-NYAP2, GGCX-F13A1, LRP1B-PDE4D, RARB-NR3C1, CCL2-SEMA6D, ROBO1-TLE1, CSNK1A1-PTK7, MYO5B-PCDH15.*

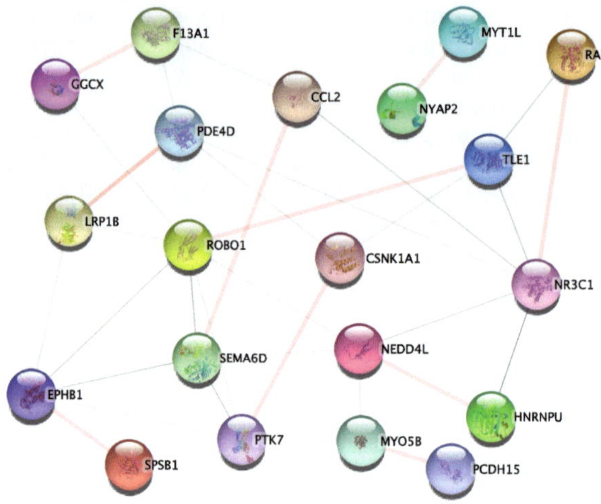

Figure 5. The PPI subnetwork of studied genes. The nodes and edges represent the proteins (genes) and their interactions, respectively. The PPI subnetwork contained 20 nodes and 33 edges. The light pink connections represent ten overlapped gene-gene interaction pairs with the PPI network.

4. Discussion

In this study, detection of genome-wide SNP-SNP interaction based on multi-GPU were performed. To our knowledge, this study is a highly comprehensive epistatic study of QT at the P-tau level. A total of 758 SNP-SNP pairs were found to be statistically significant, and highly significant SNP–SNP interactions were detected between the marginal main effect SNPs. In particular, the interaction effects were much higher than the main effects (Figure 1). As we expected, all identified interaction pairs explained a relatively high-level variance at the P-tau level (Figure 2 and Table 1), which could be helpful for explaining some part of the "missing heritability" of AD.

To identify potential genetic epistasis implicated in AD and obtain biologically meaningful explanations at gene-gene interaction level, 174 gene-gene interaction pairs, which from 1st relationship (Figure 3a) and 2nd relationship (Figure 3b) were submitted to the STRING database to perform the PPI enrichment analysis. As shown in Figure 5, the PPI sub-network containing 20 genes and 33 gene-gene interactions were identified. As a result, ten gene-gene interaction pairs overlapped with the PPI network need further discussion.

CSNK1A1 is a casein kinase which is involved in the phosphorylation state of tau [35]. Protein tyrosine kinase 7(*PTK7*) is a regulator of Wnt signaling pathways [36]. Wnt signaling is deregulated in AD, which could contribute to synapse degeneration and cognitive decline. This deficiency in Wnt signaling may further exacerbate tau hyperphosphorylation [37]. Therefore, the *CSNK1A1* and *PTK7* interaction shows strong associations with tau phosphorylation.

The *SPRY* domain-containing *SOCS* box protein 1 (*SPSB1*) is involved in the development of AD through nitric oxide (NO) pathways. To be specific, NO pathways contribute to pathogenesis of neurodegeneration in AD and other neurodegenerative dementias by involving in neuroinflammation, while SPSB1 negatively control NO production and limit cellular toxicity [38]. Ephrin type-B receptor 1 (*EphB1*) is upregulated in injured motor neurons, and then activated astrocytes [39]. Furthermore, activated astrocytes mediate neuroinflammation and neurodegeneration. Neuroinflammation induces neurodegeneration and the processes involved in neurodegeneration augments neuroinflammation [26].

Neurodegeneration is mediated by inflammatory and neurotoxic mediators such as chemokine (C-C motif) ligand 2 (*CCL2*), *CCL5*, tumor necrosis factor-alpha (TNF-α) and interleukin-6(*IL-6*) etc. The increased level of these mediators including *CCL2* lead to neurodegeneration and neuronal death in neurodegenerative diseases [26,40]. These inflammatory and neurotoxic mediators directly or indirectly through glial cells and inflammatory cells affect neuronal survival and induce neurodegeneration [26]. Moreover, *CCL2* is implicated in the pathways recruiting microglia and the development of P-tau pathology, and might be related to reducing neuroinflammation [41]. *SEMA6D* is a regulator of microglial phagocytosis and inflammatory cytokine (TNFα, IL-6 etc.) release in a TREM2-dependent manner [42,43]. Moreover, microglial phagocytosis is a disease-associated process emerging from AD genetics [44]. Excessive microglial phagocytosis of synapses can be observed in AD, leading to significant synapse loss and memory impairment [45]. And lack of microglial phagocytosis can exacerbate pathology of AD and induce memory impairment [46]. Microglia are also major players in neuroinflammation [23].

RP1B belonged to the low-density lipoprotein (LDL) receptor family, and several members of the LDL family have been implicated in cellular processes relevant to neurodegeneration, including tau uptake et al. Enhanced *LRP1B* activity can protect against the pathogenesis of AD and cognitive decline in old age. Inhibition of phosphodiesterase 4D (*PDE4D*) activity can enhance phosphorylation of tau.

MYT1L is a critical mediator of directly converting human brain vascular pericytes (HBVPs) into cholinergic neuronal cells, and the cholinergic deficit is thought to underlie progressed cognitive decline in AD. Neuronal tyrosine-phosphorylated phosphoinositide-3-kinase adapter 2 (*NYAP2*) is involved in remodeling of actin cytoskeleton [47]. Actin cytoskeleton has been described as an underlying factor of synaptic failure in AD, which could contribute to AD pathology [48].

Regulation of *HNRNPU* expression ameliorates impairments of learning and memory abilities in an AD rat model. *NEDD4L* is identified as potential nuclear enriched abundant transcript 1 (*NEAT1*) interaction proteins. upregulated NEAT1 can give rise to the amyloid accumulation and cognitive decline in AD.

In summary, *CSNK1A1-PTK7* interaction and *PDE4D* gene shows strong associations with P-tau, which is directly associated with pathogenesis of AD. Two pairs of gene-gene interactions show strong associations with AD in terms of neuroinflammation and neurodegeneration: *SPSB1-EPHB1, CCL2-SEMA6D*. *MYT1L* gene, *LRP1B* gene and *NEDD4L* gene show strong associations with AD, and leading to cognitive decline in AD. *LRP1B* gene is also associated with pathogenesis of AD. *HNRNPU* gene exerts its effects on learning and memory abilities in AD. *NYAP2* gene is involved in remodeling of actin cytoskeleton, which is indirectly associated with AD pathology. Two pairs of gene-gene interactions can affect the activation of neuronal cells and contribute to brain development: *RARB-NR3C1, ROBO1-TLE1* [49,50]. *MYO5B* gene and *GGCX* gene show strong associations with schizophrenia [51]. Neuroinflammation is well established in a subset of schizophrenia patients [52]. In addition, two genes have not yet been associated with AD pathology: *F13A1, PCDH15*, which warrant further investigation.

To our knowledge, neurodegeneration appears to be the biological mechanism most proximate to cognitive decline in AD [53]. Aβ and tau pathologies interact synergistically in the preclinical stages of AD, which contributing to faster neurodegeneration and cognitive decline [54,55]. Therefore, the identified gene-gene interactions and genes in the PPI network might be related to neuroinflammation and neurodegeneration, thereby leading to cognitive decline in AD, which is indirectly proves that accumulated P-tau may be the primary contributor to neurodegeneration during AD [27], and in turn supports the results of this study.

5. Conclusions

Aimed at performing genome-wide epistasis detection in the ADNI cohort, we used CSF P-tau as a QT. 331 AD-related gene were replicated, which are previously confirmed AD risk genes. We also replicated 10 findings in the PPI network, which are *SPSB1-EPHB1, HNRNPU-NEDD4L, MYT1L-NYAP2, GGCX-F13A1, LRP1B-PDE4D, RARB-NR3C1, CCL2-SEMA6D, ROBO1-TLE1, CSNK1A1- PTK7, MYO5B-PCDH15*. Moreover, 3 gene-gene pairs of interaction showed strong association with AD: *CSNK1A1-PTK7, SPSB1-EPHB1, CCL2-SEMA6D*. Interactions between *RARB* and *NR3C1*, between *ROBO1* and *TLE1* can affect the activation of neuronal cells and contribute to brain development. Our study also revealed two genes have not yet been associated with AD pathology: *F13A1*, *PCDH15*, which warrant further investigation. In summary, our results can provide useful clues to the aspect of inducing neuroinflammation and neurodegeneration, and show strong association with AD in terms of cognitive decline. Therefore, this study might open new avenues to complement common GWAS. Furthermore, our results may be replicated by considering other quantitative traits using different databases and methods to complement the PPI network. Biological interpretation is also a direction of future research.

Author Contributions: D.C. designed the entire study, executed data analysis, and wrote the manuscript. J.L. performed data analysis and wrote the manuscript. Q.Z. conceived and designed the study, reviewed data analysis, and critically revised the manuscript for important intellectual content. H.L.(Hong Liang) performed data analysis and revised the manuscript. H.L. (Hongwei Liu) played a major role in data analysis and biological analysis and contributed preparing figures. Y.X., X.L., C.Z., H.L. (Haoran Luo) and Y.W. performed literature search and prepared the manuscript. All authors have read and agreed to the published version of the manuscript.

Funding: This project was supported by the National Natural Science Foundation of China (grant number: 62206044).

Institutional Review Board Statement: Not applicable.

Informed Consent Statement: Not applicable.

Data Availability Statement: The data used in this project were funded by the Alzheimer's Disease Neuroimaging Initiative (ADNI). The database is the (http://adni.loni.usc.edu/, accessed on 19 June 2023) of the Alzheimer's disease neuroimaging database.

Acknowledgments: The authors acknowledge the Alzheimer's Disease Neuroimaging Initiative (ADNI) database.

Conflicts of Interest: The authors declare no conflict of interest.

References

1. Zhang, C.; Griciuc, A.; Hudry, E.; Wan, Y.; Quinti, L.; Ward, J.; Forte, A.M.; Shen, X.; Ran, C.; Elmaleh, D.R.; et al. Cromolyn reduces levels of the Alzheimer's disease-associated amyloid β-protein by promoting microglial phagocytosis. *Sci. Rep.* **2018**, *8*, 1144. [CrossRef]
2. Rusek, M.; Smith, J.; El-Khatib, K.; Aikins, K.; Czuczwar, S.J.; Pluta, R. The Role of the JAK/STAT Signaling Pathway in the Pathogenesis of Alzheimer's Disease: New Potential Treatment Target. *Int. J. Mol. Sci.* **2023**, *24*, 864. [CrossRef] [PubMed]
3. Bloom, J.S.; Ehrenreich, I.M.; Loo, W.T.; Lite, T.L.V.; Kruglyak, L. Finding the sources of missing heritability in a yeast cross. *Nature* **2013**, *494*, 234–237. [CrossRef]
4. Jansen, I.E.; Savage, J.E.; Watanabe, K.; Bryois, J.; Williams, D.M.; Steinberg, S.; Sealock, J.; Karlsson, I.K.; Hägg, S.; Athanasiu, L.; et al. Genome-wide meta-analysis identifies new loci and functional pathways influencing Alzheimer's disease risk. *Nat. Genet.* **2019**, *51*, 404–413. [CrossRef] [PubMed]
5. Simmonds, E.; Leonenko, G.; Schmidt, K.M.; Hill, M.; Myers, A.; Shoai, M.; de Rojas, I.; Tesi, N.; Holstege, H.; van der Flier, W.; et al. What does heritability of Alzheimer's disease represent? *PLoS ONE* **2023**, *18*, e0281440.
6. Andrews, S.J.; Renton, A.E.; Fulton-Howard, B.; Podlesny-Drabiniok, A.; Marcora, E.; Goate, A.M. The complex genetic architecture of Alzheimer's disease: novel insights and future directions. *eBiomedicine* **2023**, *90*, 104511. [CrossRef] [PubMed]
7. Ebbert, M.T.; Ridge, P.G.; Kauwe, J.S. Bridging the gap between statistical and biological epistasis in Alzheimer's disease. *BioMed Res. Int.* **2015**, *2015*, 870123. [CrossRef]
8. Min, A.; Thompson, E.; Basu, S. Comparing heritability estimators under alternative structures of linkage disequilibrium. *G3* **2022**, *12*, jkac134. [CrossRef]
9. KaradaĞ, Ö.; Altun, G.; AktaŞ, S. Assessment of SNP-SNP interactions by using square contingency table analysis. *An. Acad. Bras. CiÊNcias* **2020**, *92*. [CrossRef]
10. Lutz, M.W.; Sprague, D.; Chiba-Falek, O. Bioinformatics strategy to advance the interpretation of Alzheimer's disease GWAS discoveries: The roads from association to causation. *Alzheimer's Dement.* **2019**, *15*, 1048–1058. [CrossRef]
11. Goudey, B.; Rawlinson, D.; Wang, Q.; Shi, F.; Ferra, H.; Campbell, R.M.; Stern, L.; Inouye, M.T.; Ong, C.S.; Kowalczyk, A. GWIS-model-free, fast and exhaustive search for epistatic interactions in case-control GWAS. *BMC Genom.* **2013**, *14*, S10. [CrossRef] [PubMed]
12. Mackay, T.F.; Moore, J.H. Why epistasis is important for tackling complex human disease genetics. *Genome Med.* **2014**, *6*, 42. [CrossRef]

13. Mackay, T.F. Epistasis and quantitative traits: Using model organisms to study gene–gene interactions. *Nat. Rev. Genet.* **2014**, *15*, 22–33. [CrossRef]
14. Moore, J.H. A global view of epistasis. *Nat. Genet.* **2005**, *37*, 13–14. [CrossRef]
15. Chang, Y.C.; Wu, J.T.; Hong, M.Y.; Tung, Y.A.; Hsieh, P.H.; Yee, S.W.; Giacomini, K.M.; Oyang, Y.J.; Chen, C.Y. GenEpi: Gene-based epistasis discovery using machine learning. *BMC Bioinform.* **2020**, *21*, 68. [CrossRef]
16. Yang, Y.; Yao, S.; Ding, J.M.; Chen, W.; Guo, Y. Enhancer-gene interaction analyses identified the epidermal growth factor receptor as a susceptibility gene for type 2 diabetes mellitus. *Diabetes Metab. J.* **2021**, *45*, 241–250. [CrossRef] [PubMed]
17. Cordell, H.J. Detecting gene–gene interactions that underlie human diseases. *Nat. Rev. Genet.* **2009**, *10*, 392–404. [CrossRef] [PubMed]
18. Wang, H.; Yue, T.; Yang, J.; Wu, W.; Xing, E.P. Deep mixed model for marginal epistasis detection and population stratification correction in genome-wide association studies. *BMC Bioinform.* **2019**, *20*, 656. [CrossRef] [PubMed]
19. Russ, D.; Williams, J.A.; Cardoso, V.R.; Bravo-Merodio, L.; Pendleton, S.C.; Aziz, F.; Acharjee, A.; Gkoutos, G.V. Evaluating the detection ability of a range of epistasis detection methods on simulated data for pure and impure epistatic models. *PLoS ONE* **2022**, *17*, e0263390. [CrossRef] [PubMed]
20. Wang, Q.; Zhang, Q.; Meng, F.; Li, B. Objective-hierarchy based large-scale evolutionary algorithm for improving joint sparsity-compression of neural network. *Inf. Sci.* **2023**, *640*, 119095. [CrossRef]
21. Yoshida, M.; Koike, A. SNPInterForest: A new method for detecting epistatic interactions. *BMC Bioinform.* **2011**, *12*, 469. [CrossRef] [PubMed]
22. Wang, H.; Nie, F.; Huang, H.; Kim, S.; Nho, K.; Risacher, S.L.; Saykin, A.J.; Shen, L.; Initiative, A.D.N. Identifying quantitative trait loci via group-sparse multitask regression and feature selection: An imaging genetics study of the ADNI cohort. *Bioinformatics* **2012**, *28*, 229–237. [CrossRef] [PubMed]
23. Leng, F.; Edison, P. Neuroinflammation and microglial activation in Alzheimer disease: Where do we go from here? *Nat. Rev. Neurol.* **2021**, *17*, 157–172. [CrossRef] [PubMed]
24. Cruchaga, C.; Kauwe, J.S.; Harari, O.; Jin, S.C.; Cai, Y.; Karch, C.M.; Benitez, B.A.; Jeng, A.T.; Skorupa, T.; Carrell, D.; et al. GWAS of cerebrospinal fluid tau levels identifies risk variants for Alzheimer's disease. *Neuron* **2013**, *78*, 256–268. [CrossRef]
25. Guo, T.; Korman, D.; La Joie, R.; Shaw, L.M.; Trojanowski, J.Q.; Jagust, W.J.; Landau, S.M. Normalization of CSF pTau measurement by Aβ40 improves its performance as a biomarker of Alzheimer's disease. *Alzheimer's Res. Ther.* **2020**, *12*, 97. [CrossRef]
26. Kempuraj, D.; Thangavel, R.; Natteru, P.; Selvakumar, G.; Saeed, D.; Zahoor, H.; Zaheer, S.; Iyer, S.; Zaheer, A. Neuroinflammation induces neurodegeneration. *J. Neurol. Neurosurg. Spine* **2016**, *1*, 1003.
27. Dong, Y.; Yu, H.; Li, X.; Bian, K.; Zheng, Y.; Dai, M.; Feng, X.; Sun, Y.; He, Y.; Yu, B.; et al. Hyperphosphorylated tau mediates neuronal death by inducing necroptosis and inflammation in Alzheimer's disease. *J. Neuroinflamm.* **2022**, *19*, 205. [CrossRef]
28. Parhizkar, S.; Holtzman, D.M. APOE Mediated Neuroinflammation and Neurodegeneration in Alzheimer's Disease. In *Seminars in Immunology*; Elsevier: Amsterdam, The Netherlands, 2022; p. 101594.
29. Servaes, S.; Lussier, F.Z.; Therriault, J.; Tissot, C.; Bezgin, G.; Kang, M.S.; Wang, Y.T.; Stevenson, J.; Rahmouni, N.; Arias, J.F.; et al. pTau heterogeneity as a measure for disease severity in incipient Alzheimer's disease. *Alzheimer's Dement.* **2022**, *18*, e063749. [CrossRef]
30. Penke, B.; Szűcs, M.; Bogár, F. New Pathways Identify Novel Drug Targets for the Prevention and Treatment of Alzheimer's Disease. *Int. J. Mol. Sci.* **2023**, *24*, 5383. [CrossRef]
31. Arnsten, A.F.; Datta, D.; Del Tredici, K.; Braak, H. Hypothesis: Tau pathology is an initiating factor in sporadic Alzheimer's disease. *Alzheimer's Dement.* **2021**, *17*, 115–124. [CrossRef]
32. Meredith, J.E., Jr.; Sankaranarayanan, S.; Guss, V.; Lanzetti, A.J.; Berisha, F.; Neely, R.J.; Slemmon, J.R.; Portelius, E.; Zetterberg, H.; Blennow, K.; et al. Characterization of novel CSF Tau and ptau biomarkers for Alzheimer's disease. *PLoS ONE* **2013**, *8*, e76523. [CrossRef] [PubMed]
33. Price, A.L.; Patterson, N.J.; Plenge, R.M.; Weinblatt, M.E.; Shadick, N.A.; Reich, D. Principal components analysis corrects for stratification in genome-wide association studies. *Nat. Genet.* **2006**, *38*, 904–909. [CrossRef] [PubMed]
34. Zhang, Q.; Liu, H.; Ao, L.; Liang, H.; Chen, D. A GPU-based approach for detecting genome-wide SNP-SNP interactions of quantitative trait in ADNI cohorts. In Proceedings of the 2022 IEEE International Conference on Bioinformatics and Biomedicine (BIBM), Las Vegas, NV, USA, 6–8 December 2022; pp. 2564–2570.
35. Cheng, Y.; Pereira, M.; Raukar, N.; Reagan, J.L.; Quesenberry, M.; Goldberg, L.; Borgovan, T.; LaFrance Jr, W.C.; Dooner, M.; Deregibus, M.; et al. Potential biomarkers to detect traumatic brain injury by the profiling of salivary extracellular vesicles. *J. Cell. Physiol.* **2019**, *234*, 14377–14388. [CrossRef] [PubMed]
36. Berger, H.; Wodarz, A.; Borchers, A. PTK7 faces the Wnt in development and disease. *Front. Cell Dev. Biol.* **2017**, *5*, 31. [CrossRef] [PubMed]
37. Palomer, E.; Buechler, J.; Salinas, P.C. Wnt signaling deregulation in the aging and Alzheimer's brain. *Front. Cell. Neurosci.* **2019**, *13*, 227. [CrossRef]
38. Lee, E.; Giovanello, K.S.; Saykin, A.J.; Xie, F.; Kong, D.; Wang, Y.; Yang, L.; Ibrahim, J.G.; Doraiswamy, P.M.; Zhu, H.; et al. Single-nucleotide polymorphisms are associated with cognitive decline at Alzheimer's disease conversion within mild cognitive impairment patients. *Alzheimer's Dementia Diagn. Assess. Dis. Monit.* **2017**, *8*, 86–95. [CrossRef]

39. Tyzack, G.E.; Hall, C.E.; Sibley, C.R.; Cymes, T.; Forostyak, S.; Carlino, G.; Meyer, I.F.; Schiavo, G.; Zhang, S.C.; Gibbons, G.M.; et al. A neuroprotective astrocyte state is induced by neuronal signal EphB1 but fails in ALS models. *Nat. Commun.* **2017**, *8*, 1164. [CrossRef]
40. Conductier, G.; Blondeau, N.; Guyon, A.; Nahon, J.L.; Rovère, C. The role of monocyte chemoattractant protein MCP1/CCL2 in neuroinflammatory diseases. *J. Neuroimmunol.* **2010**, *224*, 93–100. [CrossRef]
41. Cherry, J.D.; Meng, G.; Daley, S.; Xia, W.; Svirsky, S.; Alvarez, V.E.; Nicks, R.; Pothast, M.; Kelley, H.; Huber, B.; et al. CCL2 is associated with microglia and macrophage recruitment in chronic traumatic encephalopathy. *J. Neuroinflamm.* **2020**, *17*, 370. [CrossRef]
42. Finan, G.M.; Kwak, C.; Hyun-Woo, R.; Kim, T.W. Sema6D Promotes TREM2-dependent Phagocytosis in Human iPSC-derived Microglia. *Alzheimer's Dement.* **2022**, *18*, e062563. [CrossRef]
43. Albanus, R.D.; Finan, G.M.; Brase, L.; Chen, S.; Guo, Q.; Kannan, A.; Acquarone, M.; You, S.F.; Novotny, B.C.; Ribeiro Pereira, P.M.; et al. Systematic characterization of brain cellular crosstalk reveals a novel role for SEMA6D in TREM2-associated microglial function in Alzheimer's disease. *bioRxiv* **2022**. [CrossRef]
44. Podleśny-Drabiniok, A.; Marcora, E.; Goate, A.M. Microglial phagocytosis: A disease-associated process emerging from Alzheimer's disease genetics. *Trends Neurosci.* **2020**, *43*, 965–979. [CrossRef] [PubMed]
45. Wang, K.; Li, J.; Zhang, Y.; Huang, Y.; Chen, D.; Shi, Z.; Smith, A.D.; Li, W.; Gao, Y. Central nervous system diseases related to pathological microglial phagocytosis. *Cns Neurosci. Ther.* **2021**, *27*, 528–539. [CrossRef] [PubMed]
46. Noda, M.; Suzumura, A. Sweepers in the CNS: Microglial migration and phagocytosis in the Alzheimer disease pathogenesis. *Int. J. Alzheimer'S Dis.* **2012**, *2012*, 891087. [CrossRef] [PubMed]
47. Yokoyama, K.; Tezuka, T.; Kotani, M.; Nakazawa, T.; Hoshina, N.; Shimoda, Y.; Kakuta, S.; Sudo, K.; Watanabe, K.; Iwakura, Y.; et al. NYAP: A phosphoprotein family that links PI3K to WAVE1 signalling in neurons. *Embo J.* **2011**, *30*, 4739–4754. [CrossRef]
48. Pelucchi, S.; Stringhi, R.; Marcello, E. Dendritic spines in Alzheimer's disease: How the actin cytoskeleton contributes to synaptic failure. *Int. J. Mol. Sci.* **2020**, *21*, 908. [CrossRef]
49. Tiwari, N.K.; Sathyanesan, M.; Schweinle, W.; Newton, S.S. Carbamoylated erythropoietin induces a neurotrophic gene profile in neuronal cells. *Prog. NeuroPsychopharmacol. Biol. Psychiatry* **2019**, *88*, 132–141. [CrossRef]
50. Hettige, N.C.; Ernst, C. FOXG1 dose in brain development. *Front. Pediatr.* **2019**, *7*, 482. [CrossRef]
51. Chen, Y.; Tian, L.; Zhang, F.; Liu, C.; Lu, T.; Ruan, Y.; Wang, L.; Yan, H.; Yan, J.; Liu, Q.; et al. Myosin Vb gene is associated with schizophrenia in Chinese Han population. *Psychiatry Res.* **2013**, *207*, 13–18. [CrossRef]
52. Murphy, C.E.; Walker, A.K.; Weickert, C.S. Neuroinflammation in schizophrenia: The role of nuclear factor kappa B. *Transl. Psychiatry* **2021**, *11*, 528. [CrossRef]
53. Merluzzi, A.P.; Vogt, N.M.; Norton, D.; Jonaitis, E.; Clark, L.R.; Carlsson, C.M.; Johnson, S.C.; Asthana, S.; Blennow, K.; Zetterberg, H.; et al. Differential effects of neurodegeneration biomarkers on subclinical cognitive decline. *Alzheimer's Dementia Transl. Res. Clin. Interv.* **2019**, *5*, 129–138. [CrossRef] [PubMed]
54. Rodriguez-Vieitez, E.; Montal, V.; Sepulcre, J.; Lois, C.; Hanseeuw, B.; Vilaplana, E.; Schultz, A.P.; Properzi, M.J.; Scott, M.R.; Amariglio, R.; et al. Association of cortical microstructure with amyloid-β and tau: Impact on cognitive decline, neurodegeneration, and clinical progression in older adults. *Mol. Psychiatry* **2021**, *26*, 7813–7822. [CrossRef] [PubMed]
55. Selkoe, D.J.; Hardy, J. The amyloid hypothesis of Alzheimer's disease at 25 years. *EMBO Mol. Med.* **2016**, *8*, 595–608. [CrossRef] [PubMed]

Disclaimer/Publisher's Note: The statements, opinions and data contained in all publications are solely those of the individual author(s) and contributor(s) and not of MDPI and/or the editor(s). MDPI and/or the editor(s) disclaim responsibility for any injury to people or property resulting from any ideas, methods, instructions or products referred to in the content.

genes MDPI

Article

A Patient with Corticobasal Syndrome and Progressive Non-Fluent Aphasia (CBS-PNFA), with Variants in *ATP7B*, *SETX*, *SORL1*, and *FOXP1* Genes

Katarzyna Gaweda-Walerych [1,*], Emilia J. Sitek [2,3,*], Małgorzata Borczyk [4], Ewa Narożańska [3], Bogna Brockhuis [5], Michał Korostyński [4], Michał Schinwelski [3,6,7], Mariusz Siemiński [8], Jarosław Sławek [3,7] and Cezary Zekanowski [1]

1 Department of Neurogenetics and Functional Genomics, Mossakowski Medical Research Institute, Polish Academy of Sciences, 02-106 Warsaw, Poland
2 Laboratory of Clinical Neuropsychology, Neurolinguistics and Neuropsychotherapy, Division of Neurological and Psychiatric Nursing, Faculty of Health Sciences, Medical University of Gdansk, 80-211 Gdansk, Poland
3 Neurology Department, St. Adalbert Hospital, Copernicus PL, 80-462 Gdansk, Poland
4 Laboratory of Pharmacogenomics, Department of Molecular Pharmacology, Maj Institute of Pharmacology Polish Academy of Sciences, 31-343 Krakow, Poland
5 Department of Nuclear Medicine, Faculty of Health Sciences, Medical University of Gdansk, 80-214 Gdansk, Poland
6 Neurocentrum Miwomed, 80-207 Gdansk, Poland
7 Division of Neurological and Psychiatric Nursing, Faculty of Health Sciences, Medical University of Gdansk, 80-211 Gdansk, Poland
8 Department of Emergency Medicine, Faculty of Health Sciences, Medical University of Gdansk, 80-210 Gdansk, Poland
* Correspondence: kgaweda@imdik.pan.pl (K.G.-W.); emilia.sitek@gumed.edu.pl (E.J.S.)

Citation: Gaweda-Walerych, K.; Sitek, E.J.; Borczyk, M.; Narożańska, E.; Brockhuis, B.; Korostyński, M.; Schinwelski, M.; Siemiński, M.; Sławek, J.; Zekanowski, C. A Patient with Corticobasal Syndrome and Progressive Non-Fluent Aphasia (CBS-PNFA), with Variants in *ATP7B*, *SETX*, *SORL1*, and *FOXP1* Genes. *Genes* **2022**, *13*, 2361. https://doi.org/10.3390/genes13122361

Academic Editor: Claudia Ricci

Received: 11 November 2022
Accepted: 9 December 2022
Published: 14 December 2022

Publisher's Note: MDPI stays neutral with regard to jurisdictional claims in published maps and institutional affiliations.

Copyright: © 2022 by the authors. Licensee MDPI, Basel, Switzerland. This article is an open access article distributed under the terms and conditions of the Creative Commons Attribution (CC BY) license (https://creativecommons.org/licenses/by/4.0/).

Abstract: Our aim was to analyze the phenotypic-genetic correlations in a patient diagnosed with early onset corticobasal syndrome with progressive non-fluent aphasia (CBS-PNFA), characterized by predominant apraxia of speech, accompanied by prominent right-sided upper-limb limb-kinetic apraxia, alien limb phenomenon, synkinesis, myoclonus, mild cortical sensory loss, and right-sided hemispatial neglect. Whole-exome sequencing (WES) identified rare single heterozygous variants in *ATP7B* (c.3207C>A), *SORL1* (c.352G>A), *SETX* (c.2385_2387delAAA), and *FOXP1* (c.1762G>A) genes. The functional analysis revealed that the deletion in the *SETX* gene changed the splicing pattern, which was accompanied by lower *SETX* mRNA levels in the patient's fibroblasts, suggesting loss-of-function as the underlying mechanism. In addition, the patient's fibroblasts demonstrated altered mitochondrial architecture with decreased connectivity, compared to the control individuals. This is the first association of the CBS-PNFA phenotype with the most common *ATP7B* pathogenic variant p.H1069Q, previously linked to Wilson's disease, and early onset Parkinson's disease. This study expands the complex clinical spectrum related to variants in well-known disease genes, such as *ATP7B*, *SORL1*, *SETX*, and *FOXP1*, corroborating the hypothesis of oligogenic inheritance. To date, the *FOXP1* gene has been linked exclusively to neurodevelopmental speech disorders, while our study highlights its possible relevance for adult-onset progressive apraxia of speech, which guarantees further study.

Keywords: corticobasal syndrome; progressive non-fluent aphasia (CBS-PNFA, CBS-NAV); whole exome sequencing (WES); *ATP7B*; *SETX*; *SORL1*; *FOXP1*; splicing analysis; mitochondrial network analysis

1. Introduction

Corticobasal syndrome (CBS) and progressive non-fluent aphasia (PNFA) are predominantly sporadic syndromes [1,2]. Only c.a. one-third of CBS cases are caused by a genetic mutation in *MAPT*, *PGRN*, *LRKK2*, *PSEN1*, or *C9orf72* genes [3–6], reviewed in [2,7,8], with a few predisposing factors, including *MAPT* H1 haplotype [9,10].

PNFA is usually associated with frontotemporal lobar degeneration (FTLD), where monogenic mutations account for about one-third of cases [11]. The PNFA phenotype has shown the strongest association with *PGRN* mutations [12–14]. Patients diagnosed clinically with distinct syndromes from the FTLD spectrum typically show symptoms of other syndromes later in the disease course. PNFA patients frequently develop CBS while individuals diagnosed initially with CBS are likely to present with non-fluent aphasia later in the disease course [15,16]. Apraxia of speech and non-fluent aphasia have (relatively recently) been acknowledged as parts of CBS [17]. CBS is characterized by strongly asymmetric presentation, comprising of at least 2 among the following motor features: (a) limb rigidity or akinesia, (b) limb dystonia, (c) limb myoclonus. Additionally, at least 2 other features should be present: (a) orobuccal or limb apraxia, (b) cortical sensory deficit, (c) alien limb phenomena [18,19]. Current criteria of corticobasal degeneration (CBD) specify the variant with non-fluent/agrammatic progressive aphasia (CBD-NAV) as one of the possible disease phenotypes [18]. This variant is associated with predominant dysfunction of the left-hemisphere and predominantly right-sided motor symptoms [18].

Of note, there have been very few genetic reports regarding the overlapping CBS-PNFA phenotype. To date, one case of symmetrical CBS with non-fluent aphasia caused by a *PGRN* mutation [20], two cases with CBS associated with speech apraxia caused by a *MAPT* mutation [21,22] and two cases sharing CBS and PPA features with a *LRRK2* mutation [6] were described. However, genetic screening for *PGRN* and *MAPT* mutations was negative in a group of nine patients with PNFA associated with CBD pathology [23], suggesting that most of these overlapping phenotypes may not be monogenic, which highlights the need to look for oligogenic etiology. The implementation of whole exome/whole genome sequencing (WES/WGS) technology highlighted a complex genomic architecture of disease phenotypes, inciting researchers to revisit the classical concepts of genetic causality and monogenic disorders [24]. The evidence that oligogenic inheritance may account for a substantial fraction of atypical and unresolved cases of neurodegenerative diseases is mounting [25–27]. Indeed, it is estimated that c.a. 50% of sporadic patients from the FTLD spectrum carry at least one rare missense variant in neurodegenerative disease-associated genes [28]. Although there are detailed clinical and research criteria for FTLD syndromes, the clinical phenotype frequently evolves in the disease course often comprising cross-disease symptoms of variable intensity. Thus, from the point of view of genetic testing, next-generation sequencing is more adequate for the detection of the genetic variants contributing to overlapping phenotypes.

Our aim was to analyze the phenotypic-genetic correlations in a patient diagnosed with early onset CBS-NAV with predominant apraxia of speech. As polygenic etiology may be of particular relevance in overlapping dementia syndromes with early onset [29], we conducted whole exome sequencing (WES).

2. Materials and Methods

2.1. Whole Exome Sequencing (WES) Data Preprocessing

WES was performed commercially by CeGaTGmbH | Paul-Ehrlich-Str. 23 | D-72076 Tübingen | Germany. The following library preparation kit was used: Twist Human Core + RefSeq + Mitochondrial Panel (Twist Bioscience); 92.37% percent of sequenced bases had a predicted quality score of 30 or more (Q30 value). Sequencing parameters were NovaSeq 6000; 2 × 100 bp. Demultiplexing of the sequencing reads was performed with Illumina bcl2fastq (2.20). Statistics of mapped reads are provided in Table 1.

Raw read files were then processed with Intelliseq Flow Annotation Pipeline (https://intelliseq.com/ accessed on 25 August 2022) built in Cromwell (https://cromwell.readthedocs.io/en/stable/ accessed on 25 August 2022). Within the pipeline, fastq file quality was assessed with FastQC. Fastq files were then aligned to the Broad Institute Hg38 Human Reference Genome with GATK 4.0.3. Duplicate reads were removed with Picard and base quality Phred scores were recalibrated using GATK's covariance recalibration. Variants were called using GATK best practices. Identified variants were assessed using

Intelliseq Flow annotation pipeline that implements the American College of Medical Genetics and Genomics (ACMG) recommendations.

Table 1. Statistics of mapped reads (DNA).

Number of Mapped Reads (in Millions)	Proportion of Sequenced Reads (in %)	PCR Duplicates (in %) *	Median Insert Size (in bp)	Average Coverage
198.435	86.7	12.44	206	213.5

(*) PCR duplicates are duplicated reads that are removed during mapping. In general, a high PCR duplicate rate correlates with low sample quality.

In short variant annotation and annotation-based filtering, 3182 genes were included in the screening based on Human Phenotype Ontology (HPO) terms (Bioinformatics, Supplementary Table S1). Databases used for annotation included gnomAD v2.1 and v3 (frequencies, coverage, constraint), 1000Genomes (frequencies), MITOMAP (frequencies, contributed diseases), ClinVar (contributed diseases, pathogenicity), Human Phenotype Ontology (HPO) terms (inheritance mode, contributed phenotypes and diseases), UCSC (repeats, PHAST conservation scores), SIFT4G (constraint), SnpEff (predicted impact on gene product), dbSNP (rsID), Ensembl (gene and transcript information), and COSMIC (somatic mutations data). Common and low-impact variants (with a max frequency threshold of 0.05 and minimal SnpEff predicted impact on gene product set as MODERATE) were then removed. Variants with the following impacts on protein and mRNA were retained: missense, nonsense, frameshift, and splice site variants. At this step, 711 variants remained (Bioinformatics, Supplementary Table S2). Annotated variants were then classified according to the ACMG criteria and prioritized, yielding 164 variants (Bioinformatics, Supplementary Table S3). Selected variants were manually evaluated for quality in IGV and confirmed in Sanger sequencing (Genetics Supplementary Figure S1 *ATP7B*, Figure S5 *SETX*).

The *C9orf72* expansion analysis was performed using previously described protocols [30] using Genetic Analyzer 3130 and SeqScape v2.5 software (Applied Biosystems, Foster City, CA, USA).

2.2. Fibroblast Cultures and Inhibition of Nonsense-Mediated Decay (NMD)

Primary skin fibroblasts (obtained from the patient and age-matched, unrelated, neurologically healthy subjects) were collected and cultured as previously described [31]. For inhibition of nonsense-mediated mRNA decay (NMD), primary fibroblast cultures were treated with an NMD inhibitor, cycloheximide (500 µg/mL), or DMSO for 10 h (Sigma Aldrich, Saint Louis, MO, USA), as previously described [31].

2.3. Reverse Transcription, mRNA Isoform Analysis, and Real-Time PCR (RT-PCR)

RNA was extracted according to standard protocol with QIAzol Lysis Reagent (Qiagen, Manchester, UK) and reverse transcribed with cDNA synthesis kit, NG dART RT (EURx Molecular Biology Products, Gdańsk, Poland; cat. no. E0801), using random hexamers primers [32]. For the PCR, AllTaq Master Mix Kit was used (Qiagen, Manchester, UK). The real-time PCR analysis was conducted with RT HS-PCR Mix SYBR (A&A BIOTECHNOLOGY, Gdańsk, Poland) using a StepOne Plus system (Applied Biosystems, Foster City, CA, USA). Changes in gene expressions were determined with the DeltaCt method using *GAPDH* levels for normalization. Similar results were obtained with *PPIB* as a housekeeping gene. Primers are listed in Supplementary Table S4. PCR products were run on 2% agarose gel, selected bands were excised from the gel and extracted with QIAquick Gel Extraction Kit (Qiagen, Manchester, UK). Sanger sequencing was performed by (https://oligo.ibb.waw.pl accessed on 2 September 2022).

2.4. Mitochondrial Network Staining and Analysis

Patient fibroblasts (P) and control fibroblasts (C1 and C2) were grown on glass coverslips in standard conditions for 24 h [31]. For mitochondrial staining, cells were washed twice with Phosphate Buffered Saline, fixed with 4% paraformaldehyde (PFA) for 10 min, blocked with Normal Goat Serum (Thermo Fisher Scientific Inc., Waltham, MA, USA; cat no. 50197Z) for one hour, incubated overnight with rabbit anti-Hsp60 (*HSPD1*, Heat Shock Protein Family D (Hsp60) Member 1) antibody (1:500; Abcam, Cat No. ab53109) to visualize mitochondria [33], followed by one hour of incubation with secondary anti-rabbit antibody Alexa Fluor−555 (Thermo Fisher Scientific Inc., USA; Cat No. A-31570) and 10-min incubation with DAPI to visualize cell nuclei. Images were obtained by Zeiss LSM 510 confocal microscopy under 63x objective. Mitochondrial morphology parameters, including Form Factor (reflecting mitochondrial length) and Aspect Ratio (reflecting both the length and the extent of mitochondrial branching), were assessed with ImageJ as described in [34–36].

2.5. Statistical Analysis

For each experiment, the relative values obtained from different biological replicates. (n = 3) were used to calculate means, medians, standard deviations (SD), and statistical significance. The Kruskal–Wallis test and two-tailed unpaired t-test with Welch's correction were used (GraphPad Prism 6.0). $p < 0.05$ was considered significant.

3. Results

3.1. Case Description

The patient was followed clinically by our team from the age of 66 until her death at the age of 72. Her medical history included type 2 diabetes, hypertension, and gout. At the age of 63, she experienced minor problems when parking her car (distance judgment) (Clinical Supplementary Figure S2. Case timeline). Later, at the age of 64, she noticed speech problems (possibly apraxia of speech) and fell from a chair on wheels. Soon afterward she left her job due to speech problems. PNFA was diagnosed at the age of 66. Her speech was very effortful and non-fluent. The pattern of speech impairment indicated the predominance of speech apraxia, but mild agrammatism was also present and more evident in writing. Spontaneous falls became more frequent. She dropped objects and presented with a strong left-sided preference when navigating the environment, suggestive of right-sided unilateral neglect. She was unable to tie shoelaces, zip her jacket, and struggled to use cutlery, which corresponded to apraxia. Her poor decision-making was evidenced by her succumbing to advertising. She had preserved insight into communication problems. Apathy, mild impulsivity, and irritability were also present. Her handwriting with her right hand was extremely effortful so she switched to writing on a keyboard using her left hand. Of note, she used to be left-handed and had been forced to write with her right hand. Moreover, she used to play the accordion and her left finger dexterity could have been above average. The patient was diagnosed with possible CBD with a non-fluent/agrammatic aphasia variant (NAV) phenotype [18].

At the age of 67, she became depressed. When she was 69, her gait deteriorated further and she demonstrated apraxia in her right lower limb. At the age of 70, she walked only with assistance, her speech was limited mainly to yes/no answers, she could not initiate movement with her right upper limb, and swallowing problems were evident.

When she was 72, she was unable to maintain a sitting position and dysphagia was severe. She died a few months later at a nursing home. Proband's father presented symptoms similar to those observed in our patient (tremors, loss of speech), at the age of 68, four years before his death.

3.1.1. Clinical Assessment

The patient underwent her first neurological examination at the age of 67. Non-fluent aphasia was the predominant syndrome with the right upper limb apraxia, involuntary mirror right limb movement, parkinsonian tremor, rigidity, and dystonia. The patient had postural instability with a tendency to fall forward. The symptoms were not levodopa responsive. With disease progression, predominantly right-sided spastic dystonia, facial dystonia, and spastic dysarthria appeared. The patient was assessed neuropsychologically four times (see: Clinical Supplementary Tables S5–S8, see also Figures 1C,D and S3). At the age of 67, the patient presented with apraxia of speech, non-fluent aphasia, aprosody, right-sided unilateral neglect, ideomotor apraxia, mirror movements, synkinesis (e.g., when the patient tried to write with her right hand, lip and right leg movements were observed), alien limb phenomenon, and mild cortical sensory loss, were accompanied by executive deficits. At this time, motor hemispatial neglect was predominant. During alternate bilateral hand movements, there was reduced amplitude, and then extinction in the right hand. The severity of deficits corresponded to mild cognitive impairment as the patient's disability was mainly due to motor impairment and aphasia. A year later, apraxia and alien limb phenomenon extended to the right lower limb and hemispatial neglect was more pronounced in visual search tasks. At the age of 68, there was a marked deterioration of language; agrammatism was present both in speech and writing. The patient was unable to perform a pegboard task with the right hand. Moreover, her tactile object recognition and finger recognition deteriorated, and cortical sensory loss was more prominent, especially when the right hand was tested. Her cognitive status progressed to mild dementia. At the last examination, when she was 69, she could hardly repeat two-syllable words and perform a simple pegboard task with her left hand due to marked hypokinesia and difficulty in moving pegs. Her visual episodic memory was still quite well preserved at this time.

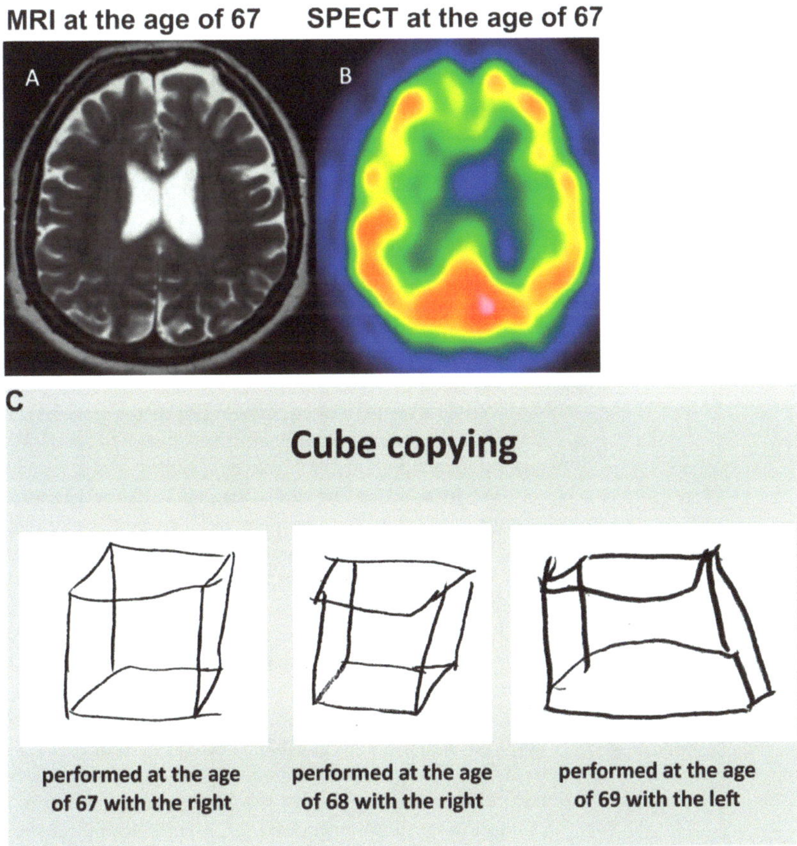

Figure 1. *Cont.*

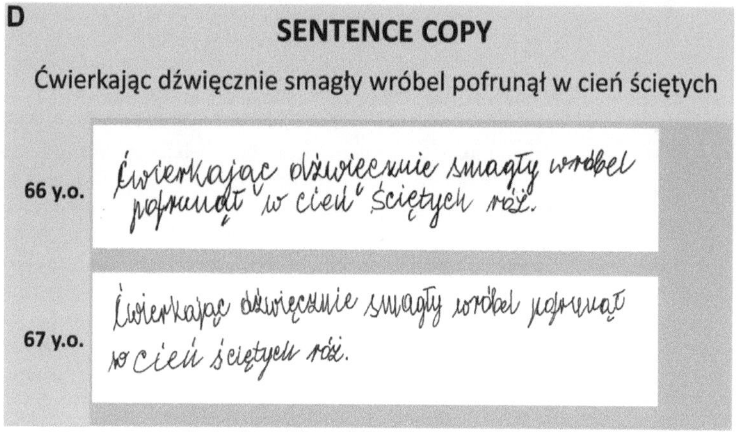

Figure 1. Neuroimaging and neuropsychology of the patient. (**A**). MRI at the age of 67 y.o.—bilateral, moderate frontal lobe atrophy; (**B**) SPECT- bilateral frontal and left parietal hypoperfusion; (**C**) cube copying task performed at 3 neuropsychological assessments. Even at the last assessment, when the patient drew with her left hand, the attempt to depict three-dimensionality is evident; (**D**) preserved sentence copying. Despite marked right upper limb rigidity, the patient was initially able to write.

3.1.2. Laboratory Testing

Repeated standard laboratory blood tests, including alanine and aspartate transferase, and an ultrasonography of the abdomen with the liver assessment were normal.

The evoked autonomic sympathetic skin potentials were within normal range, but the heart rate variability was high. Electroencephalography was normal. Electroneurography revealed no clinically significant changes.

3.1.3. Neuroimaging

Magnetic resonance imaging (MRI) of the head performed at the age of 67 demonstrated mild cortical atrophy of both frontal lobes (Figure 1A). The subtentorial atrophy did not exceed the normal range. Single photon emission computed tomography (SPECT) revealed bilateral frontal and left parietal hypoperfusion (Figure 1B).

3.1.4. Genetic Assessment

Whole exome sequencing (WES) of the patient's DNA revealed 164 rare variants classified according to the American College of Medical Genetics and Genomics (ACMG) criteria for pathogenicity (Table S3) [37].

We focused on the rare variants with the highest putative impact on the disease phenotype (Table 2). The presence of *C9orf72* expansion was excluded [30]. Among common risk variants, the patient had the *APOE* ε2, ε2 genotype.

According to the Human Splicing Finder prediction, the *SETX* variant c.2385_2387delAAA in exon 10 activates a cryptic donor site, which may lead to splicing alteration (Supplementary Material Genetics, Table S9, Scheme S1). cDNA from the patient's fibroblasts (Figure 2A, lane 6, P) had an altered pattern of exon 10 splicing with an additional aberrant isoform (band *A), visible only upon inhibition of nonsense-mediated decay mechanism with cycloheximide (CHX) treatment. Sequencing of the band *A, excised from agarose gel (Figure 2A, line 6, band *A) confirmed that band A corresponds to splicing isoform with 358 bp of exon 10 skipped (Figure 2C, Scheme S1, Supplementary Material Genetics Figure S4). The skipping of the 358 bp fragment led to the frameshift and introduced a premature termination codon (PTC). The skipped fragment had canonical splicing consensus sequences GT-AG (donor and acceptor, respectively). However, the new alternative donor site (used during the splicing event) was located downstream of the sites predicted by HSF, which highlighted the importance of performing a functional analysis since in silico predictions may not be precise.

Table 2. The rare variants of.the highest impact on the disease phenotype according to bioinformatic priotitization.

Gene	HGVS [1] DNA/Exon/Protein/rs	Predicted Effect	MAF gnomAD	CADD * Phred	ACMG Criteria
ATP7B	c.3207C>A exon 14 p.His1069Gln rs76151636 heterozygous	missense	0.001019	24	PS1, BP4, BP5, PP3 likely pathogenic
SETX	c.2385_2387delAAA exon 10 p.Ile795_Lys796delinsMet rs755971927 heterozygous	in frame deletion	0.0000329	NA	PM2, PM4, PP4 uncertain significance
SORL1	c.352G>A p.Val118Met exon 2 rs749389644 heterozygous	missense	0.00001060	28.1	PM2, PP3 uncertain significance
FOXP1	c.1762G>A p.Ala588Thr exon 20 rs202173892 heterozygous	missense	0.000318	21.7	BP5 pathogenic/likely pathogenic or of uncertain impact

[1] Variant nomenclature according to Human Genome Variation Society. *—Combined Annotation-Dependent Depletion score Phred-scaled [38].

As a positive control, to confirm correct NMD inhibition, we used mRNA transcribed from the *DSCR1* gene (the Down syndrome critical region 1), a known NMD target [39,40]. As expected, *DSCR1* levels increased upon cycloheximide treatment, showing robust NMD inhibition (Figure 2B).

Sequencing of the excised band *B (Figure 2A, lane 6) revealed that it is an artifact. The sequencing of the main band (Figure 2A, line 6) confirmed 100% identity with the expected PCR product—645 bp fragment of *SETX* exon 10 with c.2385_2387delAAA variant (Supplementary Material Genetics Figure S5, Scheme S1). In parallel, a real-time analysis showed lower levels of *SETX* mRNA (Figure 2D, see also Table S4, Scheme S1).

3.1.5. Mitochondrial Network Assessment

Since *SORL1, ATP7B, SETX*, and *FOXP1* genes have been previously linked to mitochondrial dysfunction [41–44], we performed mitochondrial staining in the patient's fibroblasts with Hsp60, a mitochondrial chaperonin essential for protein folding and assembly in the mitochondrial matrix [33]. We observed a particular network shape in the patient with significantly altered mitochondrial parameters compared to neurologically healthy controls (Figure 3A–C). The lower aspect ratio in the patient indicated decreased mitochondrial connectivity (Figure 3B), while a lower form factor pointed to shorter (and less branched) mitochondria, compared to CTRLs (Figure 3C).

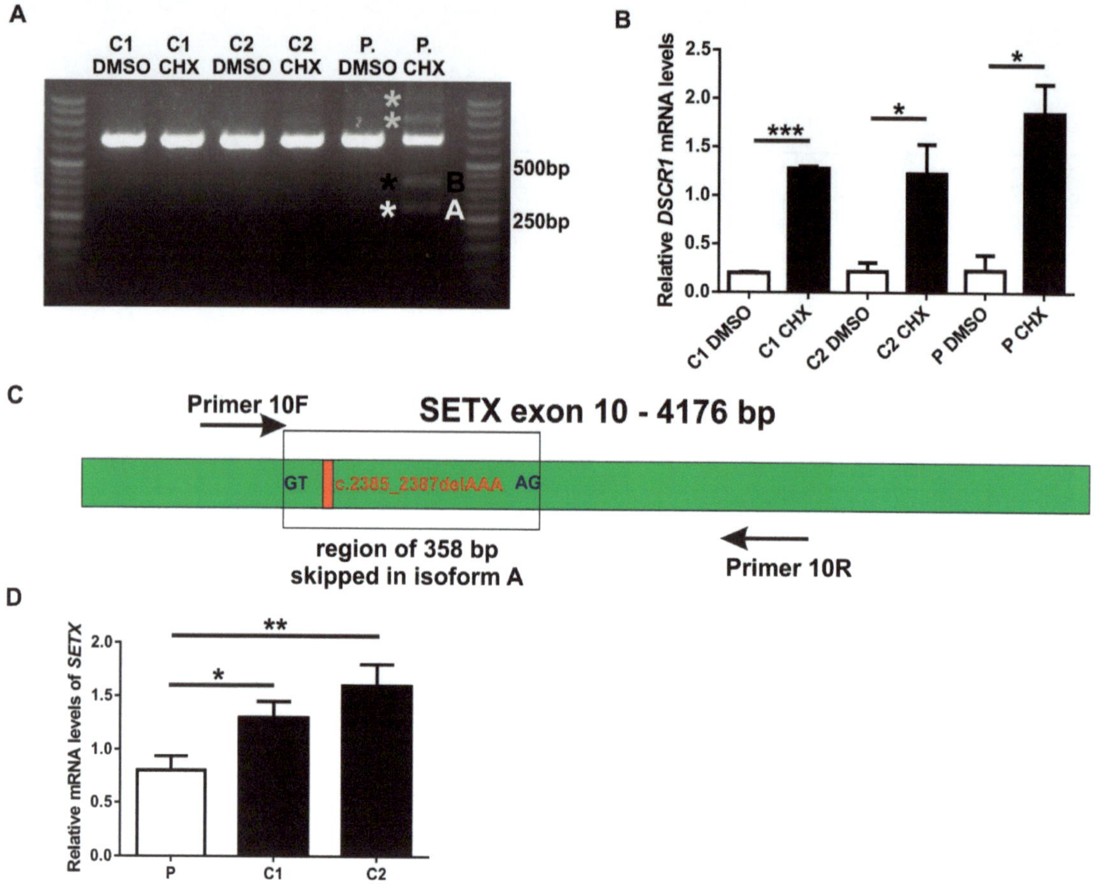

Figure 2. (**A**). Visualization of cDNA analysis on agarose gel. Fibroblasts from the patient (P) and controls (C1, C2) were treated with DMSO (lines 1, 3, 5) or cycloheximide (CHX) (line 2, 4, 6) for 10 h to block nonsense-mediated decay. White asterisk indicates aberrant isoform (*A) with partial skipping (358 bp) of *SETX* exon 10 in the patient's fibroblasts, arising due to the presence of variant c.2385_2387delAAA, as evidenced by sequencing of band A (see Figure S4). Sequencing of the excised band *B (Figure 2A, lane 6) revealed that it is an artifact; (**B**) mRNA transcribed from *DSCR1* gene (the Down syndrome critical region 1), a known NMD target, was used as a positive control to confirm NMD inhibition; (**C**) schematic representation of *SETX* exon 10, black empty square denotes 358 bp sequence skipped from exon 10 carrying c.2385_2387delAAA variant. Black arrows correspond to primers used for isoform analysis; (**D**) real-time PCR demonstrated lower levels of *SETX* mRNA in the patient compared to healthy controls (Table S4, Scheme S1). P was calculated in two-tailed *t*-test, * $p < 0.05$, ** $p < 0.01$, *** $p < 0.001$.

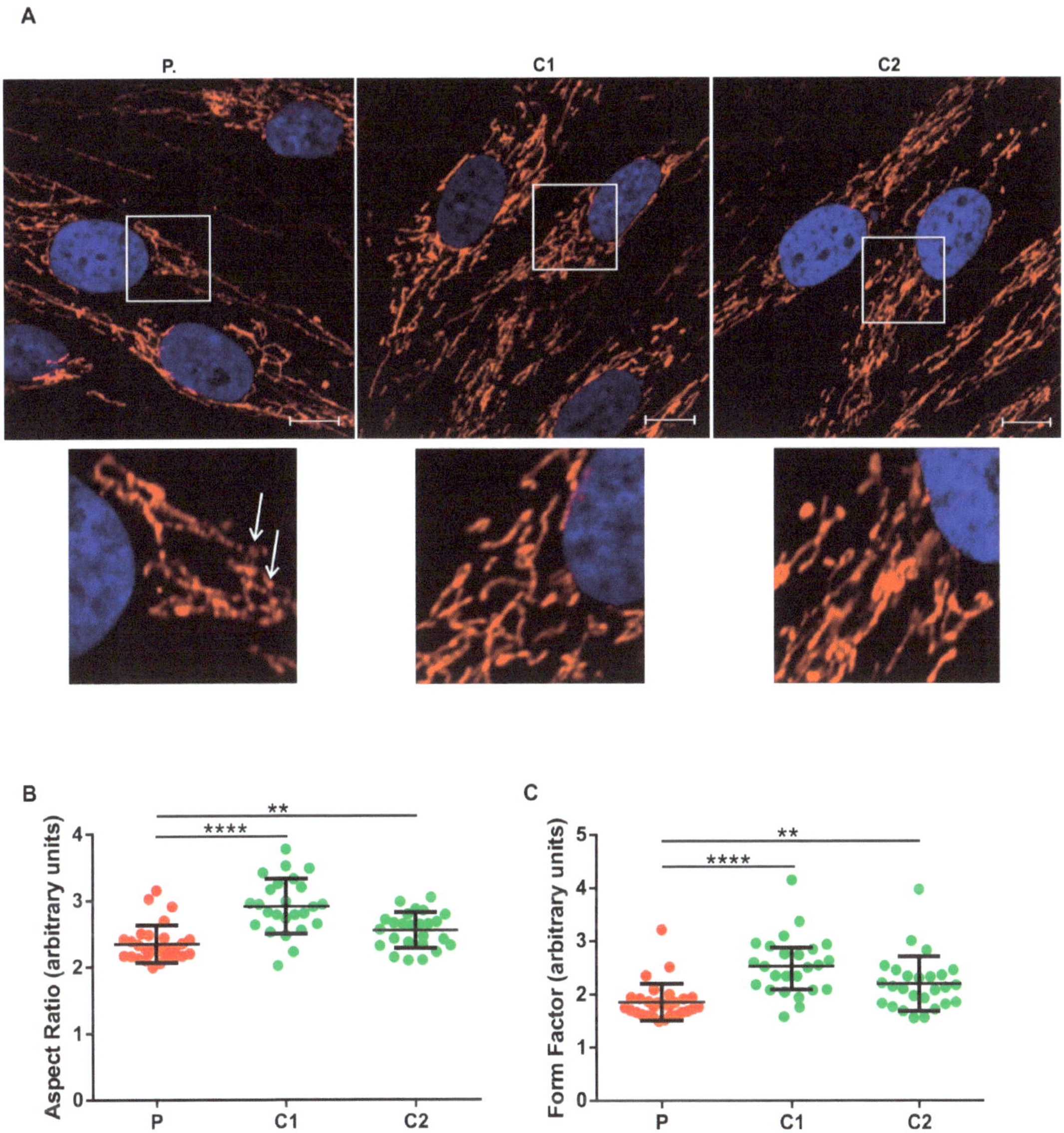

Figure 3. (**A**) The mitochondrial network in the patient's fibroblasts (P) compared to neurologically healthy, age-matched controls (C1, C2). Area delineated with the white square in the upper panel is enlarged in the lower panel. White arrows point to round, fragmented mitochondria. Red: Hsp60—mitochondria, blue: DAPI—nucleus; representative images are shown; (**B**) decreased form factor and (**C**) decreased Aspect Ratio in the patient's mitochondria compared to controls. Aspect Ratio < 0.0001, Kruskal–Wallis test; Form Factor < 0.0001, Kruskal–Wallis test; P for differences between individual cell lines: P vs. C1, P vs. C2 were calculated with unpaired t test with Welch's correction, **** < 0.0001, ** < 0.01. All images were taken at 63× magnification. Each dot corresponds to a measurement from one cell. Upper panel scale bar = 10 µm.

4. Discussion

The variability of CBS-NAV clinical phenotype suggests the important contribution of oligogenic background in disease development [24]. Detection of rare, putatively functional genetic variants linked to similar clinical phenotypes would corroborate this hypothesis.

We discuss below how the rare genetic variants detected in the patient (Table 2) could have contributed to the observed phenotype in light of the up-to-date literature and our clinical, and genetic analyses.

4.1. Complex Genetic Landscape

The *ATP7B* gene encodes a Cu-transporting P-type ATPase protein responsible for incorporating copper into apo-ceruloplasmin transporting copper in the serum. Homozygous or compound heterozygous mutations in the *ATP7B* gene caused Wilson disease (WD), characterized by excessive copper accumulation in various organs, such as the brain, liver, and cornea, leading to a wide spectrum of symptoms from hepatic to neuro-psychiatric [45–48]. The biochemical hallmarks of WD are decreased ceruloplasmin and elevated copper levels in blood serum.

In most WD patients, mutations are clustered in exon 14 of the *ATP7B* gene [49,50] that encodes part of the ATPase nucleotide-binding domain, responsible for ATP binding and enzyme activity [51]. The most common *ATP7B* pathogenic variant H1069Q in exon 14 accounts for 72% of WD cases in the Polish population [52,53]. Very rarely, carriers of single heterozygous *ATP7B* mutations (Hzgs) demonstrate clinical symptoms, and biochemical and/or brain abnormalities [54,55] (Table S10). However, recently, more disease phenotypes have been described in *ATP7B* Hzgs, such as early or late-onset Parkinson's disease (EOPD or LOPD) (summarized in Table S10). Given the mounting evidence of the possible involvement of single heterozygous *ATP7B* mutations in phenotypes with parkinsonism (Table S10), it can be speculated that p.H1069Q mutation in our CBS-NAV patient has led to the subclinical dysregulation of copper metabolism. Motor symptoms that were present in our patient included dystonia, rigidity, and rest tremor, and they initially impaired the function of the right limbs only. The marked asymmetry observed in our case was atypical for WD. No hyper- and hypointensities in the basal ganglia in T1- and T2-weighted MRI images typical for WD were seen in our patient. However, with disease progression, our patient did develop facial dystonia and dysarthria, which are common symptoms in WD.

High heart rate variability observed in our patient might be a sign of autonomic dysfunction observed in WD patients [56]. However, a comprehensive evaluation of autonomic function in WD is lacking.

Cognitive abnormalities in WD occur mainly with regard to attention, executive functions, and memory, while in our patient, despite the presence of mild executive deficits, language impairment and apraxia were predominant. Attention was not tested in detail as speech and limb apraxia made the assessment of processing speed unfeasible.

The *SETX* gene codes for senataxin, a ubiquitously expressed protein that functions as a DNA/RNA helicase, that is crucial for RNA processing, DNA damage response, neurogenesis, and autophagy [57,58]. The *SETX* gene may be considered a cross-disease gene as its mutations cause distinct neurodegenerative phenotypes. Dominant mutations in *SETX* cause juvenile-onset amyotrophic lateral sclerosis type 4 (ALS4), while recessive mutations are associated with ataxia-oculomotor apraxia 2 (AOA2) characterized by cerebellar ataxia, oculomotor apraxia, and axonal sensorimotor neuropathy [57,59]. *SETX* mutations have been detected also in Charcot Marie Tooth (CMT), distal hereditary motor neuropathy (dHMN) [60,61], childhood apraxia of speech [62], and Alzheimer's disease [63]. None of the symptoms inherent to ALS4 or AOA2 have been observed in our patient.

We observed decreased *SETX* mRNA levels in the patient's fibroblasts suggesting loss-of-function as the underlying mechanism (Figure 2D). The *SETX* variant c.2385_2387delAAA detected in our patient (Table 2) altered the splicing pattern of exon 10 (Figure 2), generating the aberrant isoform with the 358 bp fragment skipped form exon 10, which was further degraded by the nonsense-mediated decay (Figures 2A and S4). However, as evidenced by sequencing of the main band (Figure 2A, line 6; Figure S5), mRNA carrying the original 3 bp deletion was also generated. It cannot be excluded that the isoform with the original 3 bp deletion also contributes to the decreased levels of *SETX* mRNA possibly affecting the efficiency of downstream processes such as mRNA stability, maturation, transport, or translation. The production of two abnormal transcripts from the mutated allele was described previously [64,65]. For example, neurofibromatosis type 1 (NF1) mutations generated different proportions of mutated transcripts i.e., transcripts containing the exon with the original mutation and those with skipped exon [64,66]. It is worth noting that splicing of exceptionally long exons such as *SETX* exon 10 (4176 bp, Figures 2C and S2), requires precise exon definition to ensure its proper inclusion [67]. Functional variants, such as c.2385_2387delAAA may deregulate this process.

The *SORL1* (sortilin-related receptor 1) gene encodes a transmembrane protein, named sortilin-related receptor (SORLA), involved in endo-lysosomal processes, amyloid precursor protein (APP) sorting and in the degradation of amyloid-beta (Ab) peptide, responsible for Alzheimer's disease (AD) pathology [68,69]. Initially, both common and rare *SORL1* variants have been associated with AD [70–72], while recently, deleterious variants have been also detected in frontotemporal lobar degeneration (behavioral variant and primary progressive aphasia), and dementia with Lewy bodies (DLB) [28,63,73–75]. For this reason, *SORL1* is considered a cross-disease gene.

A recent functional study revealed that many *SORL1* rare variants resulted in altered maturation and cellular trafficking of the SORLA protein [76]. The V118M variant detected in our patient localized between two previously characterized mutations S114R and S124R in SORLA protein [76]. The S124R mutation showed a maturation profile similar to the wild-type protein, and the S114R mutation led to the decreased production of a mature, glycosylated form of the protein, which resulted in inefficient transport to the cell membrane [76]. It is thus plausible that also V118M variant may affect maturation defects and cellular localization of SORLA since position 118 in the protein is strongly conserved (phyloP100way).

The *FOXP1* (Forehead Box P1) gene encodes a transcription factor regulating tissue and cell-type-specific gene transcription during development and adulthood [77]. Pathogenic variants in this gene cause neurodevelopmental speech disorders with overlapping apraxic and dysarthric features, predominant motor programming problems, and poor speech intelligibility [78]. Children and adolescents with *FOXP1* syndrome present with considerable delays in language milestones [79]. Expressive deficits may be accompanied by comparable receptive problems in patients with missense variants [78] (expressive language > receptive language). Delays may also encompass motor and intellectual functions [80].

Given the frequency of the *FOXP1* c.1762G>A variant identified in our proband and low scores of various predictive algorithms, ACMG = 0.0188, SIFT = 0.142; PhastCons = 1.0; GERP = 6.17; MCAP = 0.0582, it seems to be benign. However, because our patient had predominant and pronounced speech impairment, characterized by both speech apraxia and—later in the disease course—also spastic dysarthria—it cannot be excluded that *FOXP1* rare variants, such as c.1762G>A, in a particular genetic or epigenetic context, can lead to progressive speech apraxia in adulthood. Such a hypothesis requires further studies. Of note, genetic variations in the *FOXP1* homolog and *FOXP2* gene have been previously found to modulate language performance in FTLD [81].

4.2. The significance of Motor Speech Disorders for Delineating the Patient's Complex Clinical Phenotype

Apraxia of speech (AOS), a motor speech disorder with effortful speech, impaired sequencing of articulatory gestures, and sound-based errors, distinct from both dysarthria and aphasia, is very likely to co-occur with non-fluent aphasia. It may be both neurodevelopmental (childhood-onset) or progressive. In most cases, progressive AOS accompanies PNFA or CBS.

Heterogenous criteria are used in clinical practice to diagnose AOS [82]. In the older literature, apraxia of speech may be misnamed as dysarthria since isolated primary progressive apraxia of speech (PPAOS) was recognized in the context of neurodegenerative diseases, especially CBS, relatively recently [17].

In WD, mixed dysarthria is often observed [83], but dystonic dysarthria associated with orofacial dystonia was also reported [84]. In *ATP7B* heterozygous mutation carriers dysarthria has been reported [85]. In our case, orofacial dystonia appeared late in the disease course. Initially, the patient struggled to perform vertical tongue movements which corresponded to mild oral apraxia. Progressive AOS in our patient was present in the context of full-blown progressive-non fluent aphasia, accompanied by aprosody and later in the disease course, also by spastic dysarthria. Ruggeri et al. reported the presence of dysarthria (not further specified) in 6 out of 11 cases with progressive aphasia and speech apraxia in CBS [86]. Only in 1 out of 12 cases was dysarthria unaccompanied by speech apraxia.

Detailed clinical case descriptions with very long follow-ups and neuropathology data are unfortunately rarely reported. Of note, Tetzloff et al. provided a 10-year-long observation of the PPAOS case with autopsy-confirmed CBD [87]. In this case, spastic dysarthria was added to the AOS in the 8th year of the observation period, about 1–2 years prior to the patient's death. At the same time, CBS symptoms became prominent.

In the context of *FOXP1* findings, it remains to be established if *FOXP1* abnormalities may influence the onset or pattern of progression of progressive AOS in adults.

4.3. Altered Mitochondrial Architecture

Finally, we provide evidence of an altered mitochondrial network in the patient (less branched mitochondria with decreased connectivity) compared to control individuals. As mentioned before, all four genes, *ATP7B, SETX, SORL1*, and *FOXP1*, have been associated with mitochondrial dysfunction [41–44], a common hallmark of many neurodegenerative conditions including PD and FTLD. Thus, we can hypothesize that the interaction among the four genes may contribute to altered mitochondria architecture in the patient.

4.4. Limitations of the Study

The patient's clinical diagnosis was not supported by the analysis of WD metabolic markers, cerebrospinal fluid (CSF) biomarkers, dopamine transporter SPECT [88], or amyloid/tau PET imaging, which could have helped to further elucidate the etiology [89]. Moreover, the autopsy was not performed.

As DNA from the patient's family members was not available, the co-segregation of the identified variants with the disease phenotype in the family has not been established.

5. Conclusions

The large-scale sequencing techniques have the potential to facilitate a diagnosis for intractable dementia cases with heterogenous clinical manifestations. Our results are in line with those of Ciani et al., who found that 50% of sporadic patients from the FTLD spectrum showed at least one rare missense variant in AD, PD, ALS, and LBD-associated genes [28]. Our results spoke in favor of the oligogenic inheritance in complex neurodegenerative phenotypes, such as CBS-NAV.

To our knowledge, this is the first reported case of the CBS-NAV phenotype with a single heterozygous *ATP7B* mutation. While parkinsonian phenotypes have been previously linked to *ATP7B* heterozygotes, we also described other variants of potential functional impacts in well-known disease-related genes, such as *SETX, SORL1*, and *FOXP1*. Of note, these three latter genes have been associated with speech and/or language deficits. The dissection of the molecular mechanisms underlying possible complex interactions among these genes would be a great challenge for the future.

Supplementary Materials: The following supporting information can be downloaded at: https://www.mdpi.com/article/10.3390/genes13122361/s1, **Bioinformatics Supplementary**: Table S1. Genes included in the screening; Table S2. WES Annotated; Table S3. WES ACMG classified variants; **Genetics Supplementary:** Figure S1. *ATP7B* Sanger sequencing; Table S4. Primers used in the study; Table S9. Human Splicing Finder analysis; Scheme S1. The scheme of aberrant splicing; Figure S4. Sequencing result of band *A; Figure S5. Sequencing result of the main band; **Clinical Supplementary: Figure S2 Case Timeline**; Table S5. Cognitive and behavioral screening; Figure S3. Addenbrooke's Cognitive Examination-III; Table S6. Speech and language evaluation; Table S7. Visuospatial, calculation, memory, and executive function; Table S8. Praxis, finger identification, orientation in the body schema, cortical sensory loss and tactile recognition; Table S10. Clinical phenotypes, biochemical, and brain abnormalities observed in carriers of single heterozygous *ATP7B* mutation. References [90–95] are cite in the Supplementary Materials.

Author Contributions: Conceptualization, K.G.-W., E.J.S., E.N., M.B. and M.K.; methodology, K.G.-W. and M.B.; clinical diagnosis, E.N., E.J.S., B.B., M.S. (Michał Schinwelski) and M.S. (Mariusz Siemiński); software, M.B. and M.K.; validation, K.G.-W. and M.B.; formal analysis K.G.-W. and M.B.; investigation, K.G.-W., E.J.S., M.B., E.N., B.B., M.S. (Michał Schinwelski) and M.S. (Mariusz Siemiński); resources, K.G.-W., C.Z., M.K. and J.S.; data curation, M.B.; writing—original draft preparation, K.G.-W., E.J.S., E.N. and M.B.; writing—review and editing, K.G.-W., E.J.S., M.B., E.N., B.B., M.K., M.S. (Michał Schinwelski), M.S. (Mariusz Siemiński), J.S. and C.Z.; supervision, K.G.-W., E.J.S., E.N. and M.K.; project administration, K.G.-W.; funding acquisition, K.G.-W., C.Z. and J.S. All authors have read and agreed to the published version of the manuscript.

Funding: This research received no external funding. K. Gaweda-Walerych received internal Mossakowski Medical Research Institute, Polish Academy of Sciences funding (project FBW016).

Institutional Review Board Statement: The study was conducted according to the guidelines of the Declaration of Helsinki, and approved by the Bioethics Committee of the Central Clinical Hospital of the Ministry of Interior Affairs and Administration in Warsaw (protocol code 203/2020, date of approval 2 December 2020) and the Bioethical Committee of the Medical University of Gdansk (protocol code NKB.B.N/244/2017, date of approval 29 June 2017).

Informed Consent Statement: Written informed consent was obtained from all subjects to publish this paper.

Data Availability Statement: The data presented in this study are sequencing data of human samples thus cannot be openly shared due to personal data protection.

Acknowledgments: We thank the patient and his family and other participants of the study (control subjects). We would also like to thank Emanuele Buratti (ICGEB, Italy) for the critical reading of the manuscript and providing consultation on functional analysis of the splicing. We thank Magdalena Lachwa-From for the technical assistance in confocal microscopy, which was done in the Laboratory of Advanced Microscopy Techniques at MMRI PAS, Warsaw. We thank Artur Wolny (Laboratory of Imaging Tissue Structure and Function, Nencki Institute of Experimental Biology PAS, Warsaw) for the technical assistance with the analysis of mitochondrial morphology with ImageJ.

Conflicts of Interest: The authors declare no conflict of interest.

References

1. Ramos, E.M.; Dokuru, D.R.; Van Berlo, V.; Wojta, K.; Wang, Q.; Huang, A.Y.; Miller, Z.A.; Karydas, A.M.; Bigio, E.H.; Rogalski, E.; et al. Genetic screen in a large series of patients with primary progressive aphasia. *Alzheimer's Dement.* **2019**, *15*, 553–560. [CrossRef] [PubMed]
2. Arienti, F.; Lazzeri, G.; Vizziello, M.; Monfrini, E.; Bresolin, N.; Saetti, M.C.; Picillo, M.; Franco, G.; Di Fonzo, A. Unravelling Genetic Factors Underlying Corticobasal Syndrome: A Systematic Review. *Cells* **2021**, *10*, 171. [CrossRef] [PubMed]
3. Bugiani, O.; Murrell, J.R.; Giaccone, G.; Hasegawa, M.; Ghigo, G.; Tabaton, M.; Morbin, M.; Primavera, A.; Carella, F.; Solaro, C.; et al. Frontotemporal dementia and corticobasal degeneration in a family with a P301S mutation in tau. *J. Neuropathol. Exp. Neurol.* **1999**, *58*, 667–677. [CrossRef] [PubMed]
4. Benussi, L.; Binetti, G.; Sina, E.; Gigola, L.; Bettecken, T.; Meitinger, T.; Ghidoni, R. A novel deletion in progranulin gene is associated with FTDP-17 and CBS. *Neurobiol. Aging* **2008**, *29*, 427–435. [CrossRef]

5. Lindquist, S.G.; Duno, M.; Batbayli, M.; Puschmann, A.; Braendgaard, H.; Mardosiene, S.; Svenstrup, K.; Pinborg, L.H.; Vestergaard, K.; Hjermind, L.E.; et al. Corticobasal and ataxia syndromes widen the spectrum of C9ORF72 hexanucleotide expansion disease. *Clin. Genet.* **2013**, *83*, 279–283. [CrossRef] [PubMed]
6. Chen-Plotkin, A.S.; Yuan, W.; Anderson, C.; McCarty Wood, E.; Hurtig, H.I.; Clark, C.M.; Miller, B.L.; Lee, V.M.; Trojanowski, J.Q.; Grossman, M.; et al. Corticobasal syndrome and primary progressive aphasia as manifestations of LRRK2 gene mutations. *Neurology* **2008**, *70*, 521–527. [CrossRef]
7. Constantinides, V.C.; Paraskevas, G.P.; Paraskevas, P.G.; Stefanis, L.; Kapaki, E. Corticobasal degeneration and corticobasal syndrome: A review. *Clin. Park. Relat. Disord.* **2019**, *1*, 66–71. [CrossRef]
8. Koga, S.; Josephs, K.A.; Aiba, I.; Yoshida, M.; Dickson, D.W. Neuropathology and emerging biomarkers in corticobasal syndrome. *J. Neurol. Neurosurg. Psychiatry* **2022**, *93*, 919–929. [CrossRef]
9. Houlden, H.; Baker, M.; Morris, H.R.; MacDonald, N.; Pickering-Brown, S.; Adamson, J.; Lees, A.J.; Rossor, M.N.; Quinn, N.P.; Kertesz, A.; et al. Corticobasal degeneration and progressive supranuclear palsy share a common tau haplotype. *Neurology* **2001**, *56*, 1702–1706. [CrossRef]
10. Kouri, N.; Ross, O.A.; Dombroski, B.; Younkin, C.S.; Serie, D.J.; Soto-Ortolaza, A.; Baker, M.; Finch, N.C.A.; Yoon, H.; Kim, J.; et al. Genome-wide association study of corticobasal degeneration identifies risk variants shared with progressive supranuclear palsy. *Nat. Commun.* **2015**, *6*, 7247. [CrossRef]
11. Borroni, B.; Pilotto, A.; Bianchi, M.; Gilberti, N.; Padovani, A. Genetic contributors to frontotemporal lobar degeneration: Beyond monogenic disease. *Mini Rev. Med. Chem.* **2011**, *11*, 988–1001. [CrossRef] [PubMed]
12. Snowden, J.S.; Adams, J.; Harris, J.; Thompson, J.C.; Rollinson, S.; Richardson, A.; Jones, M.; Neary, D.; Mann, D.M.; Pickering-Brown, S. Distinct clinical and pathological phenotypes in frontotemporal dementia associated with MAPT, PGRN and C9orf72 mutations. *Amyotroph. Lateral Scler. Front. Degener.* **2015**, *16*, 497–505. [CrossRef] [PubMed]
13. Gossye, H.; Van Broeckhoven, C.; Engelborghs, S. The Use of Biomarkers and Genetic Screening to Diagnose Frontotemporal Dementia: Evidence and Clinical Implications. *Front. Neurosci.* **2019**, *13*, 757. [CrossRef] [PubMed]
14. Snowden, J.S.; Pickering-Brown, S.M.; Mackenzie, I.R.; Richardson, A.M.; Varma, A.; Neary, D.; Mann, D.M. Progranulin gene mutations associated with frontotemporal dementia and progressive non-fluent aphasia. *Brain* **2006**, *129*, 3091–3102. [CrossRef] [PubMed]
15. Harciarek, M.; Sitek, E.J.; Kertesz, A. The patterns of progression in primary progressive aphasia—Implications for assessment and management. *Aphasiology* **2014**, *28*, 964–980. [CrossRef]
16. Kertesz, A.; McMonagle, P.; Blair, M.; Davidson, W.; Munoz, D.G. The evolution and pathology of frontotemporal dementia. *Brain* **2005**, *128*, 1996–2005. [CrossRef]
17. Josephs, K.A.; Duffy, J.R. Apraxia of speech and nonfluent aphasia: A new clinical marker for corticobasal degeneration and progressive supranuclear palsy. *Curr. Opin. Neurol.* **2008**, *21*, 688–692. [CrossRef]
18. Armstrong, M.J.; Litvan, I.; Lang, A.E.; Bak, T.H.; Bhatia, K.P.; Borroni, B.; Boxer, A.L.; Dickson, D.W.; Grossman, M.; Hallett, M.; et al. Criteria for the diagnosis of corticobasal degeneration. *Neurology* **2013**, *80*, 496–503. [CrossRef]
19. Wilson, D.; Le Heron, C.; Anderson, T. Corticobasal syndrome: A practical guide. *Pract. Neurol.* **2021**, *21*, 276–285. [CrossRef]
20. Dopper, E.G.; Seelaar, H.; Chiu, W.Z.; de Koning, I.; van Minkelen, R.; Baker, M.C.; Rozemuller, A.J.; Rademakers, R.; van Swieten, J.C. Symmetrical corticobasal syndrome caused by a novel C.314dup progranulin mutation. *J. Mol. Neurosci.* **2011**, *45*, 354–358. [CrossRef]
21. Marshall, C.R.; Guerreiro, R.; Thust, S.; Fletcher, P.; Rohrer, J.D.; Fox, N.C. A Novel MAPT Mutation Causing Corticobasal Syndrome Led by Progressive Apraxia of Speech. *J. Alzheimer's Dis.* **2015**, *48*, 923–926. [CrossRef] [PubMed]
22. Mazzon, G.; Menichelli, A.; Fabretto, A.; Cattaruzza, T.; Manganotti, P. A new MAPT deletion in a case of speech apraxia leading to corticobasal syndrome. *Neurocase* **2018**, *24*, 140–144. [CrossRef]
23. Santos-Santos, M.A.; Mandelli, M.L.; Binney, R.J.; Ogar, J.; Wilson, S.M.; Henry, M.L.; Hubbard, H.I.; Meese, M.; Attygalle, S.; Rosenberg, L.; et al. Features of Patients With Nonfluent/Agrammatic Primary Progressive Aphasia With Underlying Progressive Supranuclear Palsy Pathology or Corticobasal Degeneration. *JAMA Neurol.* **2016**, *73*, 733–742. [CrossRef] [PubMed]
24. Katsanis, N. The continuum of causality in human genetic disorders. *Genome Biol.* **2016**, *17*, 233. [CrossRef] [PubMed]
25. Tabuas-Pereira, M.; Guerreiro, R.; Kun-Rodrigues, C.; Almeida, M.R.; Bras, J.; Santana, I. Whole-exome sequencing reveals PSEN1 and ATP7B combined variants as a possible cause of early-onset Lewy body dementia: A case study of genotype-phenotype correlation. *Neurogenetics* **2022**, *23*, 279–283. [CrossRef]
26. Gaweda-Walerych, K.; Sitek, E.J.; Borczyk, M.; Berdynski, M.; Narozanska, E.; Brockhuis, B.; Korostynski, M.; Slawek, J.; Zekanowski, C. Two Rare Variants in PLAU and BACE1 Genes-Do They Contribute to Semantic Dementia Clinical Phenotype? *Genes* **2021**, *12*, 1806. [CrossRef]
27. Cochran, J.N.; McKinley, E.C.; Cochran, M.; Amaral, M.D.; Moyers, B.A.; Lasseigne, B.N.; Gray, D.E.; Lawlor, J.M.J.; Prokop, J.W.; Geier, E.G.; et al. Genome sequencing for early-onset or atypical dementia: High diagnostic yield and frequent observation of multiple contributory alleles. *Mol. Case Stud.* **2019**, *5*, a003491. [CrossRef]
28. Ciani, M.; Bonvicini, C.; Scassellati, C.; Carrara, M.; Maj, C.; Fostinelli, S.; Binetti, G.; Ghidoni, R.; Benussi, L. The Missing Heritability of Sporadic Frontotemporal Dementia: New Insights from Rare Variants in Neurodegenerative Candidate Genes. *Int. J. Mol. Sci.* **2019**, *20*, 3903. [CrossRef]

29. Svenningsson, P. Corticobasal degeneration: Advances in clinicopathology and biomarkers. *Curr. Opin. Neurol.* **2019**, *32*, 597–603. [CrossRef]
30. DeJesus-Hernandez, M.; Mackenzie, I.R.; Boeve, B.F.; Boxer, A.L.; Baker, M.; Rutherford, N.J.; Nicholson, A.M.; Finch, N.A.; Flynn, H.; Adamson, J.; et al. Expanded GGGGCC hexanucleotide repeat in noncoding region of C9ORF72 causes chromosome 9p-linked FTD and ALS. *Neuron* **2011**, *72*, 245–256. [CrossRef]
31. Gaweda-Walerych, K.; Sitek, E.J.; Narozanska, E.; Wezyk, M.; Brockhuis, B.; Zekanowski, C.; Slawek, J. Functional characterization of a novel progranulin mutation in a patient with progressive nonfluent aphasia. *Neurobiol. Aging* **2018**, *72*, 186-e9. [CrossRef] [PubMed]
32. Leparc, G.G.; Mitra, R.D. A sensitive procedure to detect alternatively spliced mRNA in pooled-tissue samples. *Nucleic Acids Res.* **2007**, *35*, e146. [CrossRef] [PubMed]
33. Wauters, F.; Cornelissen, T.; Imberechts, D.; Martin, S.; Koentjoro, B.; Sue, C.; Vangheluwe, P.; Vandenberghe, W. LRRK2 mutations impair depolarization-induced mitophagy through inhibition of mitochondrial accumulation of RAB10. *Autophagy* **2020**, *16*, 203–222. [CrossRef] [PubMed]
34. Merrill, K.H.F.; Strack, S. Measuring Mitochondrial Shape with ImageJ. In *Techniques to Investigate Mitochondrial Function in Neurons*; Strack, S., Usachev, Y.M., Eds.; Springer: Berlin/Heidelberg, Germany, 2017.
35. Durand, M.J.; Ait-Aissa, K.; Levchenko, V.; Staruschenko, A.; Gutterman, D.D.; Beyer, A.M. Visualization and quantification of mitochondrial structure in the endothelium of intact arteries. *Cardiovasc. Res.* **2019**, *115*, 1546–1556. [CrossRef]
36. Onesto, E.; Colombrita, C.; Gumina, V.; Borghi, M.O.; Dusi, S.; Doretti, A.; Fagiolari, G.; Invernizzi, F.; Moggio, M.; Tiranti, V.; et al. Gene-specific mitochondria dysfunctions in human TARDBP and C9ORF72 fibroblasts. *Acta Neuropathol. Commun.* **2016**, *4*, 47. [CrossRef]
37. Richards, S.; Aziz, N.; Bale, S.; Bick, D.; Das, S.; Gastier-Foster, J.; Grody, W.W.; Hegde, M.; Lyon, E.; Spector, E.; et al. Standards and guidelines for the interpretation of sequence variants: A joint consensus recommendation of the American College of Medical Genetics and Genomics and the Association for Molecular Pathology. *Genet. Med.* **2015**, *17*, 405–424. [CrossRef]
38. Rentzsch, P.; Witten, D.; Cooper, G.M.; Shendure, J.; Kircher, M. CADD: Predicting the deleteriousness of variants throughout the human genome. *Nucleic Acids Res.* **2019**, *47*, D886–D894. [CrossRef]
39. Wittmann, J.; Hol, E.M.; Jack, H.M. hUPF2 silencing identifies physiologic substrates of mammalian nonsense-mediated mRNA decay. *Mol. Cell Biol.* **2006**, *26*, 1272–1287. [CrossRef]
40. Gong, C.; Kim, Y.K.; Woeller, C.F.; Tang, Y.; Maquat, L.E. SMD and NMD are competitive pathways that contribute to myogenesis: Effects on PAX3 and myogenin mRNAs. *Genes Dev.* **2009**, *23*, 54–66. [CrossRef]
41. Barthelson, K.; Pederson, S.M.; Newman, M.; Lardelli, M. Brain transcriptome analysis reveals subtle effects on mitochondrial function and iron homeostasis of mutations in the SORL1 gene implicated in early onset familial Alzheimer's disease. *Mol. Brain* **2020**, *13*, 142. [CrossRef]
42. Renaudin, X.; Venkitaraman, A.R. A mitochondrial response to oxidative stress mediated by unscheduled RNA-DNA hybrids (R-loops). *Mol. Cell Oncol.* **2021**, *8*, 2007028. [CrossRef] [PubMed]
43. Dev, S.; Kruse, R.L.; Hamilton, J.P.; Lutsenko, S. Wilson Disease: Update on Pathophysiology and Treatment. *Front. Cell Dev. Biol.* **2022**, *10*, 871877. [CrossRef] [PubMed]
44. Wang, J.; Frohlich, H.; Torres, F.B.; Silva, R.L.; Poschet, G.; Agarwal, A.; Rappold, G.A. Mitochondrial dysfunction and oxidative stress contribute to cognitive and motor impairment in FOXP1 syndrome. *Proc. Natl. Acad. Sci. USA* **2022**, *119*, e2112852119. [CrossRef] [PubMed]
45. Roberts, E.A.; Schilsky, M.L. A practice guideline on Wilson disease. *Hepatology* **2003**, *37*, 1475–1492. [CrossRef]
46. Lutsenko, S. Modifying factors and phenotypic diversity in Wilson's disease. *Ann. N. Y. Acad. Sci.* **2014**, *1315*, 56–63. [CrossRef]
47. Bandmann, O.; Weiss, K.H.; Kaler, S.G. Wilson's disease and other neurological copper disorders. *Lancet Neurol.* **2015**, *14*, 103–113. [CrossRef]
48. Pfeiffer, R.F. Wilson Disease. *Contin. Lifelong Learn. Neurol.* **2016**, *22*, 1246–1261. [CrossRef]
49. Shah, A.B.; Chernov, I.; Zhang, H.T.; Ross, B.M.; Das, K.; Lutsenko, S.; Parano, E.; Pavone, L.; Evgrafov, O.; Ivanova-Smolenskaya, I.A.; et al. Identification and analysis of mutations in the Wilson disease gene (ATP7B): Population frequencies, genotype-phenotype correlation, and functional analyses. *Am. J. Hum. Genet.* **1997**, *61*, 317–328. [CrossRef]
50. Ivanova-Smolenskaya, I.A.; Ovchinnikov, I.V.; Karabanov, A.V.; Deineko, N.L.; Poleshchuk, V.V.; Markova, E.D.; Illarioshkin, S.N. The His1069Gln mutation in the ATP7B gene in Russian patients with Wilson disease. *J. Med. Genet.* **1999**, *36*, 174.
51. Rodriguez-Granillo, A.; Sedlak, E.; Wittung-Stafshede, P. Stability and ATP binding of the nucleotide-binding domain of the Wilson disease protein: Effect of the common H1069Q mutation. *J. Mol. Biol.* **2008**, *383*, 1097–1111. [CrossRef]
52. Gromadzka, G.; Schmidt, H.H.; Genschel, J.; Bochow, B.; Rodo, M.; Tarnacka, B.; Litwin, T.; Chabik, G.; Czlonkowska, A. Frameshift and nonsense mutations in the gene for ATPase7B are associated with severe impairment of copper metabolism and with an early clinical manifestation of Wilson's disease. *Clin. Genet.* **2005**, *68*, 524–532. [CrossRef] [PubMed]
53. Kucinskas, L.; Jeroch, J.; Vitkauskiene, A.; Sakalauskas, R.; Petrenkiene, V.; Kucinskas, V.; Naginiene, R.; Schmidt, H.; Kupcinskas, L. High frequency of the c.3207C>A (p.H1069Q) mutation in ATP7B gene of Lithuanian patients with hepatic presentation of Wilson's disease. *World J. Gastroenterol.* **2008**, *14*, 5876–5879. [CrossRef] [PubMed]
54. Tarnacka, B.; Szeszkowski, W.; Buettner, J.; Golebiowski, M.; Gromadzka, G.; Czlonkowska, A. Heterozygous carriers for Wilson's disease–magnetic spectroscopy changes in the brain. *Metab. Brain Dis.* **2009**, *24*, 463–468. [CrossRef] [PubMed]

55. Skowronska, M.; Litwin, T.; Kurkowska-Jastrzebska, I.; Czlonkowska, A. Transcranial sonography changes in heterozygotic carriers of the ATP7B gene. *Neurol. Sci.* **2020**, *41*, 2605–2612. [CrossRef] [PubMed]
56. Li, K.; Lindauer, C.; Haase, R.; Rudiger, H.; Reichmann, H.; Reuner, U.; Ziemssen, T. Autonomic Dysfunction in Wilson's Disease: A Comprehensive Evaluation during a 3-Year Follow Up. *Front. Physiol.* **2017**, *8*, 778. [CrossRef] [PubMed]
57. Groh, M.; Albulescu, L.O.; Cristini, A.; Gromak, N. Senataxin: Genome Guardian at the Interface of Transcription and Neurodegeneration. *J. Mol. Biol.* **2017**, *429*, 3181–3195. [CrossRef] [PubMed]
58. Richard, P.; Feng, S.; Tsai, Y.L.; Li, W.; Rinchetti, P.; Muhith, U.; Irizarry-Cole, J.; Stolz, K.; Sanz, L.A.; Hartono, S.; et al. SETX (senataxin), the helicase mutated in AOA2 and ALS4, functions in autophagy regulation. *Autophagy* **2021**, *17*, 1889–1906. [CrossRef]
59. Grunseich, C.; Patankar, A.; Amaya, J.; Watts, J.A.; Li, D.; Ramirez, P.; Schindler, A.B.; Fischbeck, K.H.; Cheung, V.G. Clinical and Molecular Aspects of Senataxin Mutations in Amyotrophic Lateral Sclerosis 4. *Ann. Neurol.* **2020**, *87*, 547–555. [CrossRef]
60. Hoyer, H.; Braathen, G.J.; Busk, O.L.; Holla, O.L.; Svendsen, M.; Hilmarsen, H.T.; Strand, L.; Skjelbred, C.F.; Russell, M.B. Genetic diagnosis of Charcot-Marie-Tooth disease in a population by next-generation sequencing. *Biomed Res. Int.* **2014**, *2014*, 210401. [CrossRef]
61. Dierick, I.; Baets, J.; Irobi, J.; Jacobs, A.; De Vriendt, E.; Deconinck, T.; Merlini, L.; Van den Bergh, P.; Rasic, V.M.; Robberecht, W.; et al. Relative contribution of mutations in genes for autosomal dominant distal hereditary motor neuropathies: A genotype-phenotype correlation study. *Brain* **2008**, *131*, 1217–1227. [CrossRef]
62. Worthey, E.A.; Raca, G.; Laffin, J.J.; Wilk, B.M.; Harris, J.M.; Jakielski, K.J.; Dimmock, D.P.; Strand, E.A.; Shriberg, L.D. Whole-exome sequencing supports genetic heterogeneity in childhood apraxia of speech. *J. Neurodev. Disord.* **2013**, *5*, 29. [CrossRef] [PubMed]
63. Sun, L.; Zhang, J.; Su, N.; Zhang, S.; Yan, F.; Lin, X.; Yu, J.; Li, W.; Li, X.; Xiao, S. Analysis of Genotype-Phenotype Correlations in Patients With Degenerative Dementia Through the Whole Exome Sequencing. *Front. Aging Neurosci.* **2021**, *13*, 745407. [CrossRef] [PubMed]
64. Pros, E.; Larriba, S.; Lopez, E.; Ravella, A.; Gili, M.L.; Kruyer, H.; Valls, J.; Serra, E.; Lazaro, C. NF1 mutation rather than individual genetic variability is the main determinant of the NF1-transcriptional profile of mutations affecting splicing. *Hum. Mutat.* **2006**, *27*, 1104–1114. [CrossRef] [PubMed]
65. Baralle, D.; Buratti, E. RNA splicing in human disease and in the clinic. *Clin. Sci.* **2017**, *131*, 355–368. [CrossRef]
66. Pros, E.; Gomez, C.; Martin, T.; Fabregas, P.; Serra, E.; Lazaro, C. Nature and mRNA effect of 282 different NF1 point mutations: Focus on splicing alterations. *Hum. Mutat.* **2008**, *29*, E173–E193. [CrossRef]
67. Raponi, M.; Smith, L.D.; Silipo, M.; Stuani, C.; Buratti, E.; Baralle, D. BRCA1 exon 11 a model of long exon splicing regulation. *RNA Biol.* **2014**, *11*, 351–359. [CrossRef]
68. Willnow, T.E.; Andersen, O.M. Sorting receptor SORLA–a trafficking path to avoid Alzheimer disease. *J. Cell Sci.* **2013**, *126*, 2751–2760. [CrossRef] [PubMed]
69. Schmidt, V.; Subkhangulova, A.; Willnow, T.E. Sorting receptor SORLA: Cellular mechanisms and implications for disease. *Cell Mol. Life Sci.* **2017**, *74*, 1475–1483. [CrossRef]
70. Rogaeva, E.; Meng, Y.; Lee, J.H.; Gu, Y.; Kawarai, T.; Zou, F.; Katayama, T.; Baldwin, C.T.; Cheng, R.; Hasegawa, H.; et al. The neuronal sortilin-related receptor SORL1 is genetically associated with Alzheimer disease. *Nat. Genet.* **2007**, *39*, 168–177. [CrossRef]
71. Bellenguez, C.; Charbonnier, C.; Grenier-Boley, B.; Quenez, O.; Le Guennec, K.; Nicolas, G.; Chauhan, G.; Wallon, D.; Rousseau, S.; Richard, A.C.; et al. Contribution to Alzheimer's disease risk of rare variants in TREM2, SORL1, and ABCA7 in 1779 cases and 1273 controls. *Neurobiol. Aging* **2017**, *59*, 220.e1–220.e9. [CrossRef]
72. Campion, D.; Charbonnier, C.; Nicolas, G. SORL1 genetic variants and Alzheimer disease risk: A literature review and meta-analysis of sequencing data. *Acta Neuropathol.* **2019**, *138*, 173–186. [CrossRef] [PubMed]
73. Ramos-Campoy, O.; Antonell, A.; Falgas, N.; Balasa, M.; Borrego-Ecija, S.; Rodriguez-Santiago, B.; Datta, D.; Armengol, L.; Fernandez-Villullas, G.; Bosch, B.; et al. Screening of dementia genes by whole-exome sequencing in Spanish patients with early-onset dementia: Likely pathogenic, uncertain significance and risk variants. *Neurobiol. Aging* **2020**, *93*, e1–e9. [CrossRef] [PubMed]
74. Sassi, C.; Capozzo, R.; Hammer, M.; Zecca, C.; Federoff, M.; Blauwendraat, C.; Bernstein, N.; Ding, J.; Gibbs, J.R.; Price, T.; et al. Exploring dementia and neuronal ceroid lipofuscinosis genes in 100 FTD-like patients from 6 towns and rural villages on the Adriatic Sea cost of Apulia. *Sci. Rep.* **2021**, *11*, 6353. [CrossRef] [PubMed]
75. Csaban, D.; Illes, A.; Renata, T.B.; Balicza, P.; Pentelenyi, K.; Molnar, V.; Gezsi, A.; Grosz, Z.; Gal, A.; Kovacs, T.; et al. Genetic landscape of early-onset dementia in Hungary. *Neurol. Sci.* **2022**, *43*, 5289–5300. [CrossRef]
76. Rovelet-Lecrux, A.; Feuillette, S.; Miguel, L.; Schramm, C.; Pernet, S.; Quenez, O.; Segalas-Milazzo, I.; Guilhaudis, L.; Rousseau, S.; Riou, G.; et al. Impaired SorLA maturation and trafficking as a new mechanism for SORL1 missense variants in Alzheimer disease. *Acta Neuropathol. Commun.* **2021**, *9*, 196. [CrossRef]
77. Co, M.; Anderson, A.G.; Konopka, G. FOXP transcription factors in vertebrate brain development, function, and disorders. *Wiley Interdiscip. Rev. Dev. Biol.* **2020**, *9*, e375. [CrossRef]

78. Braden, R.O.; Amor, D.J.; Fisher, S.E.; Mei, C.; Myers, C.T.; Mefford, H.; Gill, D.; Srivastava, S.; Swanson, L.C.; Goel, H.; et al. Severe speech impairment is a distinguishing feature of FOXP1-related disorder. *Dev. Med. Child Neurol.* **2021**, *63*, 1417–1426. [CrossRef]
79. Siper, P.M.; De Rubeis, S.; Trelles, M.D.P.; Durkin, A.; Di Marino, D.; Muratet, F.; Frank, Y.; Lozano, R.; Eichler, E.E.; Kelly, M.; et al. Prospective investigation of FOXP1 syndrome. *Mol. Autism* **2017**, *8*, 57. [CrossRef]
80. Bacon, C.; Rappold, G.A. The distinct and overlapping phenotypic spectra of FOXP1 and FOXP2 in cognitive disorders. *Hum. Genet.* **2012**, *131*, 1687–1698. [CrossRef]
81. Padovani, A.; Cosseddu, M.; Premi, E.; Archetti, S.; Papetti, A.; Agosti, C.; Bigni, B.; Cerini, C.; Paghera, B.; Bellelli, G.; et al. The speech and language FOXP2 gene modulates the phenotype of frontotemporal lobar degeneration. *J. Alzheimer's Dis.* **2010**, *22*, 923–931. [CrossRef]
82. Molloy, J.; Jagoe, C. Use of diverse diagnostic criteria for acquired apraxia of speech: A scoping review. *Int. J. Lang. Commun. Disord.* **2019**, *54*, 875–893. [CrossRef] [PubMed]
83. Jauer-Niworowska. A specific picture of speech disturbances in polish speaking patients with mixed dysarthria in multiple sclerosis (MS) and in Wilson's disease (WD). *Acta Neuropsychol.* **2014**, *12*, 155–166.
84. Poujois, A.; Pernon, M.; Trocello, J.M.; Woimant, F. Dystonic Dysarthria in Wilson Disease: Efficacy of Zolpidem. *Front. Neurol.* **2017**, *8*, 559. [CrossRef] [PubMed]
85. Ilyechova, E.Y.; Miliukhina, I.V.; Karpenko, M.N.; Orlov, I.A.; Puchkova, L.V.; Samsonov, S.A. Case of Early-Onset Parkinson's Disease in a Heterozygous Mutation Carrier of the ATP7B Gene. *J. Pers. Med.* **2019**, *9*, 41. [CrossRef] [PubMed]
86. Ruggeri, M.; Biagioli, C.; Ricci, M.; Gerace, C.; Blundo, C. Progressive aphasia, apraxia of speech and agraphia in corticobasal degeneration: A 12-case series clinical and neuropsychological descriptive study. *Int. J. Lang. Commun. Disord.* **2020**, *55*, 867–874. [CrossRef] [PubMed]
87. Tetzloff, K.A.; Duffy, J.R.; Strand, E.A.; Machulda, M.M.; Boland, S.M.; Utianski, R.L.; Botha, H.; Senjem, M.L.; Schwarz, C.G.; Josephs, K.A.; et al. Clinical and imaging progression over 10 years in a patient with primary progressive apraxia of speech and autopsy-confirmed corticobasal degeneration. *Neurocase* **2018**, *24*, 111–120. [CrossRef]
88. Constantinides, V.C.; Souvatzoglou, M.; Paraskevas, G.P.; Chalioti, M.; Boufidou, F.; Stefanis, L.; Kapaki, E. Dopamine transporter SPECT imaging in corticobasal syndrome: A peak into the underlying pathology? *Acta Neurol. Scand.* **2022**, *145*, 762–769. [CrossRef]
89. Nakano, Y.; Shimada, H.; Shinotoh, H.; Hirano, S.; Tagai, K.; Sano, Y.; Yamamoto, Y.; Endo, H.; Matsuoka, K.; Takahata, K.; et al. PET-based classification of corticobasal syndrome. *Park. Relat. Disord.* **2022**, *98*, 92–98. [CrossRef]
90. Chung, S.W.; Choi, B.M.; Kim, J.Y.; Lee, Y.S.; Yoon, J.P.; Oh, K.S.; Park, K.S. Altered Gene and Protein Expressions in Torn Rotator Cuff Tendon Tissues in Diabetic Patients. *Arthroscopy* **2017**, *33*, 518–526.e1. [CrossRef]
91. Lefever, S.; Vandesompele, J.; Speleman, F.; Pattyn, F. RTPrimerDB: The portal for real-time PCR primers and probes. *Nucleic Acids Res.* **2009**, *37*, D942–D945. [CrossRef]
92. Manoochehri, J.; Masoumi Dehshiri, R.; Faraji, H.; Mohammadi, S.; Dastsooz, H.; Moradi, T.; Rezaei, E.; Sadeghi, K.; Fardaei, M. Family screening for a novel ATP7B gene mutation, c.2335T>G, in the South of Iran. *Iran. J. Ped. Hematol. Oncol.* **2014**, *4*, 26–31. [PubMed]
93. Sechi, G.; Antonio Cocco, G.; Errigo, A.; Deiana, L.; Rosati, G.; Agnetti, V.; Stephen Paulus, K.; Mario Pes, G. Three sisters with very-late-onset major depression and parkinsonism. *Park. Relat. Disord.* **2007**, *13*, 122–125. [CrossRef] [PubMed]
94. Moller, J.C.; Leinweber, B.; Rissling, I.; Oertel, W.H.; Bandmann, O.; Schmidt, H.H. Prevalence of the H1069Q mutation in ATP7B in discordant pairs with early-onset Parkinson's disease. *Mov. Disord.* **2006**, *21*, 1789–1790. [CrossRef] [PubMed]
95. Zhang, Z.; Liu, J.; Zheng, W.; Hou, Q.; Zhang, L. A novel heterozygous carrier of ATP7B mutation with muscle weakness and tremor: A Chinese Case Report. *J. Musculoskelet. Neuronal Interact.* **2020**, *20*, 614–618.

genes

MDPI

Review

The Role of IgLON Cell Adhesion Molecules in Neurodegenerative Diseases

Marco Salluzzo [1], Clara Vianello [2,†], Sandra Abdullatef [3], Roberto Rimondini [2], Giovanni Piccoli [3] and Lucia Carboni [1,*]

1 Department of Pharmacy and Biotechnology, Alma Mater Studiorum University of Bologna, 40126 Bologna, Italy; marco.salluzzo@unibo.it
2 Department of Medical and Surgical Sciences, Alma Mater Studiorum University of Bologna, 40126 Bologna, Italy; clara.vianello2@unibo.it (C.V.); roberto.rimondini@unibo.it (R.R.)
3 Department of Cellular, Computational and Integrative Biology, University of Trento, 38123 Trento, Italy; sandra.abdullatef@unitn.it (S.A.); giovanni.piccoli@unitn.it (G.P.)
* Correspondence: lucia.carboni4@unibo.it; Tel.: +39-0512091793
† Present address: Department for Life Quality Studies, Alma Mater Studiorum University of Bologna, 47921 Rimini, Italy.

Abstract: In the brain, cell adhesion molecules (CAMs) are critical for neurite outgrowth, axonal fasciculation, neuronal survival and migration, and synapse formation and maintenance. Among CAMs, the IgLON family comprises five members: Opioid Binding Protein/Cell Adhesion Molecule Like (OPCML or OBCAM), Limbic System Associated Membrane Protein (LSAMP), neurotrimin (NTM), Neuronal Growth Regulator 1 (NEGR1), and IgLON5. IgLONs exhibit three N-terminal C2 immunoglobulin domains; several glycosylation sites; and a glycosylphosphatidylinositol anchoring to the membrane. Interactions as homo- or heterodimers in *cis* and in *trans*, as well as binding to other molecules, appear critical for their functions. Shedding by metalloproteases generates soluble factors interacting with cellular receptors and activating signal transduction. The aim of this review was to analyse the available data implicating a role for IgLONs in neuropsychiatric disorders. Starting from the identification of a pathological role for antibodies against IgLON5 in an autoimmune neurodegenerative disease with a poorly understood mechanism of action, accumulating evidence links IgLONs to neuropsychiatric disorders, albeit with still undefined mechanisms which will require future thorough investigations.

Keywords: IgLON5; LSAMP; NEGR1; NEUROTRIMIN; OPCML; Alzheimer's disease; Huntington's disease; Synucleinopathies; Niemann-Pick disease; psychiatric disorders

Citation: Salluzzo, M.; Vianello, C.; Abdullatef, S.; Rimondini, R.; Piccoli, G.; Carboni, L. The Role of IgLON Cell Adhesion Molecules in Neurodegenerative Diseases. *Genes* **2023**, *14*, 1886. https://doi.org/10.3390/genes14101886

Academic Editor: Xingguang Luo

Received: 1 September 2023
Revised: 26 September 2023
Accepted: 27 September 2023
Published: 28 September 2023

Copyright: © 2023 by the authors. Licensee MDPI, Basel, Switzerland. This article is an open access article distributed under the terms and conditions of the Creative Commons Attribution (CC BY) license (https://creativecommons.org/licenses/by/4.0/).

1. Introduction

Cell adhesion molecules (CAMs) placed on the cell surface perform critical functions in a number of biological processes requiring contact between cells or with the extracellular matrix, such as cell recognition, adhesion, migration, and differentiation. The principal CAM groups are integrins, selectins, cadherins, and the immunoglobulin superfamily. In the developing brain, CAMs are crucial for the correct assembly of synaptic connections as well as the interactions with supporting glial cells. In the mature brain, complex functions depend on the correct performance of CAMs in establishing contacts between neuronal bodies, axonal interactions with myelinating glial cells, axon fasciculation, and connections to non-nervous cells [1,2]. Integrins are type I transmembrane proteins consisting of a large multidomain extracellular portion, a single-pass transmembrane region, and a short cytoplasmic component. The integrin family encompasses heterodimers of α and β subunits, which can combine to form several different integrins exhibiting overlapping but non-redundant functions, with specific ligand and signalling preferences depending on the α and β subunit combinations [3,4]. Selectins consist of an N-terminal carbohydrate-recognition domain allowing for the binding to glycoconjugates, an epidermal growth factor-like domain, a series of short consensus repeats, a transmembrane region, and a short C-terminal intracellular tail. Three selectin family members exist, P-selectin, L-selectin, and E-selectin, expressed in platelet-endothelial cells, leukocytes, and endothelial cells, respectively [5,6]. In the brain, selectins expressed by endothelial cells are involved in inflammatory responses, in damage after ischemic events, and in autoimmune diseases [6–8]. Cadherins are a large number of calcium-dependent adhesion proteins. Their structure comprises a calcium-binding extracellular domain consisting of several cadherin repeats of about 100 amino acids, a transmembrane domain, and a cytoplasmic domain which interacts with signalling molecules. Cadherins can form lateral dimers (cis-dimers) as well as trans-dimers with cadherins expressed by other cells [9,10].

Immunoglobulin (Ig)-like molecules are an ancient and diverse family of proteins performing a variety of functions, such as immune and signalling molecules and CAMs. Four different subtypes of Ig-like domains exist which are named constant 1, constant 2, variable, and intermediate (C1, C2, V, and I) for their resemblance to immunoglobulin domains. The presence of at least one Ig-like domain is a requirement of this class [11,12]. Generally, the core of all Ig domains encompasses two β-sheets facing each other and an intra-chain disulfide bridge which provides stability to the structure [13]. In addition, variable numbers of fibronectin type III domains or other protein modules characterise the sub-families within this large group. Ig-like CAMs generally include type I transmembrane proteins with a large N-terminal domain, a transmembrane portion, and a cytoplasmic domain [1,14]. However, some CAMs are anchored to the membrane through a glycosylphosphatidylinositol (GPI) segment [15]. Some proteins demonstrate homophilic binding specificity, whereas others have heterophilic specificity, thus interacting with other Ig-like CAMs or with surface proteins [1,16]. Ig domains play critical roles in mediating homophilic and heterophilic interactions in *trans*, namely between CAMs on adjacent cells or CAMs localised to the extracellular environment, as well as in *cis*, i.e., with proteins located in the plasma membrane of the same cell [17,18]. The first Ig-like CAMs identified in the central nervous system (CNS) were neural cell adhesion molecule (NCAM) and L1, which, along with their isoforms, are recognised for their major functions in axon outgrowth and fasciculation, neuronal survival and migration, synapse formation, and synaptic plasticity [11,14,19]. Subsequent research led to the identification of additional sub-families, such as the nectins, MAM domain–containing GPI anchors, IGSF9, IGSF21, contactins, and IgLONs [11,20]. The IgLON subfamily, which is the focus of this review, includes five members sharing the presence of three C2 immunoglobulin domains in the N-terminus and a GPI-anchoring to the membrane (Figure 1). The first family members identified were Limbic System Associated Membrane Protein (LSAMP, initially named LAMP) [21], Opioid Binding Protein/Cell Adhesion Molecule Like (OPCML, also known as OBCAM) [22], and neurotrimin (NTM) [23]. Indeed, the name IgLON stands for Ig family containing LAMP, OBCAM, NTM. Subsequently, another family member was identified initially as KILON/neurotractin, and later re-named Neuronal Growth Regulator 1 (NEGR1) [24,25]. Lastly, antibodies against the IgLON member IgLON5 (IgLON Family Member 5) have been associated with autoimmune encephalitis [26]. Since the discovery of this family, its prominent roles in brain development, axon fasciculation, neurite extension, and synapse formation and maintenance have been identified [21–25]. Although the role in synapse regulation is well-established, further research is needed to characterise the molecular mechanisms underlying the functional outcomes [17,27–30]. A meaningful contribution to the comprehension of IgLONs' physiological functions derived from the analysis of its molecular evolution, originating from an ancestor around the emergence of Arthropods [31]. Functional analysis allowed for the identification of common as well as specific sequences for interacting partner recognition, post-translational modifications, metalloproteinase cleavage sites, and signal transduction pathways [31].

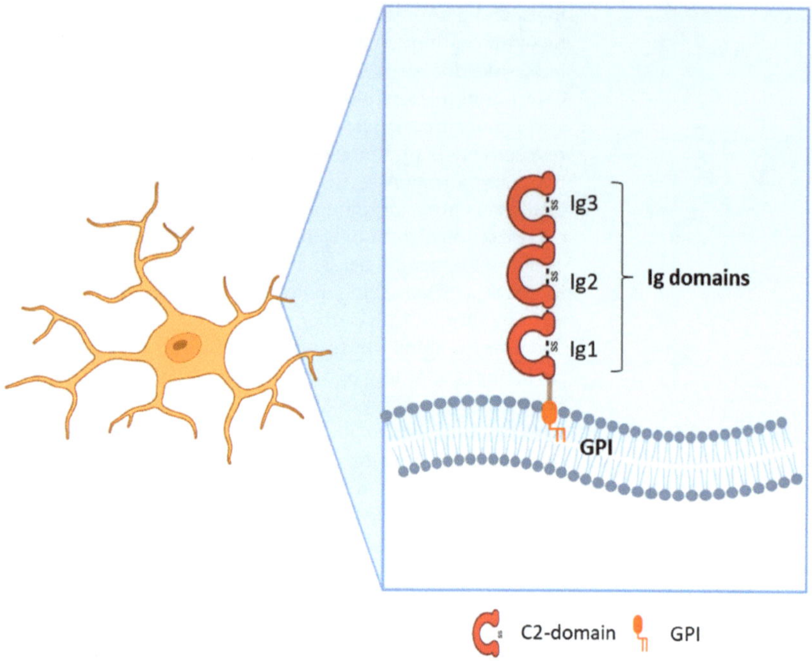

Figure 1. Structure of IgLON family members.

Available evidence supports the involvement of CAMs in the onset of neuropsychiatric disorders. In Alzheimer's disease, a role for CAMs has been suggested by genome-wide association studies (GWAS) as well as by the identification of altered levels in diseased brains [32,33]. Molecules belonging to several CAM families have been implicated in pathological manifestations through different mechanisms, involving amyloid-β metabolism, cell plasticity, and neuroinflammation [33,34]. The detection of an increase in the enzymes involved in GPI-anchoring in Alzheimer's disease brains provided further support to a relevant role of these sub-families [35]. Moreover, genetic studies have associated CAMs with a number of neuropsychiatric disorders, including CRASH syndrome, MASA syndrome, X-linked mental retardation, intellectual disability, autism spectrum disorder, schizophrenia, addiction, and bipolar disorder [15,19]. The roles of CAMs in axon growth, guidance, and fasciculation; in target recognition; and in synapse formation and maintenance might contribute to the circuit alterations characterising neuropsychiatric disorders. The main proposed mechanisms include the ability to target the formation of specific circuitries responsible for specialised brain functions through localised expression. Additionally, CAMs may impact the occurrence of neuropsychiatric diseases by modifying the balance between excitatory and inhibitory signals, thereby affecting local neural circuit activity. This influence extends to the modulation of neuromodulatory systems like monoaminergic circuits and the alteration of synaptic strength and composition [15]. Focussing more specifically on Ig-like CAMs, mutations in the genes belonging to the best-studied NCAM and L1 families, as well as changes in their expression patterns or post-translational modifications, have been associated with psychiatric and neurodegenerative disorders, including the L1 syndrome, Alzheimer's disease, schizophrenia, and bipolar disorder [36]. When further restricting the focus to GPI-anchored sub-families, their role as functional receptors controlling neurite outgrowth, synapse formation, synapse plasticity, and learning behaviours has been implicated in the pathophysiological mechanisms associated with neurodegenerative and psychiatric disorders [17]. Since a wealth of data has been recently accumulating to implicate the IgLONs, the aim of the present review is to analyse and discuss the findings indicating possible connections between this family and neurological and psychiatric disorders.

2. IgLON5 (IgLON Family Member 5)

Anti-IgLON5 disease (ORPHA:420789) is a rare disease of the CNS. Since its discovery in 2014, 60 cases have been reported worldwide [37–39]. The clinical manifestation of the disease is highly heterogeneous, encompassing unique sleep and movement disorders, bulbar dysfunction, and cognitive impairment [26]. Disease onset is around the age of 60 years with progressive symptoms that can culminate in life-threatening respiratory problems [37]. Post-mortem analysis has revealed a tauopathy restricted to neurons in the brainstem, tegmentum, hypothalamus, and hippocampus [40]. The mortality rate associated with the anti-IgLON5 syndrome is considerably high, contributing to an overall mortality rate of 34%, with no discernible link between mortality and treatment response [38]. While respiratory complications are the primary cause of death, the additional neurological symptoms (cognitive decline, sleep disturbances) significantly diminish the quality of life for patients and their families.

The role of the IgLON5 protein and the underlying mechanism of disease associated with anti-IgLON5 antibodies remain poorly understood. IgLON5 is widely expressed in the CNS, with the highest levels reached in the olfactory bulb, cortical plate, and hippocampus in mice and humans. Comprising a chain of 336 amino acids, IgLON5 shares significant structural resemblance with other proteins such as OPCML (50% similarity), NTM (48–49% similarity), LSAMP (46–47% similarity), and NEGR1 (41% similarity) (Figure 1). IgLON5 is involved in forming both homomeric and heteromeric interactions with other IgLON family members [41] (Figure 2). However, the functional implications stemming from these interactions remain poorly characterised.

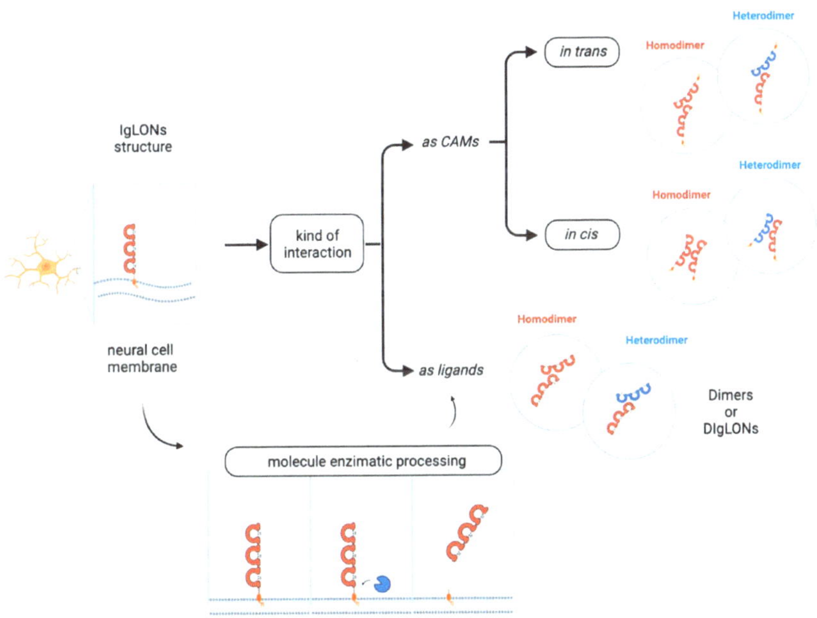

Figure 2. IgLONs can form homo- and hetero-dimers to interact in *trans* (in different cells) or in *cis* (in the same cell). They can be cleaved from the membrane by metalloproteases to form soluble interacting proteins.

The detection of anti-IgLON5 antibodies is crucial for the diagnosing of anti-IgLON5 disease. These antibodies are detectable in both serum and cerebrospinal fluid (CSF), typically coinciding with the initial diagnosis timeframe. The question of whether anti-IgLON5 antibodies directly cause neuronal dysfunction and degeneration or instead appear as a consequence of the neurodegenerative process remain unresolved. Among anti-IgLON5 antibodies, the non-complement fixing IgG4 subclass predominates over IgG1. In vitro experiments have shown that IgG4 antibodies lead to the internalization of IgLON5 in cultures of hippocampal neurons [42]. Upon exposure to anti-IgLON5 antibodies, cultured rat or human neurons showed increased neurodegenerative features such as neuronal blebbing and fragmentation [43,44]. Prolonged exposure to anti-IgLON5 IgG prompts tau hyperphosphorylation and cellular death [44].

Similar to Alzheimer's disease, the tau filaments evident in individuals with anti-IgLON5 syndrome encompass both 3R-tau and 4R-tau isoforms. However, this accumulation manifests in a distinct spatial pattern, indicative of anti-IgLON5 disease constituting a novel form of tauopathy. Recent autoptic studies have described anti-IgLON5 cases lacking hyperphosphorylated tau deposits [45,46]. These findings raise the possibility that tauopathy might arise subsequently in the disease progression, evolving as a prolonged outcome of the antibody-related effects. Thus, earlier events still depending on anti-IgLON5 antibodies may play a role in the aetiology of the disease, in addition to the proposition that a genetic predisposition to autoimmunity might exert influence, exemplified by the robust correlation with the exceptionally rare HLA-DRB1*1001 and HLA-DQB1*0501 alleles [40].

IgLONs interact and regulate receptor tyrosine kinase (RTK) trafficking and, eventually, their signalling impinging on ERK1/2 and AKT phosphorylation [28,29]. Anti-IgLON5 disease is characterised by the aggregation of hyperphosphorylated-Tau, which is phosphorylated by a variety of serine/threonine protein kinases [47]; GSK-3β is the main kinase responsible for Tau phosphorylation and precipitation [48]. An intricate signalling cascade, encompassing RTKs, insulin receptor substrate 1 (IRS-1), and AKT, has been identified as a key mechanism in curbing Tau phosphorylation by GSK-3β [49]. Hence, the hypothesis that IgLON5 may interact with and regulate a yet to be identified RTK is exceptionally intriguing (Figure 3). Recent in vitro findings put forth the proposition that antibodies against anti-IgLON5 could potentially interfere with the IgLON5 interactome [50]. It is plausible that these anti-IgLON5 antibodies might impede the IgLON5-RTK interaction, thereby exerting an influence on RTK signalling. This sequence of events could consequently trigger aberrant Tau phosphorylation and accumulation, ultimately culminating in neuronal dysfunction (Figure 3).

In order to pave the way for effective therapeutic interventions geared towards individuals afflicted with IgLON5 deficiency disease, a comprehensive grasp of both the normal and pathological roles of IgLON5, alongside the associated signalling cascade, becomes essential.

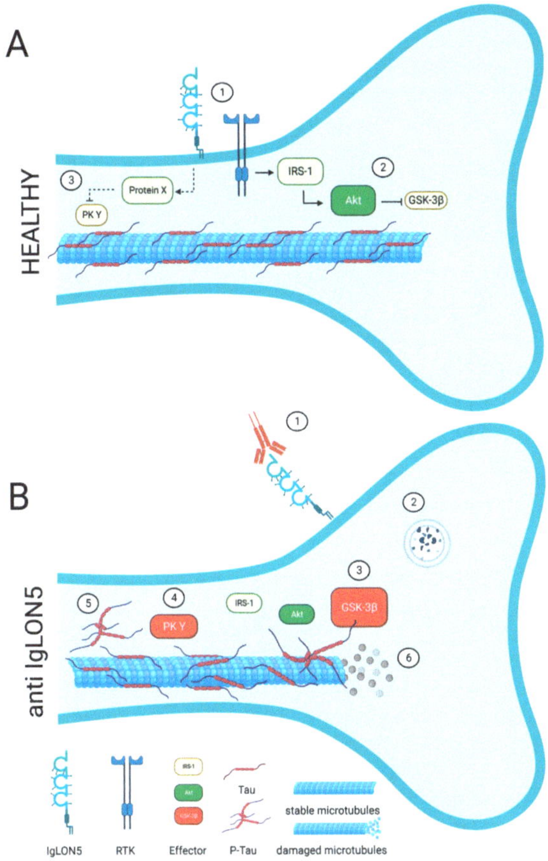

Figure 3. A potential link between IgLON5 and the aggregation of hyperphosphorylated-Tau. (**A**) In healthy conditions, IgLON5 may interact with and regulate a yet to be identified receptor tyrosine kinase (RTK, 1). A signalling cascade involving such RTK, insulin receptor substrate 1 (IRS-1), and AKT (2) prevents Tau phosphorylation by GSK-3β (3). A different RTK-triggered pathway may control Tau phosphorylation (4) and assure microtubule stability. (**B**) In patients, the anti-IgLON5 antibodies sequester IgLON5 (1) and trigger the degradation of IgLON5 together with the RTK, potentially (2). As such, RTK signalling pathways are altered (3–4). This may result in pathological tau phosphorylation and accumulation (5) and, eventually, neurotoxic microtubule disassembly (6).

3. NEGR1 (Neuronal Growth Regulator 1)

NEGR1 was first isolated and identified as a member of the IgLON family in 1999 in rats [24] and in chicken [25] as a 46–50 kDa protein; it presents the common class features of three amino-terminal C2 domains, six glycosylation sites, and GPI anchoring [24,25,51,52] (Figure 1). Homo- and heterodimerization with other IgLONs in *cis* and in *trans* across the synaptic cleft were recently thoroughly investigated [41,53] (Figure 2). Its expression in the brain is widespread, with high levels in the olfactory bulb, cerebral cortex, diencephalon, hippocampus, hypothalamus, and cerebellum [54–57], in a developmentally-regulated fashion [29,30,58,59]. NEGR1 is able to modulate neurite extension, either upon up- or down-regulation, suggesting that precise regulation is critical for correct functioning and promoting outgrowth in response to CNS injury [28,58,60]. Likewise, NEGR1 was identified as a crucial factor in determining the number of synapses in hippocampal neurons in opposite directions depending on environmental conditions [61], whereas knock-out mice showed impaired axon growth [59]. In cortical cultures, NEGR1 silencing decreased neurites' total length and number. The reduced basal dendrite arborization induced by NEGR1 downregulation was also demonstrated in vivo in the somatosensory cortex of mice exposed to miRNA-transfection in utero [62]. Ectodomain shedding through the activity of metalloproteinases is a critical mechanism for IgLONs function (Figure 4).

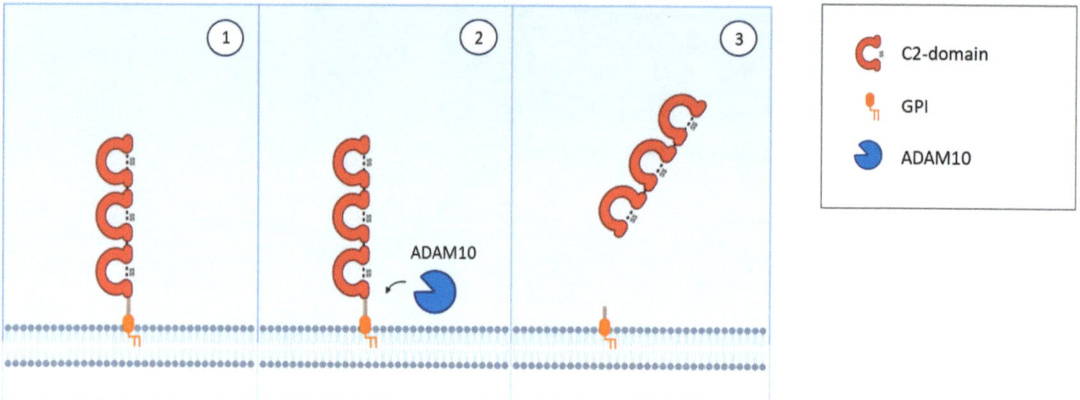

Figure 4. The metalloprotease ADAM10 can cleave IgLONs to generate soluble proteins able to bind RTKs and start intracellular signal transduction.

In particular, the Disintegrin and Metalloproteinase Domain-Containing Proteins (ADAM) were demonstrated to be responsible for shedding NEGR1 and other IgLONs from the plasma membrane with a direct impact on neurite outgrowth in cultured cortical neurons [60]. Further research demonstrated that the metalloprotease ADAM10 is specifically involved in NEGR1 shedding and that the generation of soluble NEGR1 was able to regulate neuronal morphology by binding to the FGFR2 receptor and activating its intracellular pathway [28]. Among the latter, a prominent role is exerted by ERK and AKT pathways, which decrease FGFR2 degradation from the plasma membrane [28,29] (Figure 5). NEGR1's impact on dendritic spine densities of pyramidal neurons has also been reproduced in vivo in the mouse somatosensory cortex [29]. Moreover, a role in hippocampal neurogenesis and long term potentiation are supported by the disruption of these functions in NEGR1 knock-out mice [63]. NEGR1 knock-out mice showed altered serotonergic and dopaminergic neurotransmission [64] and subtle alterations in social behaviour and in learning, as well as increased susceptibility to pentylenetetrazol-induced seizures [59]. In contrast with previous findings reporting no changes in anxiety [59], Noh et al. [63] showed that NEGR1 knock-out mice showed increased anxiety- and depressive-like behaviours, proposing the involvement of lipocalin and Leukaemia Inhibitory Factor (LIF) Receptor in the mechanism of action.

In 2009, a GWAS meta-analysis identified NEGR1 among genetic determinants of the variation in body mass index in adults and children, with a stronger influence on adiposity, which is likely to be mediated by its expression in the hypothalamic nuclei [65–69]. In line with these findings, modulation of NEGR1 in rodents affected body mass index and energy balance in complex, developmentally regulated ways [70,71]. Recently, NEGR1 has been indicated among the principal players in the genetic correlations between obesity and a number of psychiatric disorders, including major depression, schizophrenia, and anorexia nervosa [72].

Within psychiatric disorders, a statistically significant association with major depressive disorder has been discovered in recent meta-analyses of GWAS, with NEGR1 resulting among the strongest signals [73–76]. By integrating GWAS, brain expression quantitative trait loci data, and transcriptome-wide association studies, the association was confirmed and assigned the highest probability of causality; increased abundance of NEGR1 was significantly associated with an increased depression risk [77–79]. Differential regulation of transcription factor binding and the consequential altered expression has been proposed as a potential mechanism of action of the single-nucleotide polymorphisms (SNPs) associated with the increased risk of major depression [80]. Adding further evidence to the significance for depression, NEGR1 levels in the CSF of patients affected by major depression or bipolar disorder were significantly higher than in controls; this biomarker provided a major contribution to the identification of a biosignature, allowing for the correct patient group stratification [81]. In rodent models, treatment with antidepressant medications altered NEGR1 expression in brain regions [82,83], whereas genetic variants have been associated with antidepressant treatment response in obese depressed patients [84]. Supporting a possible involvement in the pathophysiology of schizophrenia, elevated levels of NEGR1 mRNA were detected in the dorsolateral prefrontal cortex of patients, [85], in agreement with previous findings in anterior pre-frontal cortex samples [86]. The association was also demonstrated at the genetic level in human patients [87]. Still within the realm of psychiatric disorders, altered NEGR1 DNA methylation has been discovered in patients affected by anorexia nervosa [88].

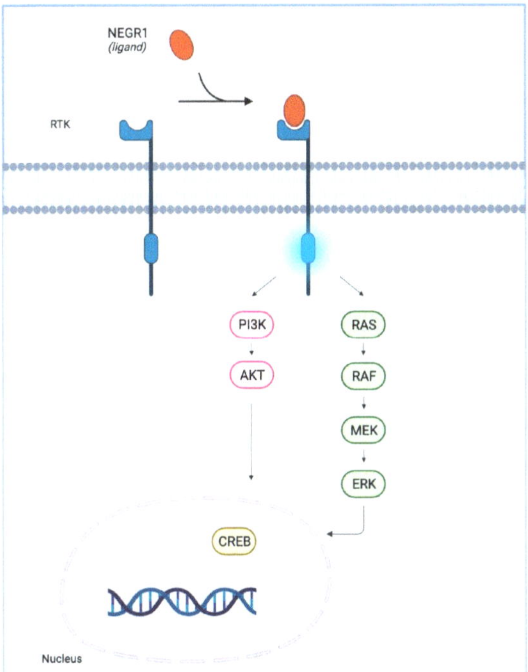

Figure 5. Soluble NEGR1 generated by shedding binds FGFR2 and activates intracellular signal transduction pathways.

Evidence to imply potential NEGR1 connections with the aetiology of neurological and neurodegenerative disorders has been accumulating recently. The existence of an association between NEGR1 and Alzheimer's disease has been investigated due to the frequent comorbidity with major depression. Ni et al. [89] reported a significant genetic association with Alzheimer's disease in Han Chinese patients, as well as altered expression in rodent models of the disease. These findings were expanded by a study identifying statistically significant associations with the protective efficacy of SNPs located in the NEGR1 gene to mediate the protective role of educational attainment in Alzheimer's disease [90]. In healthy subjects' brains, NEGR1 has been associated with white matter integrity measured as fractional anisotropy. This finding is highly suggestive, since altered white matter integrity characterises several neurodegenerative diseases, comprising Alzheimer's disease [91–93]. In synucleinopathies, including Parkinson's disease, NEGR1 has been indicated as a potential CSF biomarker in mass spectrometry studies, thus implicating relevance for diagnostic and therapeutic applications [94]. In line with these findings, NEGR1 levels in plasma were downregulated and in correlation with cognitive clinical scores and motor abilities in Parkinson's disease patients [95]. An association between NEGR1 and transmissible spongiform encephalopathies, fatal neurodegenerative diseases characterised by infectious proteins, has been discovered in rodent models in which substantially decreased levels of NEGR1 have been demonstrated. Interestingly, NEGR1 was the only glycosylated biomarker among those identified [96]. A plausible role for NEGR1 in the neurodegenerative disorder Huntington's disease has been reported by Kaltenbach et al. [97], who showed that the mutated form of huntingtin interacts with NEGR1 and that the latter is able to modify a neurodegeneration phenotype induced by treatment with a huntingtin fragment in *Drosophila*. NEGR1 is able to interact with Niemann–Pick disease type C2 protein, increase its stability, and thus modulate cholesterol accumulation and influence its homeostasis. This effect has been implicated in Niemann–Pick disease type C, an autosomal recessive disorder characterised by cholesterol accumulation leading to dementia [98].

Outside the realm of neuropsychiatric disorders, NEGR1 has been implicated in other diseases whose mechanisms may have a bearing on cognitive and intellectual disabilities that often denote neurodegenerative disorders. A possible relevance of NEGR1 in autism is supported by the observation that NEGR1 downregulation in mice reproduces brain development abnormalities resembling the cortical disorganization of neurons and abnormalities in dendritic spines characterising autism spectrum disorder features. In addition, NEGR1 and FGFR2 downregulation alters social skill-related behaviours similarly to autism spectrum disorder, in reducing ultrasound vocalizations in pups, causing sensory deficits in a hot plate test in newborns, and lowering social interactions in adults [29]. Furthermore, NEGR1 has been identified as a candidate susceptibility gene in human genetic studies [99]. An association of NEGR1 with dyslexia emerged in a copy number variation analysis in Indian families. These results suggest a potential role for NEGR1 in this neurogenetic disorder, possibly by influencing the production of dendritic postsynaptic spines in mature neurons [100]. Partial deletion of chromosome 1p31.1, only involving the NEGR1 gene, has been observed in two siblings affected by attention deficit hyperactivity disorder and language impairments, but not affected by dyslexia [101]. Another patient affected by microdeletion 1p31, including NEGR1, showed moderate intellectual disability [102]. Language impairments have been reported in a patient with an interstitial 1p31.1p31.3 deletion affecting several genes including NEGR1 [103]. A meta-analysis approach on gene expression data in multiple sclerosis patients' tissues showed a significant increase in NEGR1 levels, which was interpreted as a leading cause of impaired synaptogenesis [104]. Additional support to NEGR1 contribution to this disorder derived from a study uncovering a potential impact in the comorbidity with major depression in multiple sclerosis patients [105].

4. OPCML (Opioid Binding Protein/Cell Adhesion Molecule Like)

The IgLON member OPCML (referred also as IgLON1 or OBCAM) binds opioid receptors after shedding from the plasmatic membrane, where it functions as a CAM [53,106] (Figure 2). The OPCML gene, comprising seven exons and spanning approximately 600 Kb, is mapped to human chromosome 11, which also hosts NCAM and other brain-expressed CAMs [107]. As a result of alternative splicing, four different transcripts have been identified [108]. Originally, OPCML was purified from rat brains and presumed to be an opioid binding protein [109], and then later recognised as a member of the IgLON subgroup on the basis of its cDNA sequence [22]. In general, the primary sequence of the protein is very highly conserved between species [22]. OPCML consists of 345 amino acids, with a molecular mass of 38 kDa, harbouring six glycosylation sites at Asn44, Asn70, Asn140, Asn285, Asn293, and Asn306 [22]. Depending on the grade of glycosylation or the presence of other post-translational modifications, the molecular weight of the protein can vary [51,110]. The structure of the OPCML ectodomain is characterised by the presence of three C2 domains [22,53] and the protein is anchored to the plasma membrane through a GPI site [17,53](Figure 1). Therefore, intracellular signalling can be induced by interacting in *cis* with other transmembrane proteins and through the formation of homo- or heterodimers with other IgLONs [111]. The ability of OPCML to act independently as an opioid receptor has been debated, reaching the conclusion that its playing an accessory role seems more likely, based on primary structure similarity to IgLONs instead of G-protein coupled receptors [22,107]. OPCML transcripts have been widely detected in brain and non-brain tissues in adult and foetal animal models and humans, with brain-specific enrichment of the v2 isoform [108,112]. In line with its high expression in neurons, OPCML takes part in cell adhesion and cell–cell recognition by interacting with and modulating molecules that can promote or inhibit growth, regulating synapse number, synaptogenesis, and plasticity [111,113].

Polymorphisms in the OPCML gene have been linked to schizophrenia in GWAS studies in European patients and the association was later confirmed in a Thai population [114–116]. In a genetic study on OPCML SNPs conducted in a Han Chinese population, the association was confirmed again and reduced hippocampal expression was found in risk-allele carriers [117]. However, expression levels were not altered in the prefrontal cortex of adult patients with schizophrenia, suggesting that the control of transcription in the adult brain may not represent a prominent mechanism of the association [118]. More recently, a large meta-analysis combining data from several repositories highlighted another OPCML-associated SNP (rs2917569) with a high grade of association [119]. Although the role of OPCML in the pathophysiology of schizophrenia requires clarification, its crucial function in neurite outgrowth and spine maturation has been implicated through Eph-Cofilin signalling and F-actin polymerization [117]. Still within psychiatric disorders, a major depression study in two independent Dutch populations highlighted a significant linkage with two SNPs on chromosome 11q25 located in the intronic region 1 of the OPCML gene. The finding has been interpreted as due to the relevance of endogenous opioid neurotransmission in depressed patients [120]. Furthermore, a GWAS on anorexia nervosa detected a signal for an intronic SNP in the OPCML gene which was associated with disease risk across all discovery and replication cohorts, although it did not reach genome-wide significance [121]. A case report of two brothers with disruption in the chromosomal region encompassing OPCML and NTM suggested possible links with autistic symptoms [122].

Within neurodegenerative diseases, OPCML has been implicated in Alzheimer's disease in a GWAS carried out in patients affected by the late-onset form [123]. In line with these findings, altered OPCML mRNA levels were identified in the hippocampus of two mouse models of disease [124]. In addition, reduced OPCML glycosylation in Alzheimer's disease brains, together with other CAMs, was interpreted as a driver of altered cell adhesion and synaptic function in the disorder [125].

5. LSAMP (Limbic System Associated Membrane Protein)

The limbic system associated membrane protein (LAMP or LSAMP, as indicated in more recent works to distinguish it from the lysosomal associated membrane protein) is an IgLON family member first identified as specifically characterising neurons belonging to the rat limbic system or to areas receiving direct projections from there [21]. Subsequent purification and cloning revealed that LSAMP is an integral membrane protein with a molecular mass of 64–68 KDa containing three internal repeats typical of the Ig-like domain, with conserved pairs of cysteine residues, together with eight putative glycosylation sites [126–128] (Figure 1). Like the other IgLONs, it is a membrane protein containing a GPI anchor, which can be released from intact membranes by the action of phosphatidyl inositol-specific phospholipase C [129]. LSAMP nucleotidic and aminoacidic sequences present high sequence identity with those of OPCML and NTM, suggesting similar roles for these proteins [127]. In neurons, most protein expression is post-synaptic, located on neuronal somata and dendrites, while glial cells are not immunoreactive [126]. In the adult rat, LSAMP is expressed in regions related to the limbic system, like the hippocampus, amygdala, perirhinal cortex, anterior and lateral thalamic nuclei, preoptic area of the hypothalamus, septum, nucleus of the solitary tract, and lamina II of the dorsal horn [21,126,128,130]. The specificity of this protein for neurons belonging to the same functional system suggest its possible involvement in brain regionalization during development. Indeed, during development LSAMP mRNA is expressed in the same limbic regions as in the adult, but in a temporal fashion that relates with the time of formation of limbic axonal pathways [128,130,131]. In addition, ultramicroscopic analysis revealed that in embryonic tissue the localization of LSAMP is mainly located in the growth cone of developing axons [126,131]. The axonal staining disappeared during the second post-natal week of development, indicating a role for LSAMP in axonal growth and limbic pathway formation [130,131].

The in vivo administration of anti-LSAMP antibodies altered the formation of mossy fibres projection to pyramidal neurons during hippocampal development, suggesting that LSAMP may serve as a recognition molecule for the proper genesis of limbic connections [127]. LSAMP's crucial role in establishing limbic connections as well as limbic neural fate was further highlighted by experiments where the connections between neurons belonging to different parts of the brain were manipulated or disrupted [130,132–134]. These data suggest that the commitment to limbic fate happens early in development, and that LSAMP represents an early marker of limbic neurons determination. Thus, LSAMP acts as an attractive molecule, guiding the sprouting of limbic axons, and behaving as a repulsive signal for the growing of non-limbic axons [135]. The heterophilic binding both in *cis* and in *trans* with other IgLON members mediated by the different IgG domains has been identified as a critical mechanism of action [25,111,136–140] (Figure 2). Another important mechanism of regulation in the development of axonal growth cones involves the molecular shedding of the IgLON ectodomain from the membrane by metalloproteases, with a prominent role played by ADAM10 [60,141] (Figure 4).

In pre-clinical models of disease, evidence of alterations in the molecular expression of LSAMP is available, suggestive of implication in the neurobiological underpinnings of psychiatric disorders. Increased LSAMP expression was found in rats with low exploratory activity in the elevated plus maze test, suggesting a correlation with anxiety [142,143]. Further evidence derives from LSAMP knock-out mice, which demonstrated a heightened responsiveness to novelty in behavioural tests as well as a maladaptive response to environmental stressors and altered interaction with the serotoninergic system, associated with increased sensitivity to psychostimulants [142,144–148]. Moreover, LSAMP deletion impacts synaptogenesis and synaptic plasticity, hippocampus formation, and spatial memory [57,59,149]. In human genetic studies, SNPs in the LSAMP gene have been associated with male suicide [150] when it is well established that the most important risk factor for suicide is depression [151]. Indeed, a significant association with major depression and panic disorder has been detected [152]. In addition, altered LSAMP levels were observed in the dorsolateral prefrontal cortex of schizophrenic patients, albeit with contrasting findings in different patient cohorts [85,153]. In line with these results, significant allelic and haplotypic associations have been reported between SNPs in the LSAMP gene and schizophrenia [154].

Little is known about LSAMP involvement in neurodegenerative diseases. A study aiming at identifying biomarkers for Lewy bodies dementia, Alzheimer's disease, and Parkinson's disease in patients' CSF recognized LSAMP as a novel non-specific biomarker for neurodegeneration due to its significantly increased expression in all three diseases with respect to controls [155]. A similar study, where the relative expression of N-glycosylated proteins in Alzheimer's disease was analysed through mass spectrometry, showed that LSAMP negatively correlated with the disease, suggesting that decreased post-translational modification can occur [125]. However, neither report provided clues about the molecular mechanisms of LSAMP involvement in neurodegeneration, thus further studies are required.

6. NTM (Neurotrimin)

NTM was characterised in 1995 as a 65 kDa integral protein endowed with three C2 amino-terminal domains, seven glycosylation sites, GPI anchoring, and high sequence homology with other IgLONs [23,156] (Figure 1). Like them, NTM can bind other family members in *cis* and in *trans*, as well as form homodimers [41,53,111,136,137,157] (Figure 2). The dimers are essential for NTM function in the regulation of neurite outgrowth and cell adhesion to control neuronal connectivity and synaptogenesis [139,157–160]. Both membrane-bound and soluble forms are involved in outgrowth regulation, with sensitivity to metalloproteases, particularly ADAM10 [60,141,157,161] (Figure 4). NTM is expressed in the developing and adult brain, with high expression in the sensorimotor cortex, olfactory bulb, thalamus, hypothalamus, hippocampus, basal ganglia, and cerebellum [23,137,156]. Differently from LSAMP knock out mice, NTM deficient mice had no differences in anxiety, social interaction, or locomotor activity, only sharing a lower sensitivity to the locomotor stimulating effect of amphetamine, although to a lower extent, and displaying subtle deficits in cognitively challenging emotional learning tasks [162].

SNPs in the NTM gene have been associated with intelligence in a family-based low-density genome-wide study; these findings are noteworthy since lower cognitive function increases the risk of neuropsychiatric disorders [163]. Autism spectrum disorder patients harboured copy number variants encompassing the NTM gene and two patients with a chromosomal translocation disrupting the NTM gene displayed autistic symptoms [122,164]. Furthermore, a signal for association with NTM has been discovered for childhood aggressiveness in attention deficit hyperactivity disorder, although the signal did not reach statistical significance [165]. A GWAS in a genetically isolated population from the Netherland linked four SNPs placed within intron 1 of the NTM gene to late-onset Alzheimer's disease [123].

7. Conclusions

A wealth of data is available linking IgLON with neurodegenerative, neurological, and psychiatric disorders (Figure 6); however, since most findings derive from genetic-association studies, the underlying mechanisms of action are poorly understood.

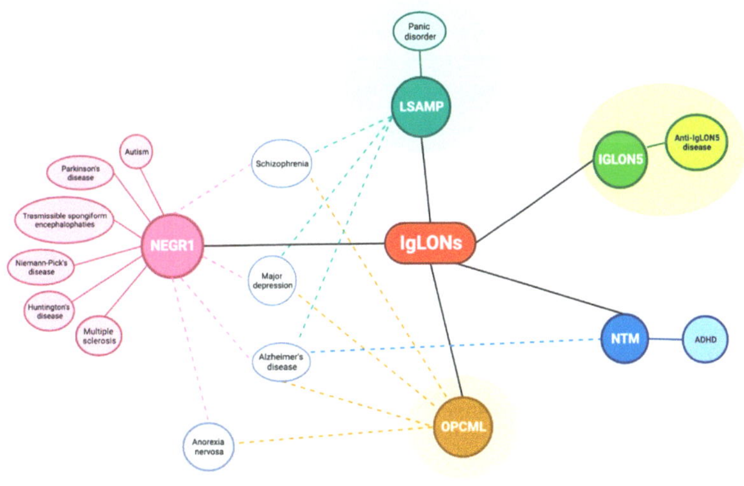

Figure 6. Associations between IgLONs and neuropsychiatric disorders.

For this family, the available evidence supports functions in nervous system development, neurite extension, synaptic formation and maintenance, blood-brain barrier structure, extracellular matrix protein recognition, and activation of signal transduction within cells able to elicit long-term responses. It is likely that the physiological role of IgLONs exerts a critical function in the pathophysiology of these disorders. However, the neurobiological underpinnings are yet to be investigated. We can speculate that the established function in synaptic plasticity and strength exerts a potential role in learning and memory formation, which can represent a common dysregulation impinging on the molecular underpinning of several disorders, with a prominent impact on neurodegenerative diseases. Dendritic spine pathology is an established hallmark of neurodegenerative diseases [166]. The accumulation of IgLONs in CSF samples gathered from patients affected by neurodegenerative disorders may reflect the spine loss. In addition, the discovery of the autoimmune response associated with anti-IgLON5 disease hints to a possible involvement of autoimmune responses as a contributing mechanism of action in other neurodegenerative and neurodevelopmental disorders associated with this family [167]. Accumulating evidence shows that IgLONs interact and regulate RTK trafficking and, eventually, their signalling impinging on ERK1/2 and Akt phosphorylation [28,29,168]. RTKs are tightly involved in the dynamics of the actin and tubulin cytoskeleton. An intriguing hypothesis is that IgLONs influence the signalling of RTKs and, consequently, cytoskeletal stability. Alteration in the cytoskeleton, such as the tau phosphorylation/accumulation observed in anti-IgLON5 disease, may eventually cause neuronal dysfunction and death.

As a caveat, it should be underlined that the existence of genetic associations does not necessarily imply a prominent role in the neurobiological underpinning of the disorder. In the same line, phenotypes associated with genetic deletions do not automatically imply a general role for the deleted gene in the pathology. Moreover, the partial overlap of disease association with IgLON family members supports the possible cross-disease alterations to be imputed to these CAMs. Further studies will elucidate the potential relevance of IgLON family members as future targets for therapeutic intervention or as biomarkers to aid diagnosis.

Author Contributions: M.S., C.V., S.A., R.R., G.P. and L.C. contributed to literature analysis, L.C. wrote the first draft of the manuscript. All authors have read and agreed to the published version of the manuscript.

Funding: This research was partially funded by Fondazione Cariplo, grant number 2019-3415 to G.P.

Institutional Review Board Statement: Not applicable.

Informed Consent Statement: Not applicable.

Data Availability Statement: Not applicable.

Acknowledgments: Figures were created with BioRender.com.

Conflicts of Interest: The authors declare no conflict of interest.

References

1. Shapiro, L.; Love, J.; Colman, D.R. Adhesion Molecules in the Nervous System: Structural Insights into Function and Diversity. *Annu. Rev. Neurosci.* **2007**, *30*, 451–474. [CrossRef] [PubMed]
2. Stachowicz, K. Physicochemical Principles of Adhesion Mechanisms in the Brain. *Int. J. Mol. Sci.* **2023**, *24*, 5070. [CrossRef] [PubMed]
3. Kadry, Y.A.; Calderwood, D.A. Chapter 22: Structural and signaling functions of integrins. *Biochim. Biophys. Acta—Biomembr.* **2020**, *1862*, 183206. [CrossRef] [PubMed]
4. Hynes, R.O. Integrins: Bidirectional, allosteric signaling machines. *Cell* **2002**, *110*, 673–687. [CrossRef] [PubMed]
5. McEver, R.P.; Zhu, C. Rolling cell adhesion. *Annu. Rev. Cell Dev. Biol.* **2010**, *26*, 363–396. [CrossRef]
6. Ley, K.; Kansas, G.S. Selectins in T-cell recruitment to non-lymphoid tissues and sites of inflammation. *Nat. Rev. Immunol.* **2004**, *4*, 325–335. [CrossRef]
7. Siew, J.J.; Chern, Y. Microglial Lectins in Health and Neurological Diseases. *Front. Mol. Neurosci.* **2018**, *11*, 158. [CrossRef]

8. Angiari, S. Selectin-mediated leukocyte trafficking during the development of autoimmune disease. *Autoimmun. Rev.* **2015**, *14*, 984–995. [CrossRef]
9. Redies, C. Cadherins in the central nervous system. *Prog. Neurobiol.* **2000**, *61*, 611–648. [CrossRef]
10. Takeichi, M. The cadherin superfamily in neuronal connections and interactions. *Nat. Rev. Neurosci.* **2007**, *8*, 11–20. [CrossRef]
11. Cameron, S.; McAllister, A.K. Immunoglobulin-Like Receptors and Their Impact on Wiring of Brain Synapses. *Annu. Rev. Genet.* **2018**, *52*, 567–590. [CrossRef] [PubMed]
12. Barclay, A.N. Membrane proteins with immunoglobulin-like domains—A master superfamily of interaction molecules. *Semin. Immunol.* **2003**, *15*, 215–223. [CrossRef] [PubMed]
13. Chothia, C.; Gelfand, I.; Kister, A. Structural determinants in the sequences of immunoglobulin variable domain. *J. Mol. Biol.* **1998**, *278*, 457–479. [CrossRef]
14. Walsh, F.S.; Doherty, P. Neural cell adhesion molecules of the immunoglobulin superfamily: Role in axon growth and guidance. *Annu. Rev. Cell Dev. Biol.* **1997**, *13*, 425–456. [CrossRef] [PubMed]
15. Sakurai, T. The role of cell adhesion molecules in brain wiring and neuropsychiatric disorders. *Mol. Cell. Neurosci.* **2017**, *81*, 4–11. [CrossRef]
16. Zinn, K.; Özkan, E. Neural immunoglobulin superfamily interaction networks. *Curr. Opin. Neurobiol.* **2017**, *45*, 99–105. [CrossRef]
17. Tan, R.P.A.; Leshchyns'ka, I.; Sytnyk, V. Glycosylphosphatidylinositol-Anchored Immunoglobulin Superfamily Cell Adhesion Molecules and Their Role in Neuronal Development and Synapse Regulation. *Front. Mol. Neurosci.* **2017**, *10*, 378. [CrossRef]
18. Leshchyns'ka, I.; Sytnyk, V. Reciprocal Interactions between Cell Adhesion Molecules of the Immunoglobulin Superfamily and the Cytoskeleton in Neurons. *Front. Cell Dev. Biol.* **2016**, *4*, 9. [CrossRef]
19. Maness, P.F.; Schachner, M. Neural recognition molecules of the immunoglobulin superfamily: Signaling transducers of axon guidance and neuronal migration. *Nat. Neurosci.* **2007**, *10*, 19–26. [CrossRef]
20. Karagogeos, D. Neural GPI-anchored cell adhesion molecules. *Front. Biosci.* **2003**, *8*, s1304–s1320. [CrossRef]
21. Levitt, P. A monoclonal antibody to limbic system neurons. *Science* **1984**, *223*, 299–301. [CrossRef]
22. Schofield, P.R.; McFarland, K.C.; Hayflick, J.S.; Wilcox, J.N.; Cho, T.M.; Roy, S.; Lee, N.M.; Loh, H.H.; Seeburg, P.H. Molecular characterization of a new immunoglobulin superfamily protein with potential roles in opioid binding and cell contact. *EMBO J.* **1989**, *8*, 489–495. [CrossRef] [PubMed]
23. Struyk, A.F.; Canoll, P.D.; Wolfgang, M.J.; Rosen, C.L.; D'Eustachio, P.; Salzer, J.L. Cloning of neurotrimin defines a new subfamily of differentially expressed neural cell adhesion molecules. *J. Neurosci.* **1995**, *15*, 2141–2156. [CrossRef]
24. Funatsu, N.; Miyata, S.; Kumanogoh, H.; Shigeta, M.; Hamada, K.; Endo, Y.; Sokawa, Y.; Maekawa, S. Characterization of a novel rat brain glycosylphosphatidylinositol- anchored protein (Kilon), a member of the IgLON cell adhesion molecule family. *J. Biol. Chem.* **1999**, *274*, 8224–8230. [CrossRef]
25. Marg, A.; Sirim, P.; Spaltmann, F.; Plagge, A.; Kauselmann, G.; Buck, F.; Rathjen, F.G.; Brümmendorf, T. Neurotractin, a novel neurite outgrowth-promoting Ig-like protein that interacts with CEPU-1 and LAMP. *J. Cell Biol.* **1999**, *145*, 865–876. [CrossRef] [PubMed]
26. Sabater, L.; Gaig, C.; Gelpi, E.; Bataller, L.; Lewerenz, J.; Torres-Vega, E.; Contreras, A.; Giometto, B.; Compta, Y.; Embid, C.; et al. A novel non-rapid-eye movement and rapid-eye-movement parasomnia with sleep breathing disorder associated with antibodies to IgLON5: A case series, characterisation of the antigen, and post-mortem study. *Lancet Neurol.* **2014**, *13*, 575–586. [CrossRef] [PubMed]
27. Takamori, S.; Holt, M.; Stenius, K.; Lemke, E.A.; Grønborg, M.; Riedel, D.; Urlaub, H.; Schenck, S.; Brügger, B.; Ringler, P.; et al. Molecular anatomy of a trafficking organelle. *Cell* **2006**, *127*, 831–846. [CrossRef] [PubMed]
28. Pischedda, F.; Piccoli, G. The IgLON Family Member Negr1 Promotes Neuronal Arborization Acting as Soluble Factor via FGFR2. *Front. Mol. Neurosci.* **2016**, *8*, 89. [CrossRef]
29. Szczurkowska, J.; Pischedda, F.; Pinto, B.; Managò, F.; Haas, C.A.; Summa, M.; Bertorelli, R.; Papaleo, F.; Schäfer, M.K.; Piccoli, G.; et al. NEGR1 and FGFR2 cooperatively regulate cortical development and core behaviours related to autism disorders in mice. *Brain* **2018**, *141*, 2772–2794. [CrossRef]
30. Fearnley, S.; Raja, R.; Cloutier, J.F. Spatiotemporal expression of IgLON family members in the developing mouse nervous system. *Sci. Rep.* **2021**, *11*, 19536. [CrossRef]
31. Kubick, N.; Brösamle, D.; Mickael, M.E. Molecular Evolution and Functional Divergence of the IgLON Family. *Evol. Bioinforma.* **2018**, *14*, 1176934318775081. [CrossRef] [PubMed]
32. Leshchyns'ka, I.; Sytnyk, V. Synaptic Cell Adhesion Molecules in Alzheimer's Disease. *Neural Plast.* **2016**, *2016*, 6427537. [CrossRef] [PubMed]
33. Wennström, M.; Nielsen, H.M. Cell adhesion molecules in Alzheimer's disease. *Degener. Neurol. Neuromuscul. Dis.* **2012**, *2*, 65–77. [CrossRef] [PubMed]
34. Leshchyns'ka, I.; Liew, H.T.; Shepherd, C.; Halliday, G.M.; Stevens, C.H.; Ke, Y.D.; Ittner, L.M.; Sytnyk, V. Aβ-dependent reduction of NCAM2-mediated synaptic adhesion contributes to synapse loss in Alzheimer's disease. *Nat. Commun.* **2015**, *6*, 8836. [CrossRef]
35. Tang, X.; Tena, J.; Di Lucente, J.; Maezawa, I.; Harvey, D.J.; Jin, L.-W.; Lebrilla, C.B.; Zivkovic, A.M. Transcriptomic and glycomic analyses highlight pathway-specific glycosylation alterations unique to Alzheimer's disease. *Sci. Rep.* **2023**, *13*, 7816. [CrossRef]

36. Sytnyk, V.; Leshchyns'ka, I.; Schachner, M. Neural Cell Adhesion Molecules of the Immunoglobulin Superfamily Regulate Synapse Formation, Maintenance, and Function. *Trends Neurosci.* **2017**, *40*, 295–308. [CrossRef]
37. Gaig, C.; Graus, F.; Compta, Y.; Högl, B.; Bataller, L.; Brüggemann, N.; Giordana, C.; Heidbreder, A.; Kotschet, K.; Lewerenz, J.; et al. Clinical manifestations of the anti-IgLON5 disease. *Neurology* **2017**, *88*, 1736–1743. [CrossRef]
38. Nissen, M.S.; Blaabjerg, M. Anti-IgLON5 Disease: A Case With 11-Year Clinical Course and Review of the Literature. *Front. Neurol.* **2019**, *10*, 1056. [CrossRef]
39. Werner, J.; Jelcic, I.; Schwarz, E.I.; Probst-Müller, E.; Nilsson, J.; Schwizer, B.; Bloch, K.E.; Lutterotti, A.; Jung, H.-H.; Schreiner, B. Anti-IgLON5 Disease: A New Bulbar-Onset Motor Neuron Mimic Syndrome. *Neurol.-Neuroimmunol. Neuroinflamm.* **2021**, *8*, e962. [CrossRef]
40. Gelpi, E.; Höftberger, R.; Graus, F.; Ling, H.; Holton, J.L.; Dawson, T.; Popovic, M.; Pretnar-Oblak, J.; Högl, B.; Schmutzhard, E.; et al. Neuropathological criteria of anti-IgLON5-related tauopathy. *Acta Neuropathol.* **2016**, *132*, 531–543. [CrossRef]
41. Ranaivoson, F.M.; Turk, L.S.; Ozgul, S.; Kakehi, S.; von Daake, S.; Lopez, N.; Trobiani, L.; De Jaco, A.; Denissova, N.; Demeler, B.; et al. A Proteomic Screen of Neuronal Cell-Surface Molecules Reveals IgLONs as Structurally Conserved Interaction Modules at the Synapse. *Structure* **2019**, *27*, 893–906.e9. [CrossRef] [PubMed]
42. Sabater, L.; Planagumà, J.; Dalmau, J.; Graus, F. Cellular investigations with human antibodies associated with the anti-IgLON5 syndrome. *J. Neuroinflamm.* **2016**, *13*, 226. [CrossRef] [PubMed]
43. Landa, J.; Gaig, C.; Plagumà, J.; Saiz, A.; Antonell, A.; Sanchez-Valle, R.; Dalmau, J.; Graus, F.; Sabater, L. Effects of IgLON5 Antibodies on Neuronal Cytoskeleton: A Link between Autoimmunity and Neurodegeneration. *Ann. Neurol.* **2020**, *88*, 1023–1027. [CrossRef] [PubMed]
44. Ryding, M.; Gamre, M.; Nissen, M.S.; Nilsson, A.C.; Okarmus, J.; Poulsen, A.A.E.; Meyer, M.; Blaabjerg, M. Neurodegeneration Induced by Anti-IgLON5 Antibodies Studied in Induced Pluripotent Stem Cell-Derived Human Neurons. *Cells* **2021**, *10*, 837. [CrossRef] [PubMed]
45. Honorat, J.A.; Komorowski, L.; Josephs, K.A.; Fechner, K.; St Louis, E.K.; Hinson, S.R.; Lederer, S.; Kumar, N.; Gadoth, A.; Lennon, V.A.; et al. IgLON5 antibody: Neurological accompaniments and outcomes in 20 patients. *Neurol.-Neuroimmunol. Neuroinflamm.* **2017**, *4*, e385. [CrossRef] [PubMed]
46. Erro, M.E.; Sabater, L.; Martínez, L.; Herrera, M.; Ostolaza, A.; García de Gurtubay, I.; Tuñón, T.; Graus, F.; Gelpi, E. Anti-IGLON5 disease: A new case without neuropathologic evidence of brainstem tauopathy. *Neurol.-Neuroimmunol. Neuroinflamm.* **2020**, *7*, e651. [CrossRef]
47. Xia, Y.; Prokop, S.; Giasson, B.I. "Don't Phos Over Tau": Recent developments in clinical biomarkers and therapies targeting tau phosphorylation in Alzheimer's disease and other tauopathies. *Mol. Neurodegener.* **2021**, *16*, 37. [CrossRef]
48. Hernandez, F.; Lucas, J.J.; Avila, J. GSK3 and tau: Two convergence points in Alzheimer's disease. *J. Alzheimer's Dis.* **2013**, *33* (Suppl. 1), S141–S144. [CrossRef]
49. Kanno, T.; Tsuchiya, A.; Tanaka, A.; Nishizaki, T. Combination of PKCε Activation and PTP1B Inhibition Effectively Suppresses Aβ-Induced GSK-3β Activation and Tau Phosphorylation. *Mol. Neurobiol.* **2016**, *53*, 4787–4797. [CrossRef]
50. Landa, J.; Serafim, A.B.; Gaig, C.; Saiz, A.; Koneczny, I.; Hoftberger, R.; Santamaria, J.; Dalmau, J.; Graus, F.; Sabater, L. Patients' IgLON5 autoantibodies interfere with IgLON5-protein interactions. *Front. Immunol.* **2023**, *14*, 1151574. [CrossRef]
51. Itoh, S.; Hachisuka, A.; Kawasaki, N.; Hashii, N.; Teshima, R.; Hayakawa, T.; Kawanishi, T.; Yamaguchi, T. Glycosylation analysis of IgLON family proteins in rat brain by liquid chromatography and multiple-stage mass spectrometry. *Biochemistry* **2008**, *47*, 10132–10154. [CrossRef] [PubMed]
52. Sim, G.; Jeong, M.; Seo, H.; Kim, J.; Lee, S. The Role of N-Glycosylation in the Intracellular Trafficking and Functionality of Neuronal Growth Regulator 1. *Cells* **2022**, *11*, 1242. [CrossRef] [PubMed]
53. Venkannagari, H.; Kasper, J.M.; Misra, A.; Rush, S.A.; Fan, S.; Lee, H.; Sun, H.; Seshadrinathan, S.; Machius, M.; Hommel, J.D.; et al. Highly Conserved Molecular Features in IgLONs Contrast Their Distinct Structural and Biological Outcomes. *J. Mol. Biol.* **2020**, *432*, 5287–5303. [CrossRef] [PubMed]
54. Miyata, S.; Funatsu, N.; Matsunaga, W.; Kiyohara, T.; Sokawa, Y.; Maekawa, S. Expression of the IgLON cell adhesion molecules Kilon and OBCAM in hypothalamic magnocellular neurons. *J. Comp. Neurol.* **2000**, *424*, 74–85. [CrossRef]
55. Miyata, S.; Matsumoto, N.; Taguchi, K.; Akagi, A.; Iino, T.; Funatsu, N.; Maekawa, S. Biochemical and ultrastructural analyses of IgLON cell adhesion molecules, Kilon and OBCAM in the rat brain. *Neuroscience* **2003**, *117*, 645–658. [CrossRef]
56. Bräuer, A.U.; Savaskan, N.E.; Plaschke, M.; Prehn, S.; Ninnemann, O.; Nitsch, R. IG-molecule kilon shows differential expression pattern from LAMP in the developing and adult rat hippocampus. *Hippocampus* **2000**, *10*, 632–644. [CrossRef]
57. Vanaveski, T.; Singh, K.; Narvik, J.; Eskla, K.-L.L.; Visnapuu, T.; Heinla, I.; Jayaram, M.; Innos, J.; Lilleväli, K.; Philips, M.-A.A.; et al. Promoter-Specific Expression and Genomic Structure of IgLON Family Genes in Mouse. *Front. Neurosci.* **2017**, *11*, 38. [CrossRef]
58. Schäfer, M.; Bräuer, A.U.; Savaskan, N.E.; Rathjen, F.G.; Brümmendorf, T. Neurotractin/kilon promotes neurite outgrowth and is expressed on reactive astrocytes after entorhinal cortex lesion. *Mol. Cell. Neurosci.* **2005**, *29*, 580–590. [CrossRef]
59. Singh, K.; Loreth, D.; Pöttker, B.; Hefti, K.; Innos, J.; Schwald, K.; Hengstler, H.; Menzel, L.; Sommer, C.J.; Radyushkin, K.; et al. Neuronal growth and behavioral alterations in mice deficient for the psychiatric disease-associated negr1 gene. *Front. Mol. Neurosci.* **2018**, *11*, 30. [CrossRef]

60. Sanz, R.; Ferraro, G.B.; Fournier, A.E. IgLON Cell Adhesion Molecules Are Shed from the Cell Surface of Cortical Neurons to Promote Neuronal Growth. *J. Biol. Chem.* **2015**, *290*, 4330–4342. [CrossRef]
61. Hashimoto, T.; Yamada, M.; Maekawa, S.; Nakashima, T.; Miyata, S. IgLON cell adhesion molecule Kilon is a crucial modulator for synapse number in hippocampal neurons. *Brain Res.* **2008**, *1224*, 1–11. [CrossRef] [PubMed]
62. Pischedda, F.; Szczurkowska, J.; Cirnaru, M.D.; Giesert, F.; Vezzoli, E.; Ueffing, M.; Sala, C.; Francolini, M.; Hauck, S.M.; Cancedda, L.; et al. A cell surface biotinylation assay to reveal membrane-associated neuronal cues: Negr1 regulates dendritic arborization. *Mol. Cell. Proteom.* **2014**, *13*, 733–748. [CrossRef] [PubMed]
63. Noh, K.; Lee, H.; Choi, T.-Y.; Joo, Y.; Kim, S.-J.; Kim, H.; Kim, J.Y.; Jahng, J.W.; Lee, S.; Choi, S.-Y.; et al. Negr1 controls adult hippocampal neurogenesis and affective behaviors. *Mol. Psychiatry* **2019**, *24*, 1189–1205. [CrossRef] [PubMed]
64. Kaare, M.; Jayaram, M.; Jagomäe, T.; Singh, K.; Kilk, K.; Mikheim, K.; Leevik, M.; Leidmaa, E.; Varul, J.; Nõmm, H.; et al. Depression-Associated Negr1 Gene-Deficiency Induces Alterations in the Monoaminergic Neurotransmission Enhancing Time-Dependent Sensitization to Amphetamine in Male Mice. *Brain Sci.* **2022**, *12*, 1696. [CrossRef]
65. Willer, C.; Speliotes, E.; Loos, R.; Li, S.; Lindgren, C.; Heid, I.; Berndt, S.; Elliott, A.; Jackson, A.; Lamina, C.; et al. Six new loci associated with body mass index highlight a neuronal influence on body weight regulation. *Nat. Genet.* **2009**, *41*, 25–34. [CrossRef]
66. Renström, F.; Payne, F.; Nordström, A.; Brito, E.C.; Rolandsson, O.; Hallmans, G.; Barroso, I.; Nordström, P.; Franks, P.W. Replication and extension of genome-wide association study results for obesity in 4923 adults from northern Sweden. *Hum. Mol. Genet.* **2009**, *18*, 1489–1496. [CrossRef]
67. Zhao, J.; Bradfield, J.P.; Li, M.; Wang, K.; Zhang, H.; Kim, C.E.; Annaiah, K.; Glessner, J.T.; Thomas, K.; Garris, M.; et al. The role of obesity-associated loci identified in genome-wide association studies in the determination of pediatric BMI. *Obesity* **2009**, *17*, 2254–2257. [CrossRef]
68. Schmid, P.M.; Heid, I.; Buechler, C.; Steege, A.; Resch, M.; Birner, C.; Endemann, D.H.; Riegger, G.A.; Luchner, A. Expression of fourteen novel obesity-related genes in zucker diabetic fatty rats. *Cardiovasc. Diabetol.* **2012**, *11*, 48. [CrossRef]
69. Boender, A.J.; Van Rozen, A.J.; Adan, R.A.H. Nutritional state affects the expression of the obesity-associated genes Etv5, faim2, Fto, and negr1. *Obesity* **2012**, *20*, 2420–2425. [CrossRef]
70. Boender, A.J.; van Gestel, M.A.; Garnerv, K.M.; Luijendijk, M.C.M.; Adan, R.A.H. The Obesity-Associated gene NEGR1 regulates aspects of energy balance in rat hypothalamic areas. *Physiol. Rep.* **2014**, *2*, e12083. [CrossRef]
71. Lee, A.W.S.; Hengstler, H.; Schwald, K.; Berriel-Diaz, M.; Loreth, D.; Kirsch, M.; Kretz, O.; Haas, C.A.; de Angelis, M.H.; Herzig, S.; et al. Functional inactivation of the genome-wide association study obesity gene neuronal growth regulator 1 in mice causes a body mass phenotype. *PLoS ONE* **2012**, *7*, e41537. [CrossRef] [PubMed]
72. Ding, H.; Ouyang, M.; Wang, J.; Xie, M.; Huang, Y.; Yuan, F.; Jia, Y.; Zhang, X.; Liu, N.; Zhang, N. Shared genetics between classes of obesity and psychiatric disorders: A large-scale genome-wide cross-trait analysis. *J. Psychosom. Res.* **2022**, *162*, 111032. [CrossRef] [PubMed]
73. Hyde, C.L.; Nagle, M.W.; Tian, C.; Chen, X.; Paciga, S.A.; Wendland, J.R.; Tung, J.Y.; Hinds, D.A.; Perlis, R.H.; Winslow, A.R. Identification of 15 genetic loci associated with risk of major depression in individuals of European descent. *Nat. Genet.* **2016**, *48*, 1031–1036. [CrossRef] [PubMed]
74. Wray, N.R.; Ripke, S.; Mattheisen, M.; Trzaskowski, M.; Byrne, E.M.; Abdellaoui, A.; Adams, M.J.; Agerbo, E.; Air, T.M.; Andlauer, T.M.F.; et al. Genome-wide association analyses identify 44 risk variants and refine the genetic architecture of major depression. *Nat. Genet.* **2018**, *50*, 668–681. [CrossRef]
75. Howard, D.M.; Adams, M.J.; Clarke, T.-K.; Hafferty, J.D.; Gibson, J.; Shirali, M.; Coleman, J.R.I.; Hagenaars, S.P.; Ward, J.; Wigmore, E.M.; et al. Genome-wide meta-analysis of depression identifies 102 independent variants and highlights the importance of the prefrontal brain regions. *Nat. Neurosci.* **2019**, *22*, 343–352. [CrossRef]
76. Levey, D.F.; Stein, M.B.; Wendt, F.R.; Pathak, G.A.; Zhou, H.; Aslan, M.; Quaden, R.; Harrington, K.M.; Nuñez, Y.Z.; Overstreet, C.; et al. Bi-ancestral depression GWAS in the Million Veteran Program and meta-analysis in >1.2 million individuals highlight new therapeutic directions. *Nat. Neurosci.* **2021**, *24*, 954–963. [CrossRef]
77. Dall'Aglio, L.; Lewis, C.M.; Pain, O. Delineating the Genetic Component of Gene Expression in Major Depression. *Biol. Psychiatry* **2021**, *89*, 627–636. [CrossRef]
78. Wang, X.; Cheng, W.; Zhu, J.; Yin, H.; Chang, S.; Yue, W.; Yu, H. Integrating genome-wide association study and expression quantitative trait loci data identifies NEGR1 as a causal risk gene of major depression disorder. *J. Affect. Disord.* **2020**, *265*, 679–686. [CrossRef]
79. Deng, Y.T.; Ou, Y.N.; Wu, B.S.; Yang, Y.X.; Jiang, Y.; Huang, Y.Y.; Liu, Y.; Tan, L.; Dong, Q.; Suckling, J.; et al. Identifying causal genes for depression via integration of the proteome and transcriptome from brain and blood. *Mol. Psychiatry* **2022**, *27*, 2849–2857. [CrossRef]
80. Li, S.; Li, Y.; Li, X.; Liu, J.; Huo, Y.; Wang, J.; Liu, Z.; Li, M.; Luo, X.-J. Regulatory mechanisms of major depressive disorder risk variants. *Mol. Psychiatry* **2020**, *25*, 1926–1945. [CrossRef]
81. Maccarrone, G.; Ditzen, C.; Yassouridis, A.; Rewerts, C.; Uhr, M.; Uhlen, M.; Holsboer, F.; Turck, C.W. Psychiatric patient stratification using biosignatures based on cerebrospinal fluid protein expression clusters. *J. Psychiatr. Res.* **2013**, *47*, 1572–1580. [CrossRef] [PubMed]

82. Tamási, V.; Petschner, P.; Adori, C.; Kirilly, E.; Ando, R.D.; Tothfalusi, L.; Juhasz, G.; Bagdy, G. Transcriptional evidence for the role of chronic venlafaxine treatment in neurotrophic signaling and neuroplasticity including also glutatmatergic- and insulin-mediated neuronal processes. *PLoS ONE* **2014**, *9*, e113662. [CrossRef] [PubMed]
83. Carboni, L.; Pischedda, F.; Piccoli, G.; Lauria, M.; Musazzi, L.; Popoli, M.; Mathé, A.A.; Domenici, E. Depression-Associated Gene Negr1-Fgfr2 Pathway Is Altered by Antidepressant Treatment. *Cells* **2020**, *9*, 1818. [CrossRef] [PubMed]
84. Amare, A.T.; Schubert, K.O.; Tekola-Ayele, F.; Hsu, Y.H.; Sangkuhl, K.; Jenkins, G.; Whaley, R.M.; Barman, P.; Batzler, A.; Altman, R.B.; et al. The association of obesity and coronary artery disease genes with response to SSRIs treatment in major depression. *J. Neural Transm.* **2019**, *126*, 35–45. [CrossRef] [PubMed]
85. Karis, K.; Eskla, K.L.; Kaare, M.; Täht, K.; Tuusov, J.; Visnapuu, T.; Innos, J.; Jayaram, M.; Timmusk, T.; Weickert, C.S.; et al. Altered expression profile of igLON family of neural cell adhesion molecules in the dorsolateral prefrontal cortex of schizophrenic patients. *Front. Mol. Neurosci.* **2018**, *11*, 8. [CrossRef]
86. Cox, D.A.; Gottschalk, M.G.; Wesseling, H.; Ernst, A.; Cooper, J.D.; Bahn, S. Proteomic systems evaluation of the molecular validity of preclinical psychosis models compared to schizophrenia brain pathology. *Schizophr. Res.* **2016**, *177*, 98–107. [CrossRef]
87. Zhang, Z.; Chen, G. A logical relationship for schizophrenia, bipolar, and major depressive disorder. Part 1: Evidence from chromosome 1 high density association screen. *J. Comp. Neurol.* **2020**, *528*, 2620–2635. [CrossRef]
88. Steiger, H.; Booij, L.; Thaler, L.; St-Hilaire, A.; Israël, M.; Casey, K.F.; Oliverio, S.; Crescenzi, O.; Lee, V.; Turecki, G.; et al. DNA methylation in people with anorexia nervosa: Epigenome-wide patterns in actively ill, long-term remitted, and healthy-eater women. *World J. Biol. Psychiatry* **2023**, *24*, 254–259. [CrossRef]
89. Ni, H.; Xu, M.; Zhan, G.-L.; Fan, Y.; Zhou, H.; Jiang, H.-Y.; Lu, W.-H.; Tan, L.; Zhang, D.-F.; Yao, Y.-G.; et al. The GWAS Risk Genes for Depression May Be Actively Involved in Alzheimer's Disease. *J. Alzheimer's Dis.* **2018**, *64*, 1149–1161. [CrossRef]
90. Raghavan, N.S.; Vardarajan, B.; Mayeux, R. Genomic variation in educational attainment modifies Alzheimer disease risk. *Neurol. Genet.* **2019**, *5*, e310. [CrossRef]
91. Harrison, J.R.; Bhatia, S.; Tan, Z.X.; Mirza-Davies, A.; Benkert, H.; Tax, C.M.W.; Jones, D.K. Imaging Alzheimer's genetic risk using diffusion MRI: A systematic review. *NeuroImage Clin.* **2020**, *27*, 102359. [CrossRef] [PubMed]
92. Li, J.P.; Pan, P.L.; Song, W.; Huang, R.; Chen, K.; Shang, H.F. A meta-analysis of diffusion tensor imaging studies in amyotrophic lateral sclerosis. *Neurobiol. Aging* **2012**, *33*, 1833–1838. [CrossRef]
93. Ciccarelli, O.; Catani, M.; Johansen-Berg, H.; Clark, C.; Thompson, A. Diffusion-based tractography in neurological disorders: Concepts, applications, and future developments. *Lancet Neurol.* **2008**, *7*, 715–727. [CrossRef] [PubMed]
94. Lachén-Montes, M.; González-Morales, A.; Fernández-Irigoyen, J.; Santamaría, E. Deployment of Label-Free Quantitative Olfactory Proteomics to Detect Cerebrospinal Fluid Biomarker Candidates in Synucleinopathies. *Methods Mol. Biol.* **2019**, *2044*, 273–289. [CrossRef] [PubMed]
95. Abdi, I.Y.; Bartl, M.; Dakna, M.; Abdesselem, H.; Majbour, N.; Trenkwalder, C.; El-Agnaf, O.; Mollenhauer, B. Cross-sectional proteomic expression in Parkinson's disease-related proteins in drug-naïve patients vs healthy controls with longitudinal clinical follow-up. *Neurobiol. Dis.* **2023**, *177*, 105997. [CrossRef] [PubMed]
96. Lamoureux, L.; Simon, S.L.R.; Waitt, B.; Knox, J.D. Proteomic Screen of Brain Glycoproteome Reveals Prion Specific Marker of Pathogenesis. *Proteomics* **2018**, *18*, 1700296. [CrossRef]
97. Kaltenbach, L.S.; Romero, E.; Becklin, R.R.; Chettier, R.; Bell, R.; Phansalkar, A.; Strand, A.; Torcassi, C.; Savage, J.; Hurlburt, A.; et al. Huntingtin interacting proteins are genetic modifiers of neurodegeneration. *PLoS Genet.* **2007**, *3*, 689–708. [CrossRef] [PubMed]
98. Kim, H.; Chun, Y.; Che, L.; Kim, J.; Lee, S.; Lee, S. The new obesity-associated protein, neuronal growth regulator 1 (NEGR1), is implicated in Niemann-Pick disease Type C (NPC2)-mediated cholesterol trafficking. *Biochem. Biophys. Res. Commun.* **2017**, *482*, 1367–1374. [CrossRef] [PubMed]
99. Marshall, C.R.; Noor, A.; Vincent, J.B.; Lionel, A.C.; Feuk, L.; Skaug, J.; Shago, M.; Moessner, R.; Pinto, D.; Ren, Y.; et al. Structural Variation of Chromosomes in Autism Spectrum Disorder. *Am. J. Hum. Genet.* **2008**, *82*, 477–488. [CrossRef]
100. Veerappa, A.M.; Saldanha, M.; Padakannaya, P.; Ramachandra, N.B. Family-based genome-wide copy number scan identifies five new genes of dyslexia involved in dendritic spinal plasticity. *J. Hum. Genet.* **2013**, *58*, 539–547. [CrossRef]
101. Genovese, A.; Cox, D.; Butler, M. Partial Deletion of Chromosome 1p31.1 Including only the Neuronal Growth Regulator 1 Gene in Two Siblings. *J. Pediatr. Genet.* **2015**, *4*, 023–028. [CrossRef]
102. Biswal, S.; Parida, P.; Dubbudu, A.; Sharawat, I.K.; Panda, P.K. Chromosome 1p31.1 Deletion Syndrome: Limited Expression. *Ann. Indian Acad. Neurol.* **2021**, *24*, 78–80. [CrossRef] [PubMed]
103. Tassano, E.; Gamucci, A.; Celle, M.E.; Ronchetto, P.; Cuoco, C.; Gimelli, G. Clinical and Molecular Cytogenetic Characterization of a de novo Interstitial 1p31.1p31.3 Deletion in a Boy with Moderate Intellectual Disability and Severe Language Impairment. *Cytogenet. Genome Res.* **2015**, *146*, 39–43. [CrossRef]
104. Katiyar, A.; Sharma, S.; Singh, T.P.; Kaur, P. Identification of shared molecular signatures indicate the susceptibility of endometriosis to multiple sclerosis. *Front. Genet.* **2018**, *9*, 42. [CrossRef]
105. Brugger, S.W.; Gardner, M.C.; Beales, J.T.; Briggs, F.; Davis, M.F. Depression in multiple sclerosis patients associated with risk variant near NEGR1. *Mult. Scler. Relat. Disord.* **2020**, *46*, 102537. [CrossRef] [PubMed]
106. Lippman, D.A.; Lee, N.M.; Loh, H.H. Opioid-binding cell adhesion molecule (OBCAM)-related clones from a rat brain cDNA library. *Gene* **1992**, *117*, 249–254. [CrossRef]

107. Shark, K.B.; Lee, N.M. Cloning, sequencing and localization to chromosome 11 of a cDNA encoding a human opioid-binding cell adhesion molecule (OBCAM). *Gene* **1995**, *155*, 213–217. [CrossRef] [PubMed]
108. Cui, Y.; Ying, Y.; van Hasselt, A.; Ng, K.M.; Yu, J.; Zhang, Q.; Jin, J.; Liu, D.; Rhim, J.S.; Rha, S.Y.; et al. OPCML is a broad tumor suppressor for multiple carcinomas and lymphomas with frequently epigenetic inactivation. *PLoS ONE* **2008**, *3*, e2990. [CrossRef]
109. Cho, T.M.; Hasegawa, J.; Ge, B.L.; Loh, H.H. Purification to apparent homogeneity of a mu-type opioid receptor from rat brain. *Proc. Natl. Acad. Sci. USA* **1986**, *83*, 4138–4142. [CrossRef]
110. Hachisuka, A.; Yamazaki, T.; Sawada, J.; Terao, T. Characterization and tissue distribution of opioid-binding cell adhesion molecule (OBCAM) using monoclonal antibodies. *Neurochem. Int.* **1996**, *28*, 373–379. [CrossRef]
111. Reed, J.; McNamee, C.; Rackstraw, S.; Jenkins, J.; Moss, D. Diglons are heterodimeric proteins composed of IgLON subunits, and Diglon-CO inhibits neurite outgrowth from cerebellar granule cells. *J. Cell Sci.* **2004**, *117*, 3961–3973. [CrossRef]
112. Hachisuka, A.; Nakajima, O.; Yamazaki, T.; Sawada, J. Developmental expression of opioid-binding cell adhesion molecule (OBCAM) in rat brain. *Dev. Brain Res.* **2000**, *122*, 183–191. [CrossRef]
113. Yamada, M.; Hashimoto, T.; Hayashi, N.; Higuchi, M.; Murakami, A.; Nakashima, T.; Maekawa, S.; Miyata, S. Synaptic adhesion molecule OBCAM; synaptogenesis and dynamic internalization. *Brain Res.* **2007**, *1165*, 5–14. [CrossRef]
114. Athanasiu, L.; Mattingsdal, M.; Kähler, A.K.; Brown, A.; Gustafsson, O.; Agartz, I.; Giegling, I.; Muglia, P.; Cichon, S.; Rietschel, M.; et al. Gene variants associated with schizophrenia in a Norwegian genome-wide study are replicated in a large European cohort. *J. Psychiatr. Res.* **2010**, *44*, 748–753. [CrossRef]
115. Panichareon, B.; Nakayama, K.; Thurakitwannakarn, W.; Iwamoto, S.; Sukhumsirichart, W. OPCML gene as a schizophrenia susceptibility locus in Thai population. *J. Mol. Neurosci.* **2012**, *46*, 373–377. [CrossRef]
116. O'Donovan, M.C.; Craddock, N.; Norton, N.; Williams, H.; Peirce, T.; Moskvina, V.; Nikolov, I.; Hamshere, M.; Carroll, L.; Georgieva, L.; et al. Identification of loci associated with schizophrenia by genome-wide association and follow-up. *Nat. Genet.* **2008**, *40*, 1053–1055. [CrossRef]
117. Zhang, Z.; Ye, M.; Li, Q.; You, Y.; Yu, H.; Ma, Y.; Mei, L.; Sun, X.; Wang, L.; Yue, W.; et al. The Schizophrenia Susceptibility Gene OPCML Regulates Spine Maturation and Cognitive Behaviors through Eph-Cofilin Signaling. *Cell Rep.* **2019**, *29*, 49–61.e7. [CrossRef]
118. Umeda-Yano, S.; Hashimoto, R.; Yamamori, H.; Weickert, C.S.; Yasuda, Y.; Ohi, K.; Fujimoto, M.; Ito, A.; Takeda, M. Expression analysis of the genes identified in GWAS of the postmortem brain tissues from patients with schizophrenia. *Neurosci. Lett.* **2014**, *568*, 12–16. [CrossRef]
119. Pardiñas, A.F.; Holmans, P.; Pocklington, A.J.; Escott-Price, V.; Ripke, S.; Carrera, N.; Legge, S.E.; Bishop, S.; Cameron, D.; Hamshere, M.L.; et al. Common schizophrenia alleles are enriched in mutation-intolerant genes and in regions under strong background selection. *Nat. Genet.* **2018**, *50*, 381–389. [CrossRef]
120. Schol-Gelok, S.; Janssens, A.C.J.W.; Tiemeier, H.; Liu, F.; Lopez-Leon, S.; Zorkoltseva, I.V.; Axenovich, T.I.; Van Swieten, J.C.; Uitterlinden, A.G.; Hofman, A.; et al. A genome-wide screen for depression in two independent dutch populations. *Biol. Psychiatry* **2010**, *68*, 187–196. [CrossRef]
121. Huckins, L.M.; Hatzikotoulas, K.; Southam, L.; Thornton, L.M.; Steinberg, J.; Aguilera-McKay, F.; Treasure, J.; Schmidt, U.; Gunasinghe, C.; Romero, A.; et al. Investigation of common, low-frequency and rare genome-wide variation in anorexia nervosa. *Mol. Psychiatry* **2018**, *23*, 1169–1180. [CrossRef] [PubMed]
122. Minhas, H.M.; Pescosolido, M.F.; Schwede, M.; Piasecka, J.; Gaitanis, J.; Tantravahi, U.; Morrow, E.M. An unbalanced translocation involving loss of 10q26.2 and gain of 11q25 in a pedigree with autism spectrum disorder and cerebellar juvenile pilocytic astrocytoma. *Am. J. Med. Genet. A* **2013**, *161A*, 787–791. [CrossRef]
123. Liu, F.; Arias-Vásquez, A.; Sleegers, K.; Aulchenko, Y.S.; Kayser, M.; Sanchez-Juan, P.; Feng, B.-J.J.; Bertoli-Avella, A.M.; van Swieten, J.; Axenovich, T.I.; et al. A genomewide screen for late-onset Alzheimer disease in a genetically isolated Dutch population. *Am. J. Hum. Genet.* **2007**, *81*, 17–31. [CrossRef]
124. Weller, A.E.; Ferraro, T.N.; Doyle, G.A.; Reiner, B.C.; Crist, R.C.; Berrettini, W.H. Single Nucleus Transcriptome Data from Alzheimer's Disease Mouse Models Yield New Insight into Pathophysiology. *J. Alzheimer's Dis.* **2022**, *90*, 1233–1247. [CrossRef]
125. Zhang, Q.; Ma, C.; Chin, L.-S.; Li, L. Integrative glycoproteomics reveals protein N-glycosylation aberrations and glycoproteomic network alterations in Alzheimer's disease. *Sci. Adv.* **2020**, *6*, eabc5802. [CrossRef]
126. Zacco, A.; Cooper, V.; Chantler, P.; Fisher-Hyland, S.; Horton, H.; Levitt, P. Isolation, biochemical characterization and ultrastructural analysis of the limbic system-associated membrane protein (LAMP), a protein expressed by neurons comprising functional neural circuits. *J. Neurosci.* **1990**, *10*, 73–90. [CrossRef]
127. Pimenta, A.F.; Zhukareva, V.; Barbe, M.F.; Reinoso, B.S.; Grimley, C.; Henzel, W.; Fischer, I.; Levitt, P. The limbic system-associated membrane protein is an Ig superfamily member that mediates selective neuronal growth and axon targeting. *Neuron* **1995**, *15*, 287–297. [CrossRef]
128. Pimenta, A.F.; Reinoso, B.S.; Levitt, P.; Pimenta, A.F.; Levitt, P. Expression of the mRNAs encoding the limbic system-associated membrane protein (LAMP): II. Fetal rat brain. *J. Comp. Neurol.* **1996**, *375*, 274–288. [CrossRef]
129. Zhukareva, V.; Levitt, P. The limbic system-associated membrane protein (LAMP) selectively mediates interactions with specific central neuron populations. *Development* **1995**, *121*, 1161–1172. [CrossRef]
130. Keller, F.; Levitt, P. Developmental and regeneration-associated regulation of the limbic system associated membrane protein in explant cultures of the rat brain. *Neuroscience* **1989**, *28*, 455–474. [CrossRef]

131. Horton, H.; Levitt, P. A unique membrane protein is expressed on early developing limbic system axons and cortical targets. *J. Neurosci.* **1988**, *8*, 4653–4661. [CrossRef] [PubMed]
132. Barbe, M.F.; Levitt, P. The early commitment of fetal neurons to the limbic cortex. *J. Neurosci.* **1991**, *11*, 519–533. [CrossRef]
133. Barbe, M.F.; Levitt, P. Attraction of specific thalamic input by cerebral grafts depends on the molecular identity of the implant. *Proc. Natl. Acad. Sci. USA* **1992**, *89*, 3706–3710. [CrossRef]
134. Ferri, R.T.; Levitt, P. Cerebral Cortical Progenitors Are Fated to Produce Region-specific Neuronal Populations. *Cereb. Cortex* **1993**, *3*, 187–198. [CrossRef]
135. Mann, F.; Zhukareva, V.; Pimenta, A.; Levitt, P.; Bolz, J. Membrane-Associated Molecules Guide Limbic and Nonlimbic Thalamocortical Projections. *J. Neurosci.* **1998**, *18*, 9409–9419. [CrossRef] [PubMed]
136. Lodge, A.P.; Howard, M.R.; McNamee, C.J.; Moss, D.J. Co-localisation, heterophilic interactions and regulated expression of IgLON family proteins in the chick nervous system. *Mol. Brain Res.* **2000**, *82*, 84–94. [CrossRef]
137. Gil, O.D.; Zhang, L.; Chen, S.; Ren, Y.Q.Q.; Pimenta, A.; Zanazzi, G.; Hillman, D.; Levitt, P.; Salzer, J.L. Complementary expression and heterophilic interactions between igLON family members neurotrimin and LAMP. *J. Neurobiol.* **2002**, *51*, 190–204. [CrossRef]
138. McNamee, C.J.; Youssef, S.; Moss, D. IgLONs form heterodimeric complexes on forebrain neurons. *Cell Biochem. Funct.* **2011**, *29*, 114–119. [CrossRef]
139. Hashimoto, T.; Maekawa, S.; Miyata, S. IgLON cell adhesion molecules regulate synaptogenesis in hippocampal neurons. *Cell Biochem. Funct.* **2009**, *27*, 496–498. [CrossRef]
140. Eagleson, K.L.; Pimenta, A.F.; Burns, M.M.; Fairfull, L.D.; Cornuet, P.K.; Zhang, L.; Levitt, P. Distinct domains of the limbic system-associated membrane protein (LAMP) mediate discrete effects on neurite outgrowth. *Mol. Cell. Neurosci.* **2003**, *24*, 725–740. [CrossRef]
141. Sanz, R.L.; Ferraro, G.B.; Girouard, M.-P.P.; Fournier, A.E. Ectodomain shedding of Limbic System-Associated Membrane Protein (LSAMP) by ADAM Metallopeptidases promotes neurite outgrowth in DRG neurons. *Sci. Rep.* **2017**, *7*, 7961. [CrossRef] [PubMed]
142. Nelovkov, A.; Philips, M.A.; Kõks, S.; Vasar, E. Rats with low exploratory activity in the elevated plus-maze have the increased expression of limbic system-associated membrane protein gene in the periaqueductal grey. *Neurosci. Lett.* **2003**, *352*, 179–182. [CrossRef]
143. Nelovkov, A.; Areda, T.; Innos, J.; Kõks, S.; Vasar, E. Rats displaying distinct exploratory activity also have different expression patterns of γ-aminobutyric acid- and cholecystokinin-related genes in brain regions. *Brain Res.* **2006**, *1100*, 21–31. [CrossRef]
144. Catania, E.H.; Pimenta, A.; Levitt, P. Genetic deletion of Lsamp causes exaggerated behavioral activation in novel environments. *Behav. Brain Res.* **2008**, *188*, 380–390. [CrossRef]
145. Innos, J.; Philips, M.-A.A.; Leidmaa, E.; Heinla, I.; Raud, S.; Reemann, P.; Plaas, M.; Nurk, K.; Kurrikoff, K.; Matto, V.; et al. Lower anxiety and a decrease in agonistic behaviour in Lsamp-deficient mice. *Behav. Brain Res.* **2011**, *217*, 21–31. [CrossRef]
146. Bregin, A.; Mazitov, T.; Aug, I.; Philips, M.-A.; Innos, J.; Vasar, E. Increased sensitivity to psychostimulants and GABAergic drugs in Lsamp-deficient mice. *Pharmacol. Biochem. Behav.* **2019**, *183*, 87–97. [CrossRef]
147. Innos, J.; Leidmaa, E.; Philips, M.-A.A.; Sütt, S.; Alttoa, A.; Harro, J.; Kõks, S.; Vasar, E. Lsamp$^{-/-}$ mice display lower sensitivity to amphetamine and have elevated 5-HT turnover. *Biochem. Biophys. Res. Commun.* **2013**, *430*, 413–418. [CrossRef]
148. Innos, J.; Philips, M.-A.; Raud, S.; Lilleväli, K.; Kõks, S.; Vasar, E. Deletion of the Lsamp gene lowers sensitivity to stressful environmental manipulations in mice. *Behav. Brain Res.* **2012**, *228*, 74–81. [CrossRef]
149. Qiu, S.; Champagne, D.L.; Peters, M.; Catania, E.H.; Weeber, E.J.; Levitt, P.; Pimenta, A.F. Loss of limbic system-associated membrane protein leads to reduced hippocampal mineralocorticoid receptor expression, impaired synaptic plasticity, and spatial memory deficit. *Biol. Psychiatry* **2010**, *68*, 197–204. [CrossRef]
150. Must, A.; Tasa, G.; Lang, A.; Vasar, E.; Kõks, S.; Maron, E.; Väli, M. Association of limbic system-associated membrane protein (LSAMP) to male completed suicide. *BMC Med. Genet.* **2008**, *9*, 34. [CrossRef]
151. Miret, M.; Ayuso-Mateos, J.L.; Sanchez-Moreno, J.; Vieta, E. Depressive disorders and suicide: Epidemiology, risk factors, and burden. *Neurosci. Biobehav. Rev.* **2013**, *37*, 2372–2374. [CrossRef] [PubMed]
152. Koido, K.; Traks, T.; Balõtšev, R.; Eller, T.; Must, A.; Koks, S.; Maron, E.; Tõru, I.; Shlik, J.; Vasar, V.; et al. Associations between LSAMP gene polymorphisms and major depressive disorder and panic disorder. *Transl. Psychiatry* **2012**, *2*, e152. [CrossRef] [PubMed]
153. Behan, A.T.; Byrne, C.; Dunn, M.J.; Cagney, G.; Cotter, D.R.; Behan, Á.; Byrne, C.; Dunn, M.J.; Cagney, G.; Cotter, D.R. Proteomic analysis of membrane microdomain-associated proteins in the dorsolateral prefrontal cortex in schizophrenia and bipolar disorder reveals alterations in LAMP, STXBP1 and BASP1 protein expression. *Mol. Psychiatry* **2009**, *14*, 601–613. [CrossRef] [PubMed]
154. Koido, K.; Janno, S.; Traks, T.; Parksepp, M.; Ljubajev, Ü.; Veiksaar, P.; Must, A.; Shlik, J.; Vasar, V.; Vasar, E. Associations between polymorphisms of LSAMP gene and schizophrenia. *Psychiatry Res.* **2014**, *215*, 797–798. [CrossRef]
155. Heywood, W.E.; Galimberti, D.; Bliss, E.; Sirka, E.; Paterson, R.W.; Magdalinou, N.K.; Carecchio, M.; Reid, E.; Heslegrave, A.; Fenoglio, C.; et al. Identification of novel CSF biomarkers for neurodegeneration and their validation by a high-throughput multiplexed targeted proteomic assay. *Mol. Neurodegener.* **2015**, *10*, 64. [CrossRef]
156. Spaltmann, F.; Brümmendorf, T. CEPU-1, a novel immunoglobulin superfamily molecule, is expressed by developing cerebellar Purkinje cells. *J. Neurosci.* **1996**, *16*, 1770–1779. [CrossRef]
157. Gil, O.D.; Zanazzi, G.; Struyk, A.F.; Salzer, J.L. Neurotrimin mediates bifunctional effects on neurite outgrowth via homophilic and heterophilic interactions. *J. Neurosci.* **1998**, *18*, 9312–9325. [CrossRef]

158. McNamee, C.J.; Reed, J.E.; Howard, M.R.; Lodge, A.P.; Moss, D.J. Promotion of neuronal cell adhesion by members of the IgLON family occurs in the absence of either support or modification of neurite outgrowth. *J. Neurochem.* **2002**, *80*, 941–948. [CrossRef]
159. Chen, S.; Gil, O.; Ren, Y.Q.; Zanazzi, G.; Salzer, J.L.; Hillman, D.E. Neurotrimin expression during cerebellar development suggests roles in axon fasciculation and synaptogenesis. *J. Neurocytol.* **2001**, *30*, 927–937. [CrossRef]
160. Singh, K.; Lilleväli, K.; Gilbert, S.F.; Bregin, A.; Narvik, J.; Jayaram, M.; Rahi, M.; Innos, J.; Kaasik, A.; Vasar, E.; et al. The combined impact of IgLON family proteins Lsamp and Neurotrimin on developing neurons and behavioral profiles in mouse. *Brain Res. Bull.* **2018**, *140*, 5–18. [CrossRef]
161. Lodge, A.P.; McNamee, C.J.; Howard, M.R.; Reed, J.E.; Moss, D.J. Identification and characterization of CEPU-Se-A secreted isoform of the IgLON family protein, CEPU-1. *Mol. Cell. Neurosci.* **2001**, *17*, 746–760. [CrossRef]
162. Mazitov, T.; Bregin, A.; Philips, M.A.; Innos, J.; Vasar, E. Deficit in emotional learning in neurotrimin knockout mice. *Behav. Brain Res.* **2017**, *317*, 311–318. [CrossRef]
163. Pan, Y.; Wang, K.S.; Aragam, N. NTM and NR3C2 polymorphisms influencing intelligence: Family-based association studies. *Prog. Neuro-Psychopharmacol. Biol. Psychiatry* **2011**, *35*, 154–160. [CrossRef] [PubMed]
164. Maruani, A.; Huguet, G.; Beggiato, A.; ElMaleh, M.; Toro, R.; Leblond, C.S.; Mathieu, A.; Amsellem, F.; Lemière, N.; Verloes, A.; et al. 11q24.2-25 micro-rearrangements in autism spectrum disorders: Relation to brain structures. *Am. J. Med. Genet. A* **2015**, *167A*, 3019–3030. [CrossRef] [PubMed]
165. Brevik, E.J.; van Donkelaar, M.M.J.; Weber, H.; Sánchez-Mora, C.; Jacob, C.; Rivero, O.; Kittel-Schneider, S.; Garcia-Martínez, I.; Aebi, M.; van Hulzen, K.; et al. Genome-wide analyses of aggressiveness in attention-deficit hyperactivity disorder. *Am. J. Med. Genet. B Neuropsychiatr. Genet.* **2016**, *171*, 733–747. [CrossRef] [PubMed]
166. Herms, J.; Dorostkar, M.M. Dendritic Spine Pathology in Neurodegenerative Diseases. *Annu. Rev. Pathol.* **2016**, *11*, 221–250. [CrossRef]
167. Kayser, M.S.; Dalmau, J. The emerging link between autoimmune disorders and neuropsychiatric disease. *J. Neuropsychiatry Clin. Neurosci.* **2011**, *23*, 90–97. [CrossRef] [PubMed]
168. McKie, A.B.; Vaughan, S.; Zanini, E.; Okon, I.S.; Louis, L.; de Sousa, C.; Greene, M.I.; Wang, Q.; Agarwal, R.; Shaposhnikov, D.; et al. The OPCML tumor suppressor functions as a cell surface repressor-adaptor, negatively regulating receptor tyrosine kinases in epithelial ovarian cancer. *Cancer Discov.* **2012**, *2*, 156–171. [CrossRef] [PubMed]

Disclaimer/Publisher's Note: The statements, opinions and data contained in all publications are solely those of the individual author(s) and contributor(s) and not of MDPI and/or the editor(s). MDPI and/or the editor(s) disclaim responsibility for any injury to people or property resulting from any ideas, methods, instructions or products referred to in the content.

genes

MDPI

Review

INPP5D/SHIP1: Expression, Regulation and Roles in Alzheimer's Disease Pathophysiology

Edward O. Olufunmilayo [1,2] and R. M. Damian Holsinger [1,3,*]

1 Laboratory of Molecular Neuroscience and Dementia, School of Medical Sciences, Faculty of Medicine and Health, The University of Sydney, Camperdown, NSW 2050, Australia; eddietobie@gmail.com
2 Department of Medicine, University College Hospital, Queen Elizabeth Road, Oritamefa, Ibadan 2002012, Nigeria
3 Neuroscience, School of Medical Sciences, Faculty of Medicine and Health, The University of Sydney, Sydney, NSW 2006, Australia
* Correspondence: damian.holsinger@sydney.edu.au

Abstract: Alzheimer's disease (AD) is the most common form of dementia, accounting for approximately 38.5 million cases of all-cause dementia. Over 60% of these individuals live in low- and middle-income countries and are the worst affected, especially by its deleterious effects on the productivity of both patients and caregivers. Numerous risk factors for the disease have been identified and our understanding of gene–environment interactions have shed light on several gene variants that contribute to the most common, sporadic form of AD. Microglial cells, the innate immune cells of the central nervous system (CNS), have long been established as guardians of the brain by providing neuroprotection and maintaining cellular homeostasis. A protein with a myriad of effects on various important signaling pathways that is expressed in microglia is the Src Homology 2 (SH2) domain-containing Inositol 5′ Phosphatase 1 (SHIP1) protein. Encoded by the *INPP5D* (Inositol Polyphosphate-5-Phosphatase D) gene, SHIP1 has diminutive effects on most microglia signaling processes. Polymorphisms of the *INPP5D* gene have been found to be associated with a significantly increased risk of AD. Several studies have elucidated mechanistic processes by which SHIP1 exerts its perturbations on signaling processes in peripheral immune cells. However, current knowledge of the controllers of *INPP5D*/SHIP1 expression and the idiosyncrasies of its influences on signaling processes in microglia and their relevance to AD pathophysiology is limited. In this review, we summarize these discoveries and discuss the potential of leveraging *INPP5D*/SHIP1 as a therapeutic target for Alzheimer's disease.

Keywords: SHIP1; Alzheimer's disease risk gene; single-nucleotide polymorphism; microglia

Citation: Olufunmilayo, E.O.; Holsinger, R.M.D. *INPP5D*/SHIP1: Expression, Regulation and Roles in Alzheimer's Disease Pathophysiology. *Genes* **2023**, *14*, 1845. https://doi.org/10.3390/genes14101845

Academic Editor: Claudia Ricci

Received: 17 August 2023
Revised: 20 September 2023
Accepted: 20 September 2023
Published: 23 September 2023

Copyright: © 2023 by the authors. Licensee MDPI, Basel, Switzerland. This article is an open access article distributed under the terms and conditions of the Creative Commons Attribution (CC BY) license (https://creativecommons.org/licenses/by/4.0/).

1. Introduction

Alzheimer's disease (AD) is a common progressive neurodegenerative disease typically affecting individuals over the age of 65. It accounts for approximately 70% of all-cause dementia. Recent statistics have revealed that approximately 55 million individuals are living with dementia worldwide [1]. Cognitive decline and neuropsychiatric symptoms cause deterioration in quality of life and the need for 24-hour care, culminating in serious personal impact on patients and caregivers. Dementia, therefore, is an enormous global health burden, and the development of effective therapies is critical to effectively managing AD.

Genetic predisposition is an important AD risk determinant, with heritability estimates of 60–80% [2]. Microglia are the predominant immune cells in the brain, and genes exclusively or highly expressed in microglia, such as TREM2, CD33, PLCG2, ABI3 [3] and INPP5D [4], have been found to exert various influences on the development of AD.

Central nervous system (CNS)-resident microglia provide innate immune defense against infection and trauma and maintain tissue homeostasis by performing synaptic pruning and clearing apoptotic cells and cellular debris. The blood–brain barrier prevents the adaptive immune system from playing a significant role in the CNS by excluding peripheral adaptive immune cells from the brain, although new evidence suggests that antigens and antigen-presenting cells originating from the CNS have been found in the peripheral lymphatic system and may stimulate adaptive and humoral immune reactions [5]. Once activated from their homeostatic or quiescent state, microglia assume a ramified morphology and efficiently clear various cellular and acellular material that has been marked for disposal. Such materials are generally produced by a variety of insults such as acute damage caused by trauma, or ongoing damage, such as that characteristic of chronic neurodegenerative diseases, including AD. In AD, these damage-related materials and signals arise from plaques, fibrils and insoluble aggregates of amyloid-β (Aβ), damaged myelin, extracellular matrix, and stressed as well as degenerating neurons [6].

As AD develops, microglia lose their innate function and form clusters surrounding amyloid plaques and other damage signals [7]. Either through local proliferation or migration, microglia alter their morphology to become more amoeboid as they transform into a reactive state and prime their protein-synthesizing machinery to express various proteins needed to mount an immune response and, importantly, promote phagocytosis [7].

In 1994, Saxton and colleagues [8] reported that a 145 kDa protein became phosphorylated and associated with the adaptor protein Shc1 following B-cell or cytokine receptor stimulation. Shortly thereafter, the 145 kDa protein was identified by a number of research groups as Src Homology 2 (SH2) domain-containing Inositol 5′ Phosphatase (SHIP) [9–11], a protein encoded by the gene *INPP5D* (a member of the inositol polyphosphate-5-phosphatase) at the locus 2q37.1. Polymorphisms of *INPP5D* have been associated with higher risk of late-onset AD (LOAD) [4]. The single-nucleotide polymorphisms (SNPs) rs35349669 and rs10933431 in the *INPP5D* gene were significantly associated with increased AD risk [12]. *INPP5D* is associated with another major AD risk gene—*TREM2* (Triggering Receptor Expressed on Myeloid cells 2). It inhibits TREM2 signaling through its interactions with DAP12 (DNAX-activating protein of 12 kDa), an adaptor protein that is important for TREM2 signaling and function [13]. Thus, SHIP1 potentially plays important roles in the pathophysiology associated with Alzheimer's disease, and therefore, numerous small-molecule inhibitors of SHIP1 protein are currently being studied as possible immune-based therapies for AD. In this review, we aim to discuss the current knowledge of processes driving *INPP5D* transcription, the actions of its gene product SHIP1 in microglia and their relevance to AD, while examining its potential as a novel therapeutic target for Alzheimer's disease.

2. Expression and Regulation of *INPP5D* in AD

INPP5D is a gene that is advancing to the forefront of genetics research in AD as a result of one of its SNPs, rs35349669, being associated as a common variant of the disease, accounting for 3.8% of all genetic risk for AD [14].

INPP5D, expressed in the brain, mostly in microglia, typically includes a 27-exon gene that encodes several isoforms. The full-length 27-exon isoform encodes an amino-terminal SH2 domain followed by a phosphatase domain, while the transcription of truncated isoforms lacking the SH2 domain originates from internal transcription start sites (TSS) [12].

The minor intron 10 allele of rs35349669 is associated with increased risk of AD, while the minor allele of rs10933431 located in intron 2 is associated with decreased AD risk [4,15]. According to Zajac et al. [12], isoforms of *INPP5D* associated with TSSs located in exon 1 and intron 14 are markedly increased in individuals with high AD neuropathology. In addition, expression of a novel variant (referred to as the D47 variant) that lacks a 47 bp segment from exon 12 was also increased in AD brains, resulting in ~13% of total *INPP5D* expression, and was found to undergo nonsense-mediated decay [12]. The missing 47 base pairs results in a frameshift and a consequent introduction of a premature termination codon (PTC), with the resulting protein being devoid of a phosphatase domain. The rs35349669 SNP correlates with an allele-specific expression of full-length *INPP5D* [12].

The identified TSS of *INPP5D* includes TSS-A in the SH2 domain, and TSS-B and TSS-C in the phosphatase domain, all of which result in the production of different SHIP1 isoforms. TSS-A is located in exon 1 and generates isoforms 201, 204 and 205, whilst TSS-B originates in exon 11, producing isoform 202, and TSS-C is located in intron 14 and generates isoform 213. They also showed that isoforms originating from TSS-A and TSS-C are upregulated in the AD brain compared to the isoform produced by TSS-B. They concluded that individuals with AD may have an altered transcription factor profile or open chromatin sites that influence access to the TSS of the *INPP5D* gene, which in turn promotes transcription of various SHIP1 isoforms, leading to the propagation of AD pathology [12]. The Ikaros family of genes encodes for five zinc-finger proteins—Ikaros (*IKZF1*), Helios (IKZF2), Aiolos (*IKZF3*), Eos (*IKZF4*) and Pegasus (*IKZF5*)—that act as transcription factors, exerting critical regulatory roles during the development and function of various immune cells [16,17]. Ikaros exerts a crucial regulatory role in B-cell receptor (BCR) signaling, as studies have found that disruption of Ikaros in the chicken B cell line DT40 results in a diminution of BCR signaling, with reduced PLCγ2 phosphorylation and impaired intracellular calcium mobilization and signaling [18,19]. Other studies have revealed that Helios exerts a directly opposite effect as its knockout resulted in elevated BCR signaling [20]. Both Ikaros and Helios bind similar DNA sequences and chromatin remodeling complexes in the upstream regulatory region of the *INPP5D* gene to exert their effects as transcription factors [21,22].

Studies by Tsai and colleagues [23] revealed significantly elevated *INPP5D* expression in human AD microglial cells, and demonstrated a direct correlation between increased *INPP5D* expression and amyloid deposition in specific brain regions, including the inferior frontal, parahippocampal and superior temporal gyri, as well as in the frontal pole, although they did not delve into mechanisms driving specific *INPP5D* gene expression. There have been attempts to explain the increased expression of *INPP5D* in AD, where it has been proposed to be related to neuroinflammation and amyloid-plaque-associated increase in the expression of the transcription factor PU.1, which has been strongly associated in AD genetics. PU1 has been shown to bind to and upregulate *INPP5D* expression as it does other AD-related genes including TYROBP, MS4As, TREM2, and CD33 [24]. However, more studies are required to characterize the mechanisms that underpin the increase in *INPP5D* expression in AD, how this might be mitigated, and if such manipulations may be of any tenable therapeutic benefits.

3. SHIP1: Structure

Saxton and colleagues [8] were studying antigen-induced B-cell signaling and Ras activation when they discovered that AgR-induced signaling stimulated phosphorylation of the adapter protein SHC and mSOS1. This in turn resulted in the formation of membrane-associated complexes composed of SHC, GRB-2, mSOS1, and a previously uncharacterized 145 kDa protein [8]. Chacko et al. [9] subsequently identified the protein to be Src homology 2 (SH2) domain-containing inositol 5′ phosphatase 1 (SHIP1). Their work revealed that phosphorylation of SHIP1 and its association with Shc required co-clustering with the Fc receptor for IgG (FcγRIIB) and not stimulation of the B-cell receptor (BCR) alone. Since it was previously discovered that co-clustering of the BCR with FcγRII reduced proliferation induced by the stimulation of the receptor, they concluded that tyrosine phosphorylation of SHIP1 and its association with Shc resulted in negative signaling through effects on phosphatidylinositol metabolism [9]. SHIP1 exerts its effects on FcγRIIB by acting at its immunoreceptor tyrosine-based inhibitory motif (ITIM) [25].

SHIP1 is a 1189-amino-acid protein with several splice variants that have multiple distinct motifs and domains that contribute to its numerous cellular functions [26]. The functionally important domains include the phosphatase domain, SRC-Homology 2 (SHC) domain, Pleckstrin Homology-related (PH-R) domain and the C-terminus [25] (Figure 1).

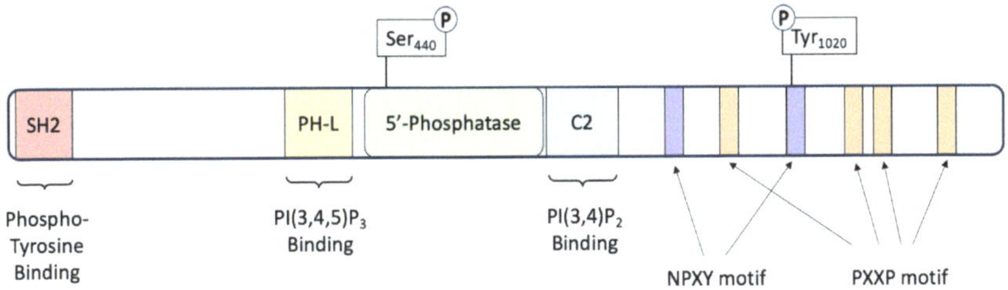

Figure 1. Schematic of the SHIP1 protein. SHIP1 is a 145 kDa protein consisting of 1189 amino acids. The protein consists of several domains that impart various functionalities. A central catalytic domain (5′-Phosphatase) is flanked by a Pleckstrin homology-like (PH-L) and a C2 domain for binding of the substrate PI(3,4,5)P3 and product PI(3,4)P2 of the enzyme. The carboxy-terminus of the protein contains four Proline-rich (PXXP) motifs, important for SRC Homology 3 (SH3) binding. The C-terminus also harbors two Asn-Pro-X-Tyr (NPXY) motifs. The protein also contains two phosphorylation sites: one within the phosphatase catalytic domain at Serine 440, and the second on a tyrosine residue at amino acid 1020 located within the second NPXY motif.

The phosphatase domain of SHIP1 shares a 64% similarity to that of SHIP2, spans amino acids 401–866 and is responsible for its catalytic activity as a lipid phosphatase [27]. There are two allosteric activation sites located on either side of the phosphatase domain; a phosphorylation site at serine 440 and a C2 domain spanning amino acids 725–863 [25]. These sites are structurally and functionally important, as works by Zhang et al. [28] showed that the phosphatase activity of SHIP1 can be enhanced by protein kinase A-mediated phosphorylation at Ser440. In addition, the activity of this same domain may be increased by the direct binding of PI(3,4)P2 (phosphatidylinositol 3,4-bisphosphate) to the C2 domain [29].

The SH2 domain of SHIP1 shares 54% homology with that of SHIP2 and spans amino acids 5–101. Interaction of the SH2 domain with the intracellular components of various cell surface receptors is a critical mechanism that governs the localization of SHIP1, and therefore controls access to its substrate, PIP3 [27]. Known targets to which the SH2 domain binds include the ITIM (Immunoreceptor Tyrosine-based Inhibitory Motif) receptors FcγRIIB, gp49B1 and CD31, as well as the ITAM (Immunoreceptor Tyrosine-based Activation Motif) receptors FcγRIIA, FcγRI-associated zeta-chain, TCR zeta chain, FcεRI and the BCR [30]. Phosphorylated cytoplasmic proteins including Shc1 and Dok-3 can also interact with the SH2 domain of SHIP1 and compete with the aforementioned receptors for SHIP1 binding, or act as co-ligands in order to relay signals downstream [31,32]. The SH2 domain also interacts with PIP3 (phosphatidylinositol (3,4,5)-trisphosphate) and PI(4,5)P_2 (phosphatidylinositol 4,5-bisphosphate) on the cell membrane, ensuring SHIP1 stability and presumably contributing to SH2-mediated recruitment of SHIP1 [33].

The PH-R domain of SHIP1 folds independently and spans amino acids 292–401, which also interacts with phosphoinositides, preferencing PIP_3 [25]. Ming-Lum and colleagues [34] found that this domain was important for localization of SHIP1 to the phagocytic cup in macrophages.

The C-Terminus of SHIP1 varies from that of SHIP2 in that it contains several protein–protein interaction domains that are predicted to contribute to the regulation of adaptor and catalytic activities associated with this protein [25]. Two tyrosine residues within this region (Y915 and Y1022) mediate interaction with Shc1 [35], Dok-1 [36] and Dok-3 [30]. Mutation of these residues was shown to significantly reduce SHIP signaling in murine cells [35], thus underpinning their importance to the functions of the protein.

4. Actions of SHIP1 in Immune Cells, Including Microglia

SHIP1 is expressed in mesenchymal cells and plays a key role in the regulation of the immune system [37], while SHIP2 is ubiquitously expressed, and is mostly known for its effects on the regulation of blood glucose and metabolism [38]. Kerr [37] noted that these disparities may be due to differences in tissue distribution that translates to association with different receptors, unique actions on other signaling molecules and kinetic differences in binding to their PIP substrates. The effects of SHIP1 on signaling in immune cells are mediated by either phosphatase-dependent or phosphatase-independent mechanisms. SHIP1, via its enzymatic phosphatase domain, acts to dephosphorylate the inositol ring of PI(3,4,5)P_3 at position 5 to generate PI(3,4)P_2, thus preventing the recruitment of PI(3,4,5)P_3-binding effector proteins such as Btk (Bruton's tyrosine kinase) and Vav (Figure 2) [25]. Binding to PIP_3 is crucial to the function of Btk [39]. Btk is a crucial regulator of PLCγ2 phosphorylation and controls Ca^{2+} fluxes in response to immune cell receptor activation [40]. SHIP1 inhibits intracellular Ca^{2+} responses by hindering the access of PIP_3 to Btk, thus functionally impairing Btk. This negative effect on Ca^{2+} flux can also impair the activation of Ca^{2+}-dependent downstream signaling effectors such as MAPK and NFκB, which have been shown by to be important downstream effectors in microglial TREM2 receptor activity [41]. Vav indirectly coordinates F-actin polymerization in immune cells by activating Rho-family GTPases and WASP, and its inhibition may potentially impair this function [42]. Loss of synaptic F-actin has been shown to be directly correlated with memory loss in AD [43]. This is a direct mechanism by which SHIP1 activity may contribute to progression in AD. However, SHIP1 also enhances the expression of adhesive integrin proteins in macrophages and microglia in a phosphatase-dependent manner, and this may prove to be an important step in the phagocytic function of these cells [44]. Contrastingly, the phosphatase activity of SHIP1 produces PI(3,4)P_2, which enhances the recruitment of distinct PI(3,4)P_2-binding effector proteins, such as TAPP1/2 and lamellipodin. Landego and co-workers [45] showed that binding of TAPP (tandem pleckstrin homology domain containing protein) adaptor proteins to PI(3,4)P_2 exerts an inhibitory effect on the serine/threonine kinase, Akt. SHIP1 has previously been reported to inhibit the kinase activity of Akt, possibly via the mechanism described above [46]. Studies by Yao et al. [47] have shown that Akt plays an important role in downstream mechanisms initiated by TREM2 receptor activation in microglia, whereby SHIP1-mediated inhibition of Akt activity, albeit indirectly, may act to reduce TREM2-induced signaling in microglia. These activities may have potentially important implications for the pathogenetic processes involved in various neurodegenerative conditions.

SHIP1 also has effects on immune cells that are initiated by phosphatase-independent mechanisms, which serve to mediate or inhibit protein–protein interactions. One mechanism involves interaction of the C-terminal domain of SHIP1 with the PTB (protein tyrosine-binding) domain of Shc [48]. This interaction interferes with the binding of Shc to the Grb2-SOS complex, an association that generally results in the activation of Ras and the MAP Kinase pathway [49]. Binding of SHIP1 to Shc thus prevents activation of the MAPK pathway and ultimately signaling in immune cells. Another phosphatase-independent mechanism controlling immune cell signaling involves binding of the SHIP1 C-terminal domain to Dok-1. This causes localization of a Dok-associated RasGAP to the membrane, resulting in inhibition of Ras-Erk signaling [36]. The inhibition of this particular signal transduction system has been shown to reduce receptor-induced TNF–α generation in macrophages and microglia [50]. In monocytes and similar cells, SHIP1 inhibits NOD-2 complex-induced NFκB activation by blocking the binding of XIAP and RIP2 to NOD-2, preventing formation of this important complex [51], which can also impair signaling in microglia.

DAP12 is an ITAM, which associates with the transmembrane domain of the TREM2 receptor on microglia and is critical to the function of this receptor and activation of downstream signaling [52]. Acting via its SH2 domain, SHIP1 binds to DAP12 and inhibits signaling intermediates downstream of the TREM2-DAP12 complex, thereby reducing cellular activation, survival, actin reorganization, Ca^{2+} flux and cellular differentiation [13].

All the mechanistic effects of SHIP1 in immune cells, including microglia that have been described above, point to a uniform conclusion that SHIP1 activity tends to inhibit various receptor-induced signaling and resultant downstream mechanisms in these cells. The implications of this phenomenon for infectious, neoplastic and neurodegenerative conditions such as Alzheimer's disease still remain to be studied and could potentially lead to the development of novel, effective therapies for these diseases.

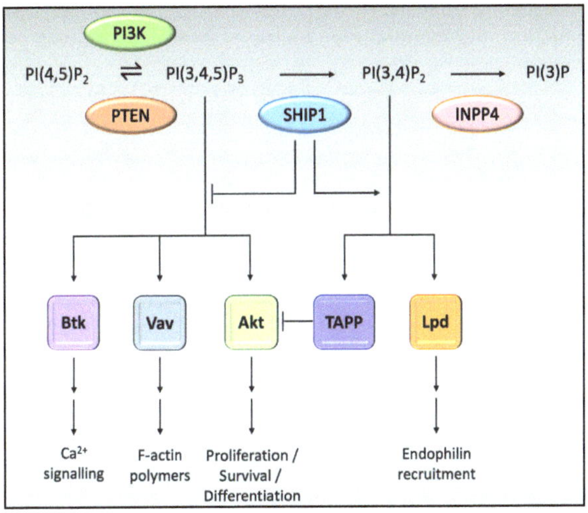

Figure 2. Phosphatase-dependent effects of SHIP1 on cell activity. PI(3,4,5)P3 is an activator of downstream signaling pathways, the most notable being the protein kinase Akt, which activates downstream signaling pathways required for cell growth, survival and proliferation. Other notable pathways include those involving Bruton's tyrosine kinase (Btk), important in calcium signaling and the Vav family of proteins, which coordinate F-actin polymerization, supporting the cell cytoskeleton and structure. SHIP1 plays a pivotal role in the dephosphorylation of PI(3,4,5)P3, which in turn has an inhibitory effect on these pathways. The product of PI(3,4,5)P3 dephosphorylation, PI(3,4)P2, binds the tandem PH domain-containing protein (TAPP) adaptor proteins, which in turn inhibits Akt, further reducing the regulatory effects of this pivotal protein kinase. PI(3,4)P2 also enhances the recruitment of Lamellipodin (Lpd), an important regulator of cyto-skeletal assembly and cell migration.

5. Roles of SHIP1 in Alzheimer's Disease Pathophysiology

The characteristic pathological changes observed in the AD-affected brain at autopsy include amyloid plaques, neurofibrillary tangles (NFTs), neuronal loss, neuronal dystrophy, amyloid angiopathy and neuroinflammation. Although the series of events leading to the final path of neurodegeneration remain unclear, the accumulation of amyloid in the form of plaques is widely accepted as a central pathological event in AD. Virtually all mutations associated with AD increase the generation of Aβ peptide. This protein is derived from the sequential cleavage of the membrane-associated amyloid precursor protein (APP) via the actions of BACE1 (β-site Amyloid Precursor Protein-cleaving Enzyme 1) and the gamma secretase complex of proteins. Aggregated species of Aβ are neurotoxic and form the core of neuritic plaques. Neuronal loss leads to neurochemical deficits that in turn lead to AD. The loss of cholinergic neurons in important regions such as the basal nucleus of Meynert, combined with the reduction in acetylcholine neurotransmitter levels, has been a well-established contributor to the cognitive deficits observed in AD. These pathological events disrupt cell-to-cell communication, result in the abnormal synthesis and accumulation of cytoskeletal proteins, contribute to the loss of synapses and exaggerated pruning of dendrites, cause neuronal damage through oxidative metabolism, and eventually lead to cell death [53]. SHIP1 has been shown to contribute via different pathways to many of these pathological events, potentially resulting in increased risk and severity of the disease.

The mutation that characterizes the Rs35349669 variant of *INPP5D* is located near an internal TSS that is critical for the transcription of the SH2 domain of SHIP1 [54]. This polymorphism results in the production of a SHIP1 transcript that lacks the SH2 domain. In *INPP5D*-haplodeficient mice models, Iguchi et al. [55] and Lin et al. [56] showed that downregulation of *INPP5D* resulted in an increased association of microglia with amyloid plaques. Additionally, they showed that *INPP5D* downregulation restored plaque compaction, and increased microglial motility towards and phagocytosis of amyloid plaques, with a resultant alleviation of cognitive deficits compared to mice expressing normal levels of *INPP5D*. This finding suggests that SHIP1 plays roles and exerts actions that generally tend to promote AD pathology. Similarly, studies by Castranio and colleagues [57] also demonstrated that downregulation of *INPP5D* increased the association of microglia with Aβ plaques, although these researchers reported that these events increased plaque burden in the animals. Further studies will be needed to address these discrepancies.

Knowledge of some of the actions of SHIP1 may be crucial to further understanding its influences in AD pathophysiology. SHIP1 has been shown to transduce inhibitory signaling of some ITIM-containing proteins and to inhibit signaling of ITAM-containing proteins such as DAP12 (Figure 3). SHIP1 thus tends to limit microglia-mediated inflammatory processes and phagocytic activities in AD [58]. SHIP1 has also been shown to display inhibitory effects on monocyte and microglia activation, partly by its action of transducing inhibitory signals from FcγRIIB and other ITIM-containing proteins [59]. Consequently, the response of microglia to detect and clear AD-associated damage-related substances such as amyloid plaques, insoluble Aβ aggregates, NFTs, extracellular matrix proteins and degenerating myelin and neurons may be grossly impaired. Also, as discussed earlier, the inhibitory effects of SHIP1 on the actions of the Vav protein in macrophages and microglial cells disrupt the processes involved in F-actin polymerization. This may result in impaired motility, microenvironment surveillance and phagocytic activity of these cells and can potentially contribute to or accelerate the molecular pathogenesis associated with AD. The depletion of PIP3 levels in these cells brought about by the actions of SHIP1 can also result in a blockage of the processes involved in microglial phagocytosis in response to the earlier mentioned AD-associated damage-related substances [60].

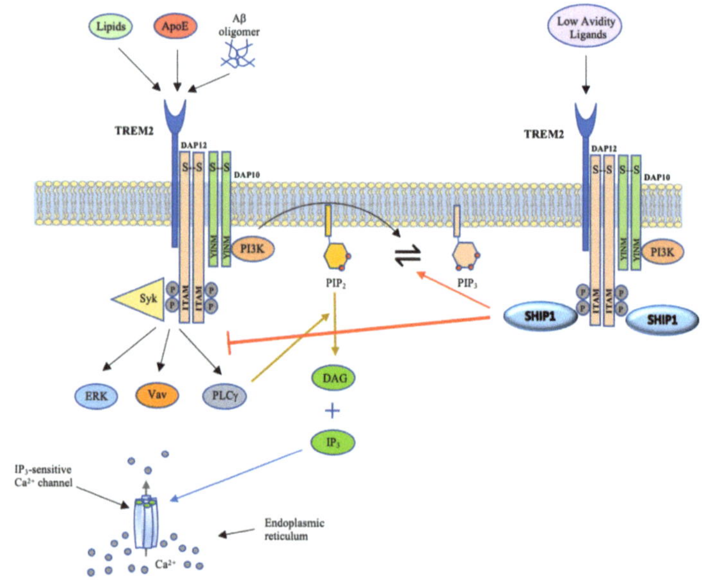

Figure 3. Schematic illustrating TREM2 signaling. The binding of ligands to TREM2 results in the

phosphorylation of tyrosine residues within ITAM of DAP12 and the YINM motif of DAP10 by Src family kinases. This in turn forms docking sites for Syk and PI3K. PI3K converts PIP_2 to PIP_3, leading to cascades of signaling events. As outlined in Figure 2, these signals increase calcium flux and activation and nuclear localization of numerous transcription factors that promote cell proliferation, survival, phagocytosis, cytokine production and cytoskeletal rearrangement. The binding of SHIP1 to phosphorylated DAP12 facilitates the dephosphorylation of PIP_3 and blocks the cascade of downstream signaling events. Image adapted from [6].

An interesting phenomenon mediated by SHIP1 that may be related to phagocytic processes is fast endophilin-mediated endocytosis (FEME) [61]. This pathway is independent of the widely known endocytotic pathway that utilizes clathrin-coated pits. Boucrot and colleagues [61] showed that in FEME, the PIP2 generated by the actions of SHIP1 on PIP3 recruits lamellipodin, which in turn recruits the protein endophilin. This process mediates ligand-triggered uptake of different receptors, cell-surface enzymes and even pathogens into the involved immune cell. The implications of this endocytotic pathway in AD pathophysiology require further investigation.

An additional function associated with SHIP1 relates to its role in regulating NF-κB (nuclear factor kappa-light-chain-enhancer of activated B cells), a protein complex that controls DNA transcription, the production of cytokines and cell survival. SHIP1 is a negative regulator of NF-κB. Exposure of postnatal day 2 astrocytes from SHIP1 knockout mice to TNF-α for 48 hours resulted in increased expression of NF-κB and a loss of the cytoplasmic inhibitor IκB alpha [62]. Similar results were observed in B and T cells from these knockout mice [63]. The ability of SHIP1 to regulate levels of NF-κB points to a critical role in the AD brain. Activation of NF-κB is known to increase expression of the AD-related β-secretase/BACE1 [64]. Whilst to our knowledge, levels of SHIP1 have not been measured in human AD brain, NF-κB levels, as measured by levels of its p65 subunit, are known to be significantly increased, particularly in the cortex, hippocampus and entorhinal cortex [65–67], frontal cortex [68] and basal forebrain [69]. Extending from these results, it could be inferred that SHIP1 levels are most probably decreased in the AD brain, indicating that, although numerous, microglial expression of SHIP1 is downregulated in AD. The reduction in SHIP1 would drive NF-κB activation, which in turn would lead to increased BACE1 in the AD brain. This results in the generation of more Aβ, which triggers further inflammation, thus setting up a vicious, positive feedback cycle as more NF-κB is produced due to inflammation.

Overall, the actions of SHIP1 in microglial cells in AD results in reduced effectiveness of the important functional processes of microglia (summarized in Table 1). These include, but are not limited to, microenvironment surveillance, detection of damage-related substances, ability to mount appropriate responses to these substances, maintenance of cytoskeletal physiology and framework and appropriate phagocytic processes, potentially leading to the accumulation of these toxic products and worsening of disease phenotype. These suggest that targeting SHIP1 with a therapeutic agent may be a viable option of novel treatment for Alzheimer's disease.

Table 1. Summary of the involvement of SHIP1 in AD pathophysiology.

INPP5D/SHIP1 Deficiency	Effect	Reference
INPP5D haploinsufficiency (Tyrobp-deficient TREM2 loss-of-function mouse)	Restored microglial association with plaques, partially restored plaque compaction and reduced phosphorylated tau(+) dystrophic neurites	[55]
INPP5D haploinsufficiency	Increased dense-core plaques, microglial association with plaques and uptake of Aβ	[56]
INPP5D downregulation	Increased microglial association with plaques, but increased plaque burden	[57]
SHIP1—transducing inhibitory signals from FcγRIIB and ITIM-containing proteins	Inhibitory effects on monocyte and microglia activation	[59]
SHIP1—NF-κB pathway	SHIP1 downregulation leads to increased NF-κB activation, and an increase in BACE1 expression in AD brain	[62–69]

6. SHIP1 as a Therapeutic Target for Alzheimer's Disease

SHIP1 inhibition in microglia, potentially via the upregulation of TREM2 signaling as well as other mechanisms that have been discussed, may lead to a reduced risk of AD. SHIP2 expression has also been shown to be highly correlated with amyloid burden and cognitive decline in AD patients [70].

Obst and colleagues [71] performed crystallographic fragment screens of 597 compounds, and noted that the SHIP1-specific inhibitor 3AC increases downstream signaling following TREM2 stimulation, as evidenced by increases in SYK phosphorylation and intracellular Ca^{2+} flux. This suggests that SHIP1 inhibitors may assist microglia in harnessing the beneficial effects triggered by TREM2 activation in microglia and may be useful as therapeutic options for AD.

The first attempt at investigating whether inhibiting the activity of a SHIP protein could be used as a therapeutic modality was performed using AS1949490, a SHIP2 inhibitor. Administration of the compound to db/db mice, a model for type 2 diabetes, significantly lowered plasma glucose and improved glucose intolerance that is characteristic of these mice [72]. Administration of AS1949490 SHIP2 transgenic mice reversed memory deficits and improved synaptic plasticity and memory formation in db/db mice [73].

Pedicone and colleagues [74] investigated the effects of a number of selective SHIP1 inhibitors, selective SHIP2 inhibitors and pan-SHIP1/2 inhibitors on microglia and in mouse models of AD. Importantly, the agents tested were shown to easily penetrate the blood–brain barrier when administered orally or via the peritoneum, an important consideration when considering drug delivery to the CNS. In particular, a pan-SHIP1/2 inhibitor, K161, was found to significantly increase microglial proliferation, facilitate microglial phagocytosis and apoptotic neurons and $A\beta_{42}$ [74]. These are major functions of microglia that are crucial for maintaining homeostasis in the CNS but are perturbed in AD. These effects were also examined using SHIP1-specific and SHIP2-specific agents, but their effects were noted to be much less significant compared to those observed using the pan-SHIP1/2 inhibitors [74]. These results suggest that a dual inhibition of the two SHIP paralogs is necessary for maximal therapeutic benefit. This may be explained by the possibility of higher $PI(3,4,5)P_3$ quantities produced by dual SHIP1/2 blockade at the plasma membrane of microglia, which in turn enhances the efficiency of phagocytosis when microglia encounter a target to be phagocytosed, as mediated by TREM2 or Dectin1 [74]. However, since studies such as those by Sala Figerio et al. [75] have emphasized that SHIP1 diminution can also be very important in AD progression, it may be reasonably inferred that SHIP1 agonism should also be considered as a possible therapeutic option for AD. It has been suggested that the stage of disease could be a determinant of whether SHIP1 antagonism or agonism should be deployed as a therapeutic strategy, with SHIP1 antagonism favored during the early stages of disease in order to enhance the homeostatic functions of microglia and agonists potentially employed late in the disease possibly to dampen the effects of dysfunctional microglia in disease progression. More recent works by Pedicone and colleagues [76] showed that K306, a SHIP1-selective agonist, suppresses induction of inflammatory cytokines and iNOS expression in microglia, reduces TNF-α production, and enhances the degradation of synaptosomes and apoptotic neurons by microglia, revealing novel SHIP1-regulated functions that may be of potential therapeutic benefits in various dementias. These studies demonstrate that the homeostatic function of microglia could be pharmacologically modulated employing small-molecule SHIP inhibitors and more recently, agonists, which could lead to potential therapies for AD.

The simultaneous inhibition of SHIP1 and -2 using pan-SHIP1/2 inhibitors may offer a unique opportunity to enhance the basal homeostatic function of microglia that could be leveraged for therapeutic purposes in AD and potentially other neurodegenerative conditions that are also characterized by similar perturbations in microglia function. Hence, pan-SHIP1/2 inhibitors represent a novel form of immunotherapy for Alzheimer's disease. However, further research is required to investigate some of these pharmacological agents, including newer lines of inquiry into SHIP1 agonists and potentially chaperone them through the drug development pipeline.

7. Conclusions and Prospects

SHIP1 is a large functional protein that is encoded by the *INPP5D* gene and has been shown to have wide-ranging mechanistic effects on cellular signaling in immune cells, particularly microglia, the immune cells of the CNS. Microglial cells are known to play critical roles in maintaining cellular homeostasis within the CNS and are intimately involved during various phases of Alzheimer's disease. *INPP5D* expression and SHIP1 activity have been shown to be important risk factors in the development of late-onset AD, as the various mechanistic actions of SHIP1 in microglia result in dampening of signaling, and consequently of some of the beneficial functions of these cells. Studies on the effects of transcriptional regulation of *INPP5D*, SHIP1 blockade and more recently SHIP1/2 agonists are currently underway, with most studies showing promise, as they point to the fact that modulating the expression of SHIP1 and its paralog, SHIP2, may significantly improve disease phenotype and eventually provide novel therapeutic options for Alzheimer's disease.

Author Contributions: Conceptualization, E.O.O. and R.M.D.H.; data curation, E.O.O.; writing—original draft preparation, E.O.O.; writing—content review and editing, R.M.D.H.; supervision, R.M.D.H. All authors have read and agreed to the published version of the manuscript.

Funding: This research received no external funding.

Institutional Review Board Statement: Not applicable.

Informed Consent Statement: Not applicable.

Data Availability Statement: All data are presented in the article.

Conflicts of Interest: The authors declare no conflict of interest.

References

1. WHO. Dementia-Social and Economic Impact. Available online: https://www.who.int/news-room/fact-sheets/detail/dementia (accessed on 18 September 2023).
2. Tanzi, R.E. The genetics of Alzheimer disease. *Cold Spring Harb. Perspect. Med.* **2012**, *2*, a006296. [CrossRef] [PubMed]
3. Wightman, D.P.; Jansen, I.E.; Savage, J.E.; Shadrin, A.A.; Bahrami, S.; Holland, D.; Rongve, A.; Borte, S.; Winsvold, B.S.; Drange, O.K.; et al. A genome-wide association study with 1,126,563 individuals identifies new risk loci for Alzheimer's disease. *Nat. Genet.* **2021**, *53*, 1276–1282. [CrossRef] [PubMed]
4. Lambert, J.C.; Ibrahim-Verbaas, C.A.; Harold, D.; Naj, A.C.; Sims, R.; Bellenguez, C.; DeStafano, A.L.; Bis, J.C.; Beecham, G.W.; Grenier-Boley, B.; et al. Meta-analysis of 74,046 individuals identifies 11 new susceptibility loci for Alzheimer's disease. *Nat. Genet.* **2013**, *45*, 1452–1458. [CrossRef]
5. Engelhardt, B.; Vajkoczy, P.; Weller, R.O. The movers and shapers in immune privilege of the CNS. *Nat. Immunol.* **2017**, *18*, 123–131. [CrossRef] [PubMed]
6. Olufunmilayo, E.O.; Holsinger, R.M.D. Variant TREM2 Signaling in Alzheimer's Disease. *J. Mol. Biol.* **2022**, *434*, 167470. [CrossRef] [PubMed]
7. Keren-Shaul, H.; Spinrad, A.; Weiner, A.; Matcovitch-Natan, O.; Dvir-Szternfeld, R.; Ulland, T.K.; David, E.; Baruch, K.; Lara-Astaiso, D.; Toth, B.; et al. A Unique Microglia Type Associated with Restricting Development of Alzheimer's Disease. *Cell* **2017**, *169*, 1276–1290.e17. [CrossRef]
8. Saxton, T.M.; van Oostveen, I.; Bowtell, D.; Aebersold, R.; Gold, M.R. B cell antigen receptor cross-linking induces phosphorylation of the p21ras oncoprotein activators SHC and mSOS1 as well as assembly of complexes containing SHC, GRB-2, mSOS1, and a 145-kDa tyrosine-phosphorylated protein. *J. Immunol.* **1994**, *153*, 623–636. [CrossRef]
9. Chacko, G.W.; Tridandapani, S.; Damen, J.E.; Liu, L.; Krystal, G.; Coggeshall, K.M. Negative signaling in B lymphocytes induces tyrosine phosphorylation of the 145-kDa inositol polyphosphate 5-phosphatase, SHIP. *J. Immunol.* **1996**, *157*, 2234–2238. [CrossRef]
10. Damen, J.E.; Liu, L.; Rosten, P.; Humphries, R.K.; Jefferson, A.B.; Majerus, P.W.; Krystal, G. The 145-kDa protein induced to associate with Shc by multiple cytokines is an inositol tetraphosphate and phosphatidylinositol 3,4,5-triphosphate 5-phosphatase. *Proc. Natl. Acad. Sci. USA* **1996**, *93*, 1689–1693. [CrossRef]
11. Ware, M.D.; Rosten, P.; Damen, J.E.; Liu, L.; Humphries, R.K.; Krystal, G. Cloning and characterization of human SHIP, the 145-kD inositol 5-phosphatase that associates with SHC after cytokine stimulation. *Blood* **1996**, *88*, 2833–2840. [CrossRef]

12. Zajac, D.J.; Simpson, J.; Zhang, E.; Parikh, I.; Estus, S. Expression of INPP5D Isoforms in Human Brain: Impact of Alzheimer's Disease Neuropathology and Genetics. *Genes* **2023**, *14*, 763. [CrossRef] [PubMed]
13. Peng, Q.; Malhotra, S.; Torchia, J.A.; Kerr, W.G.; Coggeshall, K.M.; Humphrey, M.B. TREM2- and DAP12-dependent activation of PI3K requires DAP10 and is inhibited by SHIP1. *Sci. Signal* **2010**, *3*, ra38. [CrossRef] [PubMed]
14. Van Cauwenberghe, C.; Van Broeckhoven, C.; Sleegers, K. The genetic landscape of Alzheimer disease: Clinical implications and perspectives. *Genet. Med.* **2016**, *18*, 421–430. [CrossRef] [PubMed]
15. Jansen, I.E.; Savage, J.E.; Watanabe, K.; Bryois, J.; Williams, D.M.; Steinberg, S.; Sealock, J.; Karlsson, I.K.; Hagg, S.; Athanasiu, L.; et al. Genome-wide meta-analysis identifies new loci and functional pathways influencing Alzheimer's disease risk. *Nat. Genet.* **2019**, *51*, 404–413. [CrossRef] [PubMed]
16. Thompson, E.C.; Cobb, B.S.; Sabbattini, P.; Meixlsperger, S.; Parelho, V.; Liberg, D.; Taylor, B.; Dillon, N.; Georgopoulos, K.; Jumaa, H.; et al. Ikaros DNA-binding proteins as integral components of B cell developmental-stage-specific regulatory circuits. *Immunity* **2007**, *26*, 335–344. [CrossRef]
17. Read, K.A.; Jones, D.M.; Freud, A.G.; Oestreich, K.J. Established and emergent roles for Ikaros transcription factors in lymphoid cell development and function. *Immunol. Rev.* **2021**, *300*, 82–99. [CrossRef]
18. Nera, K.P.; Alinikula, J.; Terho, P.; Narvi, E.; Tornquist, K.; Kurosaki, T.; Buerstedde, J.M.; Lassila, O. Ikaros has a crucial role in regulation of B cell receptor signaling. *Eur. J. Immunol.* **2006**, *36*, 516–525. [CrossRef]
19. Dhanyamraju, P.K.; Iyer, S.; Smink, G.; Bamme, Y.; Bhadauria, P.; Payne, J.L.; Dovat, E.; Klink, M.; Ding, Y. Transcriptional Regulation of Genes by Ikaros Tumor Suppressor in Acute Lymphoblastic Leukemia. *Int. J. Mol. Sci.* **2020**, *21*, 1377. [CrossRef]
20. Alinikula, J.; Kohonen, P.; Nera, K.P.; Lassila, O. Concerted action of Helios and Ikaros controls the expression of the inositol 5-phosphatase SHIP. *Eur. J. Immunol.* **2010**, *40*, 2599–2607. [CrossRef]
21. Hahm, K.; Cobb, B.S.; McCarty, A.S.; Brown, K.E.; Klug, C.A.; Lee, R.; Akashi, K.; Weissman, I.L.; Fisher, A.G.; Smale, S.T. Helios, a T cell-restricted Ikaros family member that quantitatively associates with Ikaros at centromeric heterochromatin. *Genes Dev.* **1998**, *12*, 782–796. [CrossRef]
22. Xia, R.; Cheng, Y.; Han, X.; Wei, Y.; Wei, X. Ikaros Proteins in Tumor: Current Perspectives and New Developments. *Front. Mol. Biosci.* **2021**, *8*, 788440. [CrossRef] [PubMed]
23. Tsai, A.P.; Lin, P.B.; Dong, C.; Moutinho, M.; Casali, B.T.; Liu, Y.; Lamb, B.T.; Landreth, G.E.; Oblak, A.L.; Nho, K. INPP5D expression is associated with risk for Alzheimer's disease and induced by plaque-associated microglia. *Neurobiol. Dis.* **2021**, *153*, 105303. [CrossRef]
24. Huang, K.L.; Marcora, E.; Pimenova, A.A.; Di Narzo, A.F.; Kapoor, M.; Jin, S.C.; Harari, O.; Bertelsen, S.; Fairfax, B.P.; Czajkowski, J.; et al. A common haplotype lowers PU.1 expression in myeloid cells and delays onset of Alzheimer's disease. *Nat. Neurosci.* **2017**, *20*, 1052–1061. [CrossRef] [PubMed]
25. Pauls, S.D.; Marshall, A.J. Regulation of immune cell signaling by SHIP1: A phosphatase, scaffold protein, and potential therapeutic target. *Eur. J. Immunol.* **2017**, *47*, 932–945. [CrossRef] [PubMed]
26. March, M.E.; Ravichandran, K. Regulation of the immune response by SHIP. *Semin. Immunol.* **2002**, *14*, 37–47. [CrossRef] [PubMed]
27. Aman, M.J.; Walk, S.F.; March, M.E.; Su, H.P.; Carver, D.J.; Ravichandran, K.S. Essential role for the C-terminal noncatalytic region of SHIP in FcgammaRIIB1-mediated inhibitory signaling. *Mol. Cell Biol.* **2000**, *20*, 3576–3589. [CrossRef]
28. Zhang, J.; Ravichandran, K.S.; Garrison, J.C. A key role for the phosphorylation of Ser440 by the cyclic AMP-dependent protein kinase in regulating the activity of the Src homology 2 domain-containing Inositol 5'-phosphatase (SHIP1). *J. Biol. Chem.* **2010**, *285*, 34839–34849. [CrossRef]
29. Ong, C.J.; Ming-Lum, A.; Nodwell, M.; Ghanipour, A.; Yang, L.; Williams, D.E.; Kim, J.; Demirjian, L.; Qasimi, P.; Ruschmann, J.; et al. Small-molecule agonists of SHIP1 inhibit the phosphoinositide 3-kinase pathway in hematopoietic cells. *Blood* **2007**, *110*, 1942–1949. [CrossRef]
30. Manno, B.; Oellerich, T.; Schnyder, T.; Corso, J.; Losing, M.; Neumann, K.; Urlaub, H.; Batista, F.D.; Engelke, M.; Wienands, J. The Dok-3/Grb2 adaptor module promotes inducible association of the lipid phosphatase SHIP with the BCR in a coreceptor-independent manner. *Eur. J. Immunol.* **2016**, *46*, 2520–2530. [CrossRef]
31. D'Ambrosio, D.; Hippen, K.L.; Cambier, J.C. Distinct mechanisms mediate SHC association with the activated and resting B cell antigen receptor. *Eur. J. Immunol.* **1996**, *26*, 1960–1965. [CrossRef]
32. Lemay, S.; Davidson, D.; Latour, S.; Veillette, A. Dok-3, a novel adapter molecule involved in the negative regulation of immunoreceptor signaling. *Mol. Cell Biol.* **2000**, *20*, 2743–2754. [CrossRef] [PubMed]
33. Park, M.J.; Sheng, R.; Silkov, A.; Jung, D.J.; Wang, Z.G.; Xin, Y.; Kim, H.; Thiagarajan-Rosenkranz, P.; Song, S.; Yoon, Y.; et al. SH2 Domains Serve as Lipid-Binding Modules for pTyr-Signaling Proteins. *Mol. Cell* **2016**, *62*, 7–20. [CrossRef] [PubMed]
34. Ming-Lum, A.; Shojania, S.; So, E.; McCarrell, E.; Shaw, E.; Vu, D.; Wang, I.; McIntosh, L.P.; Mui, A.L. A pleckstrin homology-related domain in SHIP1 mediates membrane localization during Fcgamma receptor-induced phagocytosis. *FASEB J.* **2012**, *26*, 3163–3177. [CrossRef]
35. Lamkin, T.D.; Walk, S.F.; Liu, L.; Damen, J.E.; Krystal, G.; Ravichandran, K.S. Shc interaction with Src homology 2 domain containing inositol phosphatase (SHIP) in vivo requires the Shc-phosphotyrosine binding domain and two specific phosphotyrosines on SHIP. *J. Biol. Chem.* **1997**, *272*, 10396–10401. [CrossRef]

36. Tamir, I.; Stolpa, J.C.; Helgason, C.D.; Nakamura, K.; Bruhns, P.; Daeron, M.; Cambier, J.C. The RasGAP-binding protein p62dok is a mediator of inhibitory FcgammaRIIB signals in B cells. *Immunity* **2000**, *12*, 347–358. [CrossRef]
37. Kerr, W.G. Inhibitor and activator: Dual functions for SHIP in immunity and cancer. *Ann. N. Y Acad. Sci.* **2011**, *1217*, 1–17. [CrossRef] [PubMed]
38. Clement, S.; Krause, U.; Desmedt, F.; Tanti, J.F.; Behrends, J.; Pesesse, X.; Sasaki, T.; Penninger, J.; Doherty, M.; Malaisse, W.; et al. The lipid phosphatase SHIP2 controls insulin sensitivity. *Nature* **2001**, *409*, 92–97. [CrossRef] [PubMed]
39. Salim, K.; Bottomley, M.J.; Querfurth, E.; Zvelebil, M.J.; Gout, I.; Scaife, R.; Margolis, R.L.; Gigg, R.; Smith, C.I.; Driscoll, P.C.; et al. Distinct specificity in the recognition of phosphoinositides by the pleckstrin homology domains of dynamin and Bruton's tyrosine kinase. *EMBO J.* **1996**, *15*, 6241–6250. [CrossRef]
40. Scharenberg, A.M.; El-Hillal, O.; Fruman, D.A.; Beitz, L.O.; Li, Z.; Lin, S.; Gout, I.; Cantley, L.C.; Rawlings, D.J.; Kinet, J.P. Phosphatidylinositol-3,4,5-trisphosphate (PtdIns-3,4,5-P3)/Tec kinase-dependent calcium signaling pathway: A target for SHIP-mediated inhibitory signals. *EMBO J.* **1998**, *17*, 1961–1972. [CrossRef]
41. Konishi, H.; Kiyama, H. Microglial TREM2/DAP12 Signaling: A Double-Edged Sword in Neural Diseases. *Front. Cell Neurosci.* **2018**, *12*, 206. [CrossRef]
42. Han, J.; Luby-Phelps, K.; Das, B.; Shu, X.; Xia, Y.; Mosteller, R.D.; Krishna, U.M.; Falck, J.R.; White, M.A.; Broek, D. Role of substrates and products of PI 3-kinase in regulating activation of Rac-related guanosine triphosphatases by Vav. *Science* **1998**, *279*, 558–560. [CrossRef] [PubMed]
43. Kommaddi, R.P.; Das, D.; Karunakaran, S.; Nanguneri, S.; Bapat, D.; Ray, A.; Shaw, E.; Bennett, D.A.; Nair, D.; Ravindranath, V. Abeta mediates F-actin disassembly in dendritic spines leading to cognitive deficits in Alzheimer's disease. *J. Neurosci.* **2018**, *38*, 1085–1099. [CrossRef] [PubMed]
44. Rey-Ladino, J.A.; Huber, M.; Liu, L.; Damen, J.E.; Krystal, G.; Takei, F. The SH2-containing inositol-5′-phosphatase enhances LFA-1-mediated cell adhesion and defines two signaling pathways for LFA-1 activation. *J. Immunol.* **1999**, *162*, 5792–5799. [CrossRef] [PubMed]
45. Landego, I.; Jayachandran, N.; Wullschleger, S.; Zhang, T.T.; Gibson, I.W.; Miller, A.; Alessi, D.R.; Marshall, A.J. Interaction of TAPP adapter proteins with phosphatidylinositol (3,4)-bisphosphate regulates B-cell activation and autoantibody production. *Eur. J. Immunol.* **2012**, *42*, 2760–2770. [CrossRef] [PubMed]
46. Aman, M.J.; Lamkin, T.D.; Okada, H.; Kurosaki, T.; Ravichandran, K.S. The inositol phosphatase SHIP inhibits Akt/PKB activation in B cells. *J. Biol. Chem.* **1998**, *273*, 33922–33928. [CrossRef]
47. Yao, H.; Coppola, K.; Schweig, J.E.; Crawford, F.; Mullan, M.; Paris, D. Distinct Signaling Pathways Regulate TREM2 Phagocytic and NFkappaB Antagonistic Activities. *Front. Cell Neurosci.* **2019**, *13*, 457. [CrossRef]
48. Pradhan, M.; Coggeshall, K.M. Activation-induced bi-dentate interaction of SHIP and Shc in B lymphocytes. *J. Cell. Biochem.* **1997**, *67*, 32–42. [CrossRef]
49. Tridandapani, S.; Chacko, G.W.; Van Brocklyn, J.R.; Coggeshall, K.M. Negative signaling in B cells causes reduced Ras activity by reducing Shc-Grb2 interactions. *J. Immunol.* **1997**, *158*, 1125–1132. [CrossRef]
50. An, H.; Xu, H.; Zhang, M.; Zhou, J.; Feng, T.; Qian, C.; Qi, R.; Cao, X. Src homology 2 domain-containing inositol-5-phosphatase 1 (SHIP1) negatively regulates TLR4-mediated LPS response primarily through a phosphatase activity- and PI-3K-independent mechanism. *Blood* **2005**, *105*, 4685–4692. [CrossRef]
51. Conde, C.; Rambout, X.; Lebrun, M.; Lecat, A.; Di Valentin, E.; Dequiedt, F.; Piette, J.; Gloire, G.; Legrand, S. The inositol phosphatase SHIP-1 inhibits NOD2-induced NF-kappaB activation by disturbing the interaction of XIAP with RIP2. *PLoS ONE* **2012**, *7*, e41005. [CrossRef]
52. Colonna, M.; Wang, Y. TREM2 variants: New keys to decipher Alzheimer disease pathogenesis. *Nat. Rev. Neurosci.* **2016**, *17*, 201–207. [CrossRef] [PubMed]
53. Breijyeh, Z.; Karaman, R. Comprehensive Review on Alzheimer's Disease: Causes and Treatment. *Molecules* **2020**, *25*, 5789. [CrossRef] [PubMed]
54. Cunningham, F.; Amode, M.R.; Barrell, D.; Beal, K.; Billis, K.; Brent, S.; Carvalho-Silva, D.; Clapham, P.; Coates, G.; Fitzgerald, S.; et al. Ensembl 2015. *Nucleic Acids Res* **2015**, *43*, D662–D669. [CrossRef]
55. Iguchi, A.; Takatori, S.; Kimura, S.; Muneto, H.; Wang, K.; Etani, H.; Ito, G.; Sato, H.; Hori, Y.; Sasaki, J.; et al. INPP5D modulates TREM2 loss-of-function phenotypes in a β-amyloidosis mouse model. *iScience* **2023**, *26*, 106375. [CrossRef]
56. Lin, P.B.; Tsai, A.P.; Soni, D.; Lee-Gosselin, A.; Moutinho, M.; Puntambekar, S.S.; Landreth, G.E.; Lamb, B.T.; Oblak, A.L. INPP5D deficiency attenuates amyloid pathology in a mouse model of Alzheimer's disease. *Alzheimers Dement.* **2023**, *19*, 2528–2537. [CrossRef] [PubMed]
57. Castranio, E.L.; Hasel, P.; Haure-Mirande, J.V.; Ramirez Jimenez, A.V.; Hamilton, B.W.; Kim, R.D.; Glabe, C.G.; Wang, M.; Zhang, B.; Gandy, S.; et al. Microglial INPP5D limits plaque formation and glial reactivity in the PSAPP mouse model of Alzheimer's disease. *Alzheimers Dement.* **2023**, *19*, 2239–2252. [CrossRef]
58. Malik, M.; Parikh, I.; Vasquez, J.B.; Smith, C.; Tai, L.; Bu, G.; LaDu, M.J.; Fardo, D.W.; Rebeck, G.W.; Estus, S. Genetics ignite focus on microglial inflammation in Alzheimer's disease. *Mol. Neurodegener.* **2015**, *10*, 52. [CrossRef]
59. Huang, Z.Y.; Hunter, S.; Kim, M.K.; Indik, Z.K.; Schreiber, A.D. The effect of phosphatases SHP-1 and SHIP-1 on signaling by the ITIM- and ITAM-containing Fcgamma receptors FcgammaRIIB and FcgammaRIIA. *J. Leukoc. Biol.* **2003**, *73*, 823–829. [CrossRef]

60. Hodges, A.K.; Piers, T.M.; Collier, D.; Cousins, O.; Pocock, J.M. Pathways linking Alzheimer's disease risk genes expressed highly in microglia. *Neuroimmunol. Neuroinflam* **2021**, *8*, 245. [CrossRef]
61. Boucrot, E.; Ferreira, A.P.; Almeida-Souza, L.; Debard, S.; Vallis, Y.; Howard, G.; Bertot, L.; Sauvonnet, N.; McMahon, H.T. Endophilin marks and controls a clathrin-independent endocytic pathway. *Nature* **2015**, *517*, 460–465. [CrossRef]
62. Massa, P.T.; Wu, C. Increased inducible activation of NF-kappaB and responsive genes in astrocytes deficient in the protein tyrosine phosphatase SHP-1. *J. Interferon Cytokine Res.* **1998**, *18*, 499–507. [CrossRef] [PubMed]
63. Khaled, A.R.; Butfiloski, E.J.; Sobel, E.S.; Schiffenbauer, J. Functional consequences of the SHP-1 defect in motheaten viable mice: Role of NF-kappa B. *Cell Immunol.* **1998**, *185*, 49–58. [CrossRef] [PubMed]
64. Bourne, K.Z.; Ferrari, D.C.; Lange-Dohna, C.; Rossner, S.; Wood, T.G.; Perez-Polo, J.R. Differential regulation of BACE1 promoter activity by nuclear factor-kappaB in neurons and glia upon exposure to β-amyloid peptides. *J. Neurosci. Res.* **2007**, *85*, 1194–1204. [CrossRef]
65. Terai, K.; Matsuo, A.; McGeer, P.L. Enhancement of immunoreactivity for NF-kappa B in the hippocampal formation and cerebral cortex of Alzheimer's disease. *Brain Res.* **1996**, *735*, 159–168. [CrossRef]
66. Yan, S.D.; Yan, S.F.; Chen, X.; Fu, J.; Chen, M.; Kuppusamy, P.; Smith, M.A.; Perry, G.; Godman, G.C.; Nawroth, P.; et al. Non-enzymatically glycated tau in Alzheimer's disease induces neuronal oxidant stress resulting in cytokine gene expression and release of amyloid β-peptide. *Nat. Med.* **1995**, *1*, 693–699. [CrossRef] [PubMed]
67. Ferrer, I.; Marti, E.; Lopez, E.; Tortosa, A. NF-kB immunoreactivity is observed in association with β A4 diffuse plaques in patients with Alzheimer's disease. *Neuropathol. Appl. Neurobiol.* **1998**, *24*, 271–277. [CrossRef] [PubMed]
68. Chen, C.H.; Zhou, W.; Liu, S.; Deng, Y.; Cai, F.; Tone, M.; Tone, Y.; Tong, Y.; Song, W. Increased NF-kappaB signalling up-regulates BACE1 expression and its therapeutic potential in Alzheimer's disease. *Int. J. Neuropsychopharmacol.* **2012**, *15*, 77–90. [CrossRef]
69. Kitamura, Y.; Shimohama, S.; Ota, T.; Matsuoka, Y.; Nomura, Y.; Taniguchi, T. Alteration of transcription factors NF-kappaB and STAT1 in Alzheimer's disease brains. *Neurosci. Lett.* **1997**, *237*, 17–20. [CrossRef]
70. Mostafavi, S.; Gaiteri, C.; Sullivan, S.E.; White, C.C.; Tasaki, S.; Xu, J.; Taga, M.; Klein, H.U.; Patrick, E.; Komashko, V.; et al. A molecular network of the aging human brain provides insights into the pathology and cognitive decline of Alzheimer's disease. *Nat. Neurosci.* **2018**, *21*, 811–819. [CrossRef]
71. Obst, J.; Bradshaw, W.; Roberts, H.H.; Priestley, R.; Jimenez-Antunez, C.; Cowley, S.A.; Gileadi, O.; Mead, E.; Daniel, E.D.; Davis, J.B. Targeting SHIP1 for therapeutic intervention in Alzheimer's disease. *Alzheimers Dement.* **2020**, *16*, e045839. [CrossRef]
72. Suwa, A.; Yamamoto, T.; Sawada, A.; Minoura, K.; Hosogai, N.; Tahara, A.; Kurama, T.; Shimokawa, T.; Aramori, I. Discovery and functional characterization of a novel small molecule inhibitor of the intracellular phosphatase, SHIP2. *Br. J. Pharmacol.* **2009**, *158*, 879–887. [CrossRef] [PubMed]
73. Soeda, Y.; Tsuneki, H.; Muranaka, H.; Mori, N.; Hosoh, S.; Ichihara, Y.; Kagawa, S.; Wang, X.; Toyooka, N.; Takamura, Y.; et al. The inositol phosphatase SHIP2 negatively regulates insulin/IGF-I actions implicated in neuroprotection and memory function in mouse brain. *Mol. Endocrinol.* **2010**, *24*, 1965–1977. [CrossRef] [PubMed]
74. Pedicone, C.; Fernandes, S.; Dungan, O.M.; Dormann, S.M.; Viernes, D.R.; Adhikari, A.A.; Choi, L.B.; De Jong, E.P.; Chisholm, J.D.; Kerr, W.G. Pan-SHIP1/2 inhibitors promote microglia effector functions essential for CNS homeostasis. *J. Cell Sci.* **2020**, *133*, jcs238030. [CrossRef] [PubMed]
75. Sala Frigerio, C.; Wolfs, L.; Fattorelli, N.; Thrupp, N.; Voytyuk, I.; Schmidt, I.; Mancuso, R.; Chen, W.T.; Woodbury, M.E.; Srivastava, G.; et al. The Major Risk Factors for Alzheimer's Disease: Age, Sex, and Genes Modulate the Microglia Response to Abeta Plaques. *Cell Rep.* **2019**, *27*, 1293–1306 e1296. [CrossRef] [PubMed]
76. Pedicone, C.; Fernandes, S.; Matera, A.; Meyer, S.T.; Loh, S.; Ha, J.H.; Bernard, D.; Chisholm, J.D.; Paolicelli, R.C.; Kerr, W.G. Discovery of a novel SHIP1 agonist that promotes degradation of lipid-laden phagocytic cargo by microglia. *iScience* **2022**, *25*, 104170. [CrossRef] [PubMed]

Disclaimer/Publisher's Note: The statements, opinions and data contained in all publications are solely those of the individual author(s) and contributor(s) and not of MDPI and/or the editor(s). MDPI and/or the editor(s) disclaim responsibility for any injury to people or property resulting from any ideas, methods, instructions or products referred to in the content.

GCAT TACG GCAT genes

MDPI

Review

Outcome Measures and Biomarkers for Clinical Trials in Hereditary Spastic Paraplegia: A Scoping Review

Sue-Faye Siow [1,2,*], Dennis Yeow [1,3,4,5,6], Laura I. Rudaks [1,2,4,5], Fangzhi Jia [1], Gautam Wali [1,3,†], Carolyn M. Sue [1,3,6,7,†] and Kishore R. Kumar [1,4,5,7,†]

1 Sydney Medical School, University of Sydney, Camperdown 2050, Australia; c.sue@neura.edu.au (C.M.S.)
2 Department of Clinical Genetics, Royal North Shore Hospital, St Leonards 2065, Australia
3 Neuroscience Research Australia, University of New South Wales, Randwick 2031, Australia
4 Rare Disease Program, Garvan Institute of Medical Research, Darlinghurst 2010, Australia
5 Translational Neurogenomics Group, Molecular Medicine Laboratory and Department of Neurology, Concord Hospital, Concord 2139, Australia
6 Neurodegenerative Service, Prince of Wales Hospital, Randwick 2031, Australia
7 School of Clinical Medicine, UNSW Medicine & Health, University of New South Wales, Kensington 2052, Australia
* Correspondence: suefaye.siow@health.nsw.gov.au
† These authors share senior authorship.

Abstract: Hereditary spastic paraplegia (HSP) is characterized by progressive lower limb spasticity. There is no disease-modifying treatment currently available. Therefore, standardized, validated outcome measures to facilitate clinical trials are urgently needed. We performed a scoping review of outcome measures and biomarkers for HSP to provide recommendations for future studies and identify areas for further research. We searched Embase, Medline, Scopus, Web of Science, and the Central Cochrane database. Seventy studies met the inclusion criteria, and eighty-three outcome measures were identified. The Spastic Paraplegia Rating Scale (SPRS) was the most widely used (27 studies), followed by the modified Ashworth Scale (18 studies) and magnetic resonance imaging (17 studies). Patient-reported outcome measures (PROMs) were infrequently used to assess treatment outcomes (28% of interventional studies). Diffusion tensor imaging, gait analysis and neurofilament light chain levels were the most promising biomarkers in terms of being able to differentiate patients from controls and correlate with clinical disease severity. Overall, we found variability and inconsistencies in use of outcome measures with a paucity of longitudinal data. We highlight the need for (1) a standardized set of core outcome measures, (2) validation of existing biomarkers, and (3) inclusion of PROMs in HSP clinical trials.

Keywords: hereditary spastic paraplegia; outcome measures; biomarkers; clinical trials; scoping review

Citation: Siow, S.-F.; Yeow, D.; Rudaks, L.I.; Jia, F.; Wali, G.; Sue, C.M.; Kumar, K.R. Outcome Measures and Biomarkers for Clinical Trials in Hereditary Spastic Paraplegia: A Scoping Review. *Genes* **2023**, *14*, 1756. https://doi.org/10.3390/genes14091756

Academic Editor: Claudia Ricci

Received: 9 August 2023
Revised: 30 August 2023
Accepted: 30 August 2023
Published: 3 September 2023

Copyright: © 2023 by the authors. Licensee MDPI, Basel, Switzerland. This article is an open access article distributed under the terms and conditions of the Creative Commons Attribution (CC BY) license (https://creativecommons.org/licenses/by/4.0/).

1. Introduction

Hereditary Spastic Paraplegia (HSP) refers to a group of inherited neurodegenerative conditions characterized by lower limb spasticity and weakness. HSP is rare, with a prevalence of 0.3 to 5.5 per 100,000 people, depending on country [1–4]. HSP is associated with significant disability and a negative impact on quality of life [5]. There are over 80 recognized HSP-associated genes with broad phenotypic variability [6–8]. Clinically, HSP can be categorized into pure HSP—symptoms limited to weakness, spasticity, impaired vibration sense in the lower limbs, and bladder dysfunction; and complex HSP—where there are additional neurological and non-neurological manifestations [9]. The phenotypic and genotypic heterogeneity of HSP and the rarity of the condition pose a challenge to the development of suitable outcome measures and biomarkers.

Advances in genetic testing have led to the rapid discovery of genes associated with HSP [10], accelerating the discovery of therapeutic targets. Patient-derived stem cell and animal models have identified potential drug treatment candidates targeting underlying pathogenesis for specific HSP genotypes, such as noscapine for HSP-*SPAST* [11–15]. Currently, the rate of drug discovery is far outpaced by the rate of gene discovery for HSP and a shift in HSP research to developing treatment options is required [10,16,17]. Outcome measures and biomarkers that can measure the efficacy of therapeutic interventions in clinical trials are needed to facilitate this.

The choice of appropriate outcome measures in interventional trials is critical to demonstrating a meaningful treatment effect [18]. Initiatives such as COnsensus-based Standards for the selection of health Measurement INstruments (COSMIN) and Core Outcome Measures in Effectiveness Trials (COMET) aim to guide the selection of appropriate outcome measures [19]. There is currently a lack of standardized outcome measures used in HSP clinical trials, with recent reviews of interventional trials in HSP showing heterogeneity of outcome measures [17,20]. Inconsistency of outcome measures leads to an increased risk of reporting bias due to post hoc selection of outcomes based on results rather than the use of pre-specified primary outcomes [21]. Furthermore, use of different outcome measures limits comparison and meta-analysis of results from different studies [21].

There are no current consensus guidelines for outcome measures in HSP clinical trials. To address this, we performed a scoping review of outcome measures and biomarkers in HSP to identify suitable outcome measures, provide recommendations for future HSP clinical trials, and identify areas for further research.

2. Materials and Methods

2.1. Search Strategy

This study was conducted according to published guidelines for conducting a systematic scoping review [22,23]. Under the guidance of an academic librarian, we performed a search in Embase, Medline, Scopus, Web of Science and the Central Cochrane databases using the search terms "Hereditary Spastic Paraplegia", "biomarker", "outcome measure", and "patient reported outcome measure" (Appendix A). Additional studies were identified by searching the references of the included articles and relevant review articles.

2.2. Selection Criteria

We kept the selection criteria broad to capture all possible outcome measures and biomarkers. The inclusion criteria included studies involving patients with HSP of any age and gender, and that included a description of the outcome measures or biomarkers for HSP. Abstracts were included if novel outcome measures or biomarkers were described.

We excluded review articles, single-case reports, trial protocols with no published results, abstracts with no novel outcome measures/biomarkers, and articles that did not involve humans or human samples, did not include outcome measures/biomarkers, or were not in English.

2.3. Screening of Search Results

The screening and data extraction process was conducted using Covidence, a web-based collaboration software platform that streamlines the production of systematic and other literature reviews [24]. The first author (S.F.S.) excluded all irrelevant results—studies unrelated to HSP or not in English and duplicate studies. Authors S.F.S. and D.Y. independently reviewed the abstracts according to the selection criteria for the first 50 results. Authors S.F.S. and D.Y. independently reviewed the remaining abstracts, applying the finalized exclusion criteria.

2.4. Data Extraction and Analysis

Authors S.F.S., D.Y., L.R., and F.J. reviewed two full texts each to test the data extraction template. The template was modified by discussion at a team meeting. The team then reviewed six full texts, each with each author reviewing the same three articles as two other authors (~30% overlap) and met to resolve any discrepancies and standardize the extraction process. The rest of the studies were reviewed by one reviewer (D.Y., L.R., or F.J.) and verified by a second reviewer (S.F.S.). At each stage, discrepancies were resolved through a team meeting. The data were extracted using a standardized data extraction template, including information on study characteristics, aim of study, participant characteristics, characteristics of interventions, outcome measures, and ability of outcome measure to (i) distinguish patients versus controls, (ii) demonstrate change over time; (iii) show response to the intervention; and (iv) correlate with other measures. Study quality analysis was not performed, as is usual for scoping reviews [22].

Data were analyzed descriptively to provide an overview of study characteristics and outcome measures. Outcome measures were grouped according to the types of clinical outcome assessments as defined by the U.S. Food and Drug Administration [25]:

1. Clinician-reported outcome measures (CROM): measurement of clinical signs or findings performed by a health professional.
2. Performance outcome measures (PerfOM): measurement with a standardized task, either administered by a trained individual or undertaken by the patient without assistance.
3. Patient-reported outcome measures (PROM): measurement of patient-reported health status.

Biomarkers were grouped according to the assessed modality:

1. Laboratory-based biomarkers;
2. Neuroimaging biomarkers;
3. Neurophysiology biomarkers;
4. Other biomarkers.

As other groups have previously published reviews of non-randomized interventional clinical trials [17,20], we chose to perform further analysis of randomized controlled trials (RCTs) to compare the outcome measures used, as RCTs are the study type with the highest level of evidence for treatment effectiveness [26].

3. Results

3.1. Search Results

A total of 1930 search results were identified, and 1437 abstracts were screened after duplicates were removed, of which 1222 were identified as irrelevant. The full texts of the remaining results ($n = 215$) were assessed according to inclusion criteria. Ten more studies were identified by searching the references of the included articles (Figure 1).

3.2. Study Characteristics

A total of 70 studies that met the inclusion criteria were identified. These studies were published between 1991 and 2022; date limits were not set to include as many studies as possible.

The majority (78.6%) of the included studies were observational studies, and only a quarter (25.7%) of studies were interventional studies. Most studies did not include longitudinal data (75.7%), and over half did not include a control group (51.4%). Participant genotype was predominantly mixed or unknown (65.7%), and sample sizes were small (mean of 36.99 participants) (Table 1). A list of all studies with relevant details is included as Supplementary Material S1.

We assessed each outcome measure for (1) the ability to differentiate patients versus controls if a control group was included, (2) the ability to demonstrate disease progression if longitudinal data were included, (3) the ability to demonstrate response to the intervention if an intervention was assessed, and (4) any correlation with other biomarkers or outcome measures. Although we report the number of studies that fulfilled the criteria for (1), (2) and (3) to illustrate the available evidence for each outcome measure, it is important to note that (1) in studies with multiple outcome measures, the control group was compared to the patient group for some but not all outcome measures, (2) not all longitudinal studies were designed to assess the ability of an outcome measure to measure disease progression, (3) an outcome measure may not demonstrate a response to the intervention for many reasons including efficacy of intervention, duration of trial, and timing of the outcome measure relative to the intervention.

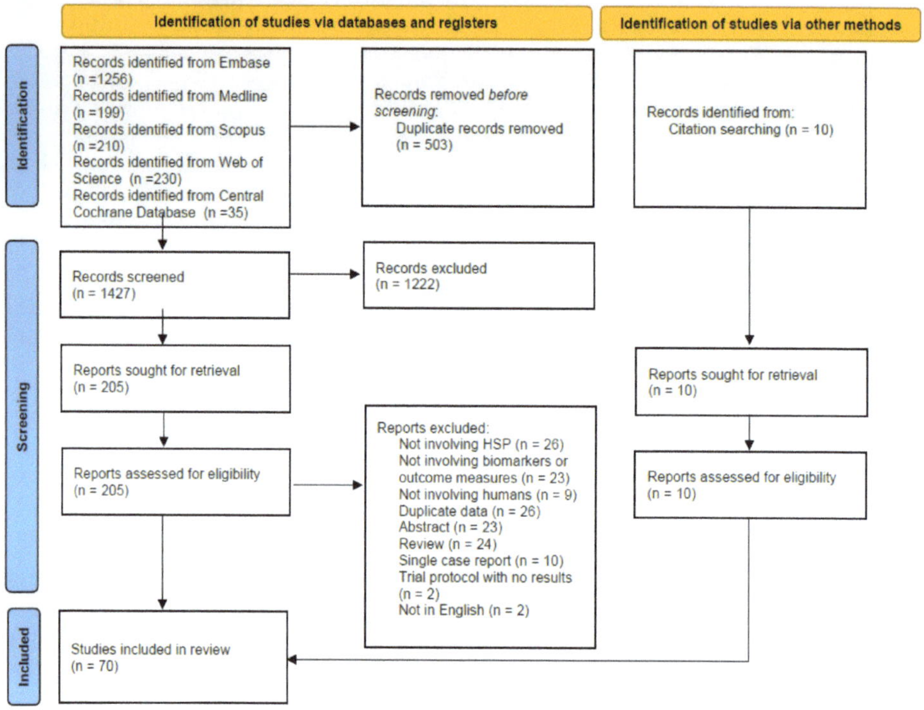

Figure 1. PRISMA flow chart [27].

Table 1. Descriptive findings.

Study Type	Number of Studies (%)
Observational (total)	55 (78.6)
• Case series	9
• Cohort	14
• Case–control	2
• Cross-sectional	30
Diagnostic test accuracy	1 (1.4)
Non-randomized interventional studies	7 (10)
Randomized controlled trials	6 (8.6)
Study Characteristics	**N (%)**
Longitudinal data presented	17 (24.3)
Control group included	34 (48.6)
Intervention	18 (25.7)
Participant Genotype	**N (%)**
Single genotype	24 (34.3)
Mixed known and unknown genotypes	35 (50)
Unknown genotype	11 (15.7)
Participant Characteristics	**N (%)**
Number of participants	37.0 people (range 2–239, SD 42.2)
Mean of mean ages (n = 64)	39.9 years (range 4.8–62, SD 14.9)

3.3. Clinical Outcome Assessments

3.3.1. Clinician Reported Outcome Measures

The Spastic Paraplegia Rating Scale (SPRS) was the most widely used CROM, reported in twenty-seven studies (Table 2). Although 12/27 included studies control groups, a comparison of SPRS scores in patients versus controls was performed in only two studies, both showing significant differences [28,29]. It is important to note that SPRS values from healthy controls are most relevant when comparing to pre-symptomatic HSP carriers rather than for comparison to individuals with symptomatic HSP. 8/27 studies were longitudinal, and only six of those studies showed disease progression over time with the SPRS (median follow-up time 12–31 months) [30–35]. The SPRS was used as an outcome measure in 4/18 interventional studies and showed a response to the intervention in only one of the four studies [36]. The SPRS was commonly used to correlate with other biomarkers or outcome measures (21/27 studies) (see Supplementary Material S2).

Table 2. Clinician-reported outcome measures.

Outcome Measure	Studies [1]	Control Group (Patient vs. Control Difference) [2]	Longitudinal (Disease Progression) [3]	Intervention (Response to Intervention) [4]
		HSP-specific CROM		
SPRS	27	12 (2)	8 (6)	5 (1)
SPATAX-EUROSPA disability score	4	1 (0)	1 (1)	1 (0)
		CROM for other neurological disorders		
SARA	3	1 (1)	1 (1)	0 (0)
ALSFRS-R	2	1 (0)	2 (1)	0 (0)
Unified Huntington's Disease Rating Scale Part IV	1	1 (0)	0 (0)	0 (0)
Multiple Sclerosis Impairment Scale	1	0 (0)	0 (0)	1 (0)
		Generic CROM		
Functional questionnaire score	1	0 (0)	0 (0)	1 (0)
Modified Rankin Scale	1	1 (0)	1 (0)	0 (0)
Disability Score (DIS)	1	0 (0)	0 (0)	1 (0)
Functional Independence Measure (FIM)	1	0 (0)	0 (0)	0 (0)

[1] Total number of studies; [2] Number of studies that included a control group (Studies where outcome measure showed a difference between patients and controls); [3] Number of studies that had longitudinal data (studies where outcome measure was able to show disease progression); [4] Number of studies that included an intervention (Studies where outcome measure showed a response to the intervention). Details of the included studies are in Supplementary Material S2.

The SPATAX-EUROSPA disability score, another HSP-specific CROM, was less commonly used than the SPRS (Table 2). There were no control data and change over time was studied in only one study showing disease progression in 3/31 patients included in the study [32]. It was used in one interventional study [37] and did not show any significant change with intervention.

CROMs developed for other neurological conditions and generic functional CROMs were also used to assess patients with HSP. The Scale for Assessment and Rating of Ataxia (SARA) was used in three studies and showed a difference between patient versus controls in 1/3 studies [38] and longitudinal change in 1/3 studies [39]. The Amyotrophic Lateral Sclerosis rating scale revised (ALSFRS-R) was used in two studies [40,41], showing disease progression in one but not the other, and was shown to correlate with serum and CSF neurofilament heavy chain [40,41], though one paper included patients with ALS in their analysis (Supplementary Material S2).

3.3.2. Performance Outcome Measures

PerfOMs assessing motor function were widely used in the included studies, with twenty-one different outcome measures identified. The most commonly reported motor PerfOMs were the modified Ashworth Scale (MAS) (18 studies) and the 10 m walk test (10MWT), 6 min walk test (6MWT) and their variations (14 studies) (Table 3). Compared to CROMs, motor PerfOMs were more commonly used as outcome measures for interventional studies—MAS in 13/18 interventional studies, 10MWT/6MWT/variations in 9/18 interventional studies. Conversely, motor PerfOMs were less commonly used to correlate with other outcome measures or biomarkers, 2/18 studies for MAS and 3/14 studies for 10MWT and its variations (Supplementary Materials S2). Similar to the CROMs, there were very little longitudinal data—3/18 for MAS, 1/14 for 10MWT and variations.

Table 3. Performance outcome measures.

Outcome Measure	Studies [1]	Control Group (Patient vs. Control Difference) [2]	Longitudinal (Disease Progression) [3]	Intervention (Response to Intervention) [4]
		Motor Function PerfOMS		
Modified Ashworth Scale	18	2 (0)	3 (1)	13 (11)
10MWT, 6MWT, 2MWT, 5MWT, 20MWT, 3-min endurance walk	14	3 (2)	1 (1)	9 (3)
Timed Up-and-Go test (TUG)	6	3 (2)	2 (2)	3 (1)
Medical Research Council muscle strength	4	0 (0)	0 (0)	2 (1)
Gross motor function measure (GMFM-66, GMFM-88)	2	0 (0)	0 (0)	2 (1)
Gross motor function classification score (GMFCS)	2	1 (0)	0 (0)	1 (0)
Falls Efficacy Scale-International (FES-I)	2	2 (1)	1 (0)	0 (0)
Berg balance scale	2	0 (0)	0 (0)	2 (1)
Composite cerebellar functional severity score (CCFSw)	2	1 (1)	1 (0)	0 (0)
Nine-hole pegboard test, click test, writing test, tapping test	2	1 (0)	0 (0)	1 (1)
Physiological Cost Index	2	1 (0)	1 (0)	2 (0)
Four-stage functional scale of motor disability (4SMD or 4FMS)	2	0 (0)	0 (0)	0 (0)
Activities specific Balance Confidence scale (ABC)	1	0 (0)	0 (0)	1 (0)
Modified version of Gillette Functional Assessment Questionnaire	1	0 (0)	0 (0)	1 (1)
Walking Handicap scale	1	0 (0)	0 (0)	1 (1)
Lower extremity subclass of Fugl–Meyer assessment	1	0 (0)	0 (0)	1 (0)
Ambulatory score (AMB)	1	0 (0)	0 (0)	1 (0)
AMBUS	1	0 (0)	1 (0)	0 (0)
Spasm Frequency Scale	1	0 (0)	1 (1)	1 (1)
Strength with microFET 2 hand-held dynamometer	1	0 (0)	0 (0)	1 (1)

Table 3. *Cont.*

Outcome Measure	Studies [1]	Control Group (Patient vs. Control Difference) [2]	Longitudinal (Disease Progression) [3]	Intervention (Response to Intervention) [4]
		Motor Function PerfOMS		
Walking ability (landmarks of disability)	1	0 (0)	0 (0)	0 (0)
		Cognitive Function PerfOMs		
Montreal Cognitive Assessment	4	4 (0)	1 (0)	0 (0)
Wechsler Adult Intelligence Scale—revised	3	1 (0)	1 (1)	0 (0)
Mini-Mental State Exam	2	2 (0)	0 (0)	0 (0)
Addenbrooke's Cognitive Exam	1	1 (0)	1 (0)	0 (0)
CANTAB assessment	1	1 (0)	0 (0)	0 (0)
VLMT, FWIT, TMT A/B, FAB, d2-R, RWT *	1	0 (0)	1 (1)	0 (0)

[1] Total number of studies; [2] Number of studies that included a control group (Studies where outcome measure showed a difference between patients and controls); [3] Studies that had longitudinal data (studies where outcome measure was able to show disease progression); [4] Studies that included an intervention (Studies where outcome measure showed a response to the intervention); * VLMT—Verbal Learning and Memory Test, FWIT—Farbe-Wort-Interferenz Test, TMT A/B—Trail Making Test Part A and B, FAB—Frontal Assessment Battery, d2-R—revised d2 Test of attention, RWT—Regensburg Word Fluency Test. Details of the included studies are in Supplementary Material S2.

PerfOMs measuring cognition were used in ten studies for descriptive purposes only rather than to measure outcomes in interventional studies.

3.3.3. Patient Reported Outcome Measures

Most PROMs used assessed quality of life (12/17), while others assessed fatigue, pain, and autonomic symptoms (Table 4). The most used PROM was the Short Form Health Survey-36 (SF-36) and its derivative, SF-12, in seven studies, followed by EuroQoL-5 Dimensions (EQ-5D) in three studies. PROMs were not commonly used in interventional studies (5/18 interventional studies). There was no longitudinal data for PROMs in the included studies. The SF-36 and its derivatives, EQ-5D, Becks Depression Inventory (BDI-V), Zung depression score, Brief Pain inventory, Modified Fatigue Impact Scale and multidimensional fatigue inventory showed differences between patients and controls in some but not all studies.

Table 4. Patient-Reported Outcome Measures.

Outcome Measure	Studies [1]	Control Group (Patient vs. Control Difference) [2]	Longitudinal (Disease Progression) [3]	Intervention (Response to Intervention) [4]
		Quality of Life PROMs		
SF-36, SF-12, RAND 36-Item Health Survey	7	3 (2)	1 (0)	2 (1)
EQ-5D	3	3 (1)	0 (0)	0 (0)
Becks Depression Inventory (BDI-V)	2	2 (1)	0 (0)	0 (0)
Visual analogue score	2	0 (0)	0 (0)	2 (1)
Patient Health Questionnaire (PHQ-9)	1	1 (0)	0 (0)	0 (0)
Modified Goal Attainment Scale (mGAS)	1	0 (0)	0 (0)	1 (1)
International Consultation of Incontinence Questionnaire (ICIQ)—LUTSqol	1	0 (0)	0 (0)	0 (0)

Table 4. *Cont.*

Outcome Measure	Studies [1]	Control Group (Patient vs. Control Difference) [2]	Longitudinal (Disease Progression) [3]	Intervention (Response to Intervention) [4]
Cerebral Palsy QoL questionnaire (CPQoL)	1	0 (0)	0 (0)	1 (1)
Caregiver Priorities and Child Health Index of Life with Disabilities (CPCHILD)	1	0 (0)	0 (0)	0 (0)
Zung depression score	1	1 (1)	0 (0)	0 (0)
ICIQ-Short Form	1	0 (0)	0 (0)	0 (0)
Hospital Anxiety and Depression Scale	1	0 (0)	0 (0)	1 (1)
Other PROMs				
Brief pain inventory	3	2 (1)	0 (0)	1 (0)
Modified Fatigue Impact Scale (MFI)	2	1 (1)	0 (0)	1 (0)
Multidimensional fatigue inventory	1	1 (0)	0 (0)	0 (0)
Scale for Outcomes in Parkinson's Disease for Autonomic Symptoms (SCOPA-AUT)	1	0 (0)	0 (0)	0 (0)
Numeric rating scale for pain	1	0 (0)	0 (0)	1 (1)

[1] Total number of studies; [2] Number of studies that included a control group (Studies where outcome measure showed a difference between patients and controls); [3] Number of studies that had longitudinal data (Studies where outcome measure was able to show disease progression); [4] studies that included an intervention (Studies where outcome measure showed a response to the intervention). Details of the included studies are in Supplementary Material S2.

3.4. *Biomarkers*

3.4.1. Laboratory-Based Biomarkers

Serum and cerebrospinal fluid (CSF) neurofilament light chain (NfL) levels, and serum and CSF 25- and 27-hydroxycholesterol (25- and 27-OHC) levels were the two most studied biochemical biomarkers—included in six and four studies, respectively (Table 5). Serum and CSF NfL levels were able to differentiate patient vs. control groups or historical control values in all studies. Serum and CSF NfL levels were shown to correlate with SPRS scores in two studies [29,42] but this correlation was not present in the other two studies [32,43] (Supplementary Material S2). Plasma and CSF 25- and 27-OHC were significantly elevated in individuals with HSP-*CYP7B1* (SPG5) compared to controls or reported normative values in all four studies [35,43–45]. In the few longitudinal studies, there was no change over time in NfL levels [32] or 25- and 27-OHC levels [35,44]. NfL levels were not used as outcome measures in any interventional studies; however, 25- and 27-OHC levels showed a response to treatment with statins in interventional studies [35,44]. Other biochemical markers reported, such as lipidomics, amino acid levels, mitochondrial DNA levels and cell morphomics, were more commonly used as diagnostic biomarkers rather than to measure disease progression or response to the intervention.

3.4.2. Neuroimaging Biomarkers

Magnetic resonance imaging (MRI) of the brain and spine, with or without volumetric analysis, was used in seventeen studies, diffusion tensor imaging (DTI) in seven studies and magnetic resonance spectroscopy (MRS) in four studies (Table 6). MRI brain and spine differentiated patients vs. controls in six studies and showed longitudinal change in only one study [46]. No imaging parameters were used as outcome measures for interventional studies. MRI findings were found to correlate with SPRS scores in 3/4 of the studies, and DTI findings correlated with SPRS scores in 2/3 of the studies (Supplementary Materials S2).

Table 5. Biochemical Biomarkers.

Biomarker	Studies [1]	Control Group (Patient vs. Control Difference) [2]	Longitudinal (Disease Progression) [3]	Intervention (Response to Intervention) [4]
Serum and CSF NfL	6	6 (5)	1 (0)	0 (0)
Serum and CSF 25-OHC and 27-OHC	4	3 (2)	2 (0)	2 (2)
Blood(plasma) and CSF amino acids	2	1 (1)	0 (0)	1 (0)
Lipidomics: fibroblast and plasma	1	1 (1)	0 (0)	0 (0)
Citrulline	1	0 (0)	0 (0)	0 (0)
Glycosylceramide profile	1	0 (0)	0 (0)	0 (0)
Neurofilament heavy chain: CSF and serum	2	1 (1)	1 (0)	0 (0)
Autophagy-related protein (ATG9A) ratio	1	1 (1)	0 (0)	0 (0)
CSF Aβ 1–42, total tau, phospho tau	1	1 (0)	0 (0)	0 (0)
Mitochondrial DNA levels; Muscle biopsy	1	1 (1)	0 (0)	0 (0)
Cell morphomics	1	1 (1)	0 (0)	1 (1)
Scanning electron microscopy of hair shafts	1	1 (1)	0 (0)	0 (0)

[1] Total number of studies; [2] Number of studies that included a control group (Studies where outcome measure showed a difference between patients and controls); [3] Studies that had longitudinal data (Studies where outcome measure was able to show disease progression); [4] Studies that included an intervention (Studies where outcome measure showed a response to the intervention). Details of the included studies are in Supplementary Material S2.

Table 6. Neuroimaging Biomarkers.

Biomarker	Studies [1]	Control Group (Patient vs. Control Difference) [2]	Longitudinal (Disease Progression) [3]	Intervention (Response to Intervention) [4]
MRI brain and spine	17	8 (5)	4 (1)	1 (0)
DTI	7	7 (7)	2 (1)	0 (0)
MRS	4	3 (2)	1 (0)	0 (0)
Ioflupane Single Photon Emission Computed Tomography (SPECT)	1	0 (0)	0 (0)	0 (0)

[1] Total number of studies; [2] Number of studies that included a control group (Studies where outcome measure showed a difference between patients and controls); [3] Studies that had longitudinal data (studies where outcome measure was able to show disease progression); [4] Studies that included an intervention (Studies where outcome measure showed a response to the intervention). Details of the included studies are in Supplementary Material S2.

3.4.3. Neurophysiology Biomarkers

Motor evoked potentials (MEPs) and nerve conduction studies/electromyography (NCS/EMG) were the most widely used neurophysiological markers, being used in ten and nine studies, respectively (Table 7). Although MEPs were used as outcome measures in three interventional studies, no changes in MEP results were demonstrated in response to any of the interventions. NCS/EMG, somatosensory evoked potentials (SSEPs), visual evoked potentials (VEPs) and brainstem auditory evoked potentials (BAEPs) were mostly used as descriptive measures with few studies using these measures to compare patients versus controls, study longitudinal cohorts, or to evaluate interventions.

3.4.4. Other Biomarkers

We identified eight studies that utilized gait analysis—seven were laboratory-based, and one was a mobile system [34] (Table 8). Four studies described the use of an infrared multi-camera motion analysis system [28,47–49], one used pressure sensors [50], and another two did not provide details of the gait analysis system used [51,52]

(Supplementary Material S1). Four interventional studies used gait analysis as an outcome measure, with only two of those studies showing a response to the intervention [49,50]. Gait analysis was able to differentiate patients with HSP from patients with spastic diplegia [47], healthy controls [34,48] and pre-symptomatic HSP-*SPAST* carriers [28]. Gait parameters also correlate with SPRS scores [28,34] (Supplementary Table S2).

Table 7. Neurophysiology Biomarkers.

Biomarker	Studies [1]	Control Group (Patient vs. Control Difference) [2]	Longitudinal (Disease Progression) [3]	Intervention (Response to Intervention) [4]
MEPs	10	2 (1)	2 (0)	3 (0)
NCS/EMG	9	2 (0)	2 (0)	1 (0)
SSEP	5	1 (1)	0 (0)	0 (0)
VEP	2	0 (0)	0 (0)	0 (0)
BAEP	1	0 (0)	0 (0)	0 (0)

[1] Total number of studies; [2] Number of studies that included a control group (Studies where outcome measure showed a difference between patients and controls); [3] Studies that had longitudinal data (Studies where outcome measure was able to show disease progression); [4] Studies that included an intervention (Studies where outcome measure showed a response to the intervention). Details of the included studies are in Supplementary Material S2.

Table 8. Other Biomarkers.

Biomarker	Studies [1]	Control Group (Patient vs. Control Difference) [2]	Longitudinal (Disease Progression) [3]	Intervention (Response to Intervention) [4]
Laboratory and mobile gait analysis	8	4 (4)	1 (1)	4 (2)
Spectral-domain optical coherence tomography (SD-OCT)	4	0 (0)	2 (0)	0 (0)
Video-oculography and rotational chair testing	1	1 (1)	0 (0)	0 (0)
Goniometer	2	0 (0)	0 (0)	1 (1)
Instrumented dynamic balance assessment	1	0 (0)	0 (0)	1 (1)
Urodynamic assessment	1	0 (0)	0 (0)	0 (0)
Video supported posturography	1	0 (0)	1 (0)	0 (0)
Protocol for Evaluation of Acquired Speech Disorders (PADAF)	1	1 (1)	0 (0)	0 (0)

[1] Total number of studies; [2] Number of studies that included a control group (Studies where outcome measure showed a difference between patients and controls); [3] Studies that had longitudinal data (Studies where outcome measure was able to show disease progression); [4] Studies that included an intervention (Studies where outcome measure showed a response to the intervention). Details of the included studies are in Supplementary Material S2.

SD-OCT was investigated in four studies, with one study showing statistically significant retinal nerve fiber layer (RNFL) thinning in patients compared to normative values but no change over time (Table 8).

Video-oculography and rotational chair testing and speech assessment showed differences between patients vs. controls [53,54].

3.4.5. Genotype-Specific Biomarkers in HSP

We identified biomarkers that were specific to particular HSP genotypes (Table 9). The most widely studied were serum and CSF oxysterols, 25- and 27-hydroxycholesterol (25- and 27-OHC) levels, in HSP-*CYP7B1* or SPG5. 25- and 27-OHC levels were significantly elevated in the serum and CSF of patients compared to healthy controls. Plasma and serum oxysterol levels were used as outcome measures in two clinical trials showing reduced levels with atorvastatin treatment but no clinical correlation [35,44]. Most genotype-specific biomarkers were reported in single studies, with some studies (citrulline, lipidomics, glycosylceramide profile, and scanning electron microscopy of hair shafts) having small patient numbers ($n \leq 5$).

Table 9. Genotype-specific biomarkers.

Biomarker	Studies	Genotype	Finding
Serum hydroxycholesterols [35,43–45]	4	HSP-*CYP7B1*	Higher 25- and 27-OHC levels in patients compared to controls
Plasma citrulline [55,56]	2	HSP-*ALDH18A1*	Low citrulline levels in patients (n = 3 and n = 4)
Lipidomics [57]	1	HSP-*PCYT2*	Accumulation of plasma phosphatidylcholine [O] etherphospholipids in patients (n = 3) compared to controls (n = 20)
Glycosylceramide profile [58]	1	HSP-*GBA2*	Elevated glycosylceramide levels in affected patient and carrier parent
ATG9A ratio (automated high-throughput imaging) [59]	1	HSP-*AP-4*	Increase in ATG9A ratio (intracellular distribution of ATG9A in trans-Golgi network compared to the remainder of the cell) in patient fibroblasts (n = 18) compared to asymptomatic carriers (n = 14)
mtDNA levels [60]	1	HSP-*SPG7*	Reduced mtDNA levels from whole blood in patients (n = 27) and carriers (n = 5) compared to controls (n = 17).
Scanning electron microscopy of hair shafts [61]	1	HSP-*FAHN*	Subtle to pronounced longitudinal grooves in hair shafts from 4/4 patients and adhesive plaques in 3/4 patients compared to controls.

3.5. Randomized Controlled Trials in HSP

We identified eight randomized controlled trials published over two decades (Supplementary Material S3). Of note, 5/8 RCTs we reviewed were not included in previously published reviews.

Six studies used a crossover design, while two studies were parallel randomized trials. Only two studies recruited patients with a single specific HSP genotype; the other six included mixed and unknown genotypes. All studies had small sample sizes (range 8–49). Two studies did not define a primary outcome measure [37,62]. Four studies showed a positive response to treatment. The outcome measures were heterogeneous, with no consistency between studies despite six studies aiming to evaluate spasticity in response to the intervention. The most common outcome measures used were the 10MWT and modified Ashworth score, with two positive studies showing a significant improvement in MAS with treatment [62,63]. Study durations were generally short—only one study had a duration of more than 6 months [64]. There was no clear association between the duration of the trial and response to the intervention.

4. Discussion

In this scoping review, we compiled a comprehensive list of outcome measures used in HSP (Figure 2) and categorized them according to construct (Tables 2–8), with information on key measurement properties. This is an important resource to inform the choice of outcome measures for future clinical trials in HSP. We identified eighty-three outcome measures highlighting the heterogeneity and inconsistency of outcome measures used. We identify a need for standardized outcome measures and recommendations for use in clinical trials, such as core outcome sets (COS) developed for other neurological conditions [19,65]. We identify common limitations of the included studies in this review (Figure 3) and list the advantages and disadvantages of commonly used outcome measures (Table 10).

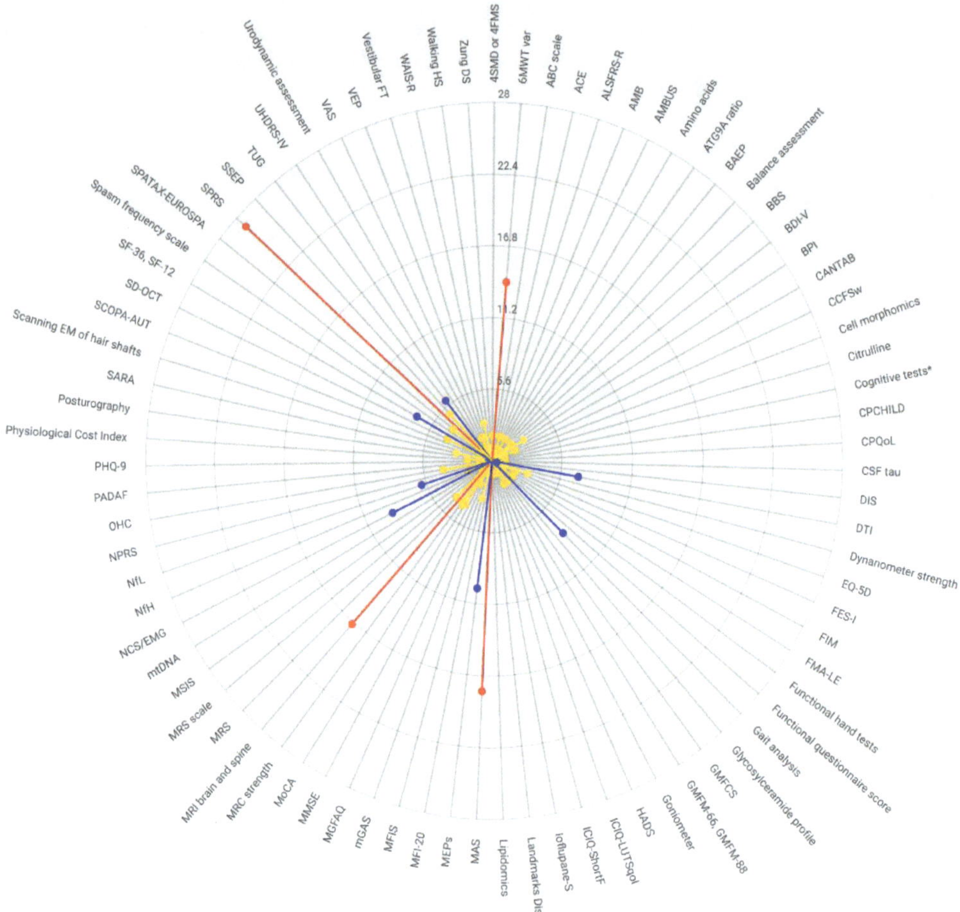

Figure 2. Outcome measures for hereditary spastic paraplegia according to frequency reported (Yellow 1–5 studies, Blue 6–10 studies, Red >10 studies). 4SMD or 4FMS—Four-stage functional scale of motor disability; 6MWT var—Six minute walk test and variations; ABC scale—Activities-specific Balance Confidence Scale; ACE—Addenbrooke's Cognitive Exam; ALSFRS-R—Amyotrophic Lateral Sclerosis Rating Scale Revised; AMB—Ambulatory score; AMBUS—Distance walked in meters walked in 5 s without help; ATG9A ratio—Autophagy-related protein 9A ratio; BAEP—brainstem auditory evoked potentials; BBS—Berg Balance Scale; BDI-V—Becks Depression Inventory; BPI—Brief Pain Inventory; CANTAB—CANTAB cognitive assessment; CCFSw—Composite Cerebellar Functional Severity Score; Cognitive tests*—Verbal Learning and Memory Test, Farbe-Wort-Interferenz Test, Trail Making Test Part A and B, Frontal Assessment Battery, revised d2 Test of attention, Regensburg Word Fluency Test; CPCHILD—Caregiver Priorities and Child Health Index of Life with Disabilities; CPQoL—Cerebral Palsy quality of life questionnaire; DIS—Disability score; DTI—diffusion tensor imaging; EQ-5D—EuroQoL 5 Dimensions; FES-I—Falls Efficacy Scale-International; FIM—Functional Independence Measure; FMA-LE—Lower extremity subclass of Fugl–Meyer assessment; GMFCS—Gross Motor Function Classification Score; GMFM—Gross Motor Function Measure; HADS—Hospital Anxiety and Depression Scale; ICIQ-LUTSqol—International Consultation of Incontinence Questionnaire lower urinary tract symptoms quality of life; ICIQ-ShortF—International Consultation of Incontinence Questionnaire Short Form; Ioflupane-S—Ioflupane-single photon emission computed tomography (SPECT); Landmarks

Dis—Landmarks of Disability; MAS—Modified Ashworth Scale; MEPs—motor evoked potentials; MFI-20—Multidimensional Fatigue Inventory; MFIS—Modified Fatigue Impact Scale; mGAS—Modified Goal Attainment Scale; MGFAQ—Modified version of Gillette Functional Assessment Questionnaire; MMSE—Mini-Mental State Exam; MoCA—Montreal Cognitive Assessment; MRC—Medical Research Council muscle strength; MRI—magnetic resonance imaging; MRS—magnetic resonance spectroscopy; MRScale—Modified Rankin Scale; MSIS—Multiple Sclerosis Impairment Scale; mtDNA—mitochondrial DNA load; NCS/EMG—nerve conduction studies/electromyography; NfH—Neurofilament heavy chain; NfL—Neurofilament light chain; NPRS—Numeric rating scale for pain; OHC—25 and 27 hydroxycholesterol; PADAF—protocol for evaluation of acquired speech disorder; PHQ-9—Patient Health Questionnaire; SARA—Scale for Assessment and Rating of Ataxia; scanning EM of hair shafts—scanning electron microscopy of hair shafts; SCOPA-AUT—Scale for Outcomes in Parkinson's Disease for Autonomic Symptoms; SD-OCT—spectral domain optical coherence tomography; SF-36, SF-12—Short Form 36, Short Form 12; SPRS—Spastic Paraplegia Rating Scale; SSEP—Somatosensory evoked potentials; TUG—Timed Up-and-Go test; UHDRS-IV—Unified Huntington's Disease Rating Scale Part IV; VAS—Visual Analogue Score; VEP—visual evoked potentials; Vestibular FT—video-oculography and rotational chair testing; WAIS-R—Wechsler Adult Intelligence Scale-revisited; Walking HS—Walking Handicap Scale; Zung DS—Zung Depression Scale.

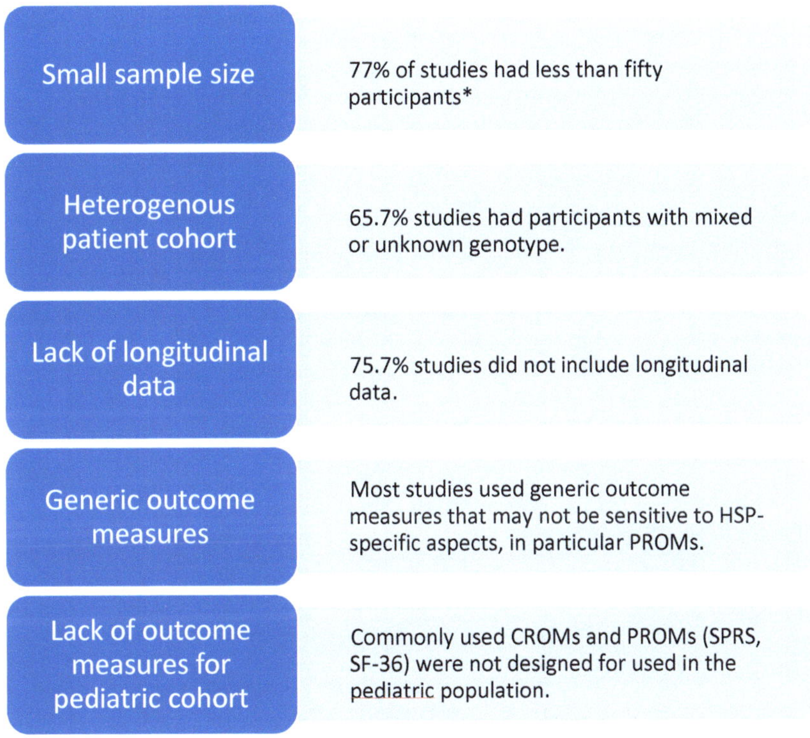

Figure 3. Common limitations of studies of HSP outcome measures. * [66] A sample size of less than fifty participants is defined as small in the COSMIN checklist. However, a smaller sample size may be justified in the study design. For example, a randomized controlled trial of 14 patients with *HSP-CYP7B1* calculated an estimate of the effect size of treatment using a pre-specified biomarker to determine the sample size required for adequate power [35].

Table 10. Advantages and disadvantages of commonly used outcome measures for hereditary spastic paraplegia.

Outcome Measure	Advantages	Disadvantages
SPRS	HSP-specific. Validated against other measures of disability. Cross-cultural validation [67,68]. Longitudinal measurement [30,33,35,69].	Not validated in a pediatric cohort. Inter-rater reliability only measured between two raters from same center [67]. Requires 10 m space and stairs to measure walking time.
SPATAX-EUROSPA disability score	HSP-specific.	Not validated. Does not measure pain, bladder or bowel function.
Modified Ashworth scale	Accessible. Moderate reliability [70].	Potential for inter-rater variation particularly for lower limb assessment [70]. Variation depends on when test is performed (e.g., the timing of anti-spasticity medication).
6MWT, 10MWT, TUG and variations	Accessible.	Not suitable for patients who are unable to mobilize. Potential variation depending on hallway lengths used due to time taken to change directions [71].
SF-36	Validated [72]. Published population norms [73]. Some versions easily accessible. Cross-cultural validity [73,74]. Demonstrated worse QoL in patients with HSP compared to controls [74,75].	Generic and does not address HSP-specific aspects of QoL.
EQ-5D	Validated [76]. Easily accessible. Cross-cultural validity [76]. Published population norms.	Generic and does not address HSP-specific aspects of QoL. Not as widely used as SF-36 in HSP population.
Serum and CSF neurofilament light chain (NfL)	Able to distinguish patients vs. controls, pre-symptomatic and symptomatic patients [29,77,78].	Non-specific, elevated in other conditions such as ALS, Alzheimer's dementia, Parkinson's Disease [79]. CSF collection requires an invasive procedure with potential adverse effects, such as a low-pressure headache.
27 and 25 hydroxycholesterol (OHC)	Specific to SPG5 (HSP-*CYP7B1*). Decrease in serum and plasma 27-OHC levels in response to atorvastatin [35,44]. Good diagnostic biomarker.	Biochemical response to treatment not reflected in clinical benefit as measured with SPRS. Role as prognostic or monitoring biomarker yet to be determined.
MRI brain and spine	Accessible. Volumetric analysis showed atrophy in certain parts of the brain or spine in patients with HSP [29,31,43,80].	Heterogenous findings between and within genotypes [81].
DTI	Abnormal in patients vs. controls (see results section). Correlate with other outcome measures [82–84]. Imaging can be performed with most MRI machines. Able to identify axonal damage not seen on MRI.	No longitudinal data. Requires analysis by experienced personnel.
MEPs	Able to measure upper motor neuron abnormalities seen in HSP [85]. Lower limb CMCT abnormal in 78% patients with HSP [86].	Inconsistent findings across various studies [86]. Requires specialized equipment and technical expertise. Some patients may find MEP studies uncomfortable.

Table 10. *Cont.*

Outcome Measure	Advantages	Disadvantages
NCS/EMG	Majority of patients with HSP-*SPG11* have axonal neuropathy [31,33].	Inconsistent findings across various studies.
Gait analysis	Able to differentiate patients vs. controls (see results). Able to demonstrate response to intervention [49,50]. Correlate with SPRS scores [28,34]. Mobile gait analysis system able to show disease progression over time [87].	Require expensive equipment. Data analysis can be complex depending on parameters used. Not suitable for participants who are unable to walk.
Retinal nerve fiber layer with OCT	Abnormal in 39% of patients with HSP [88].	No change in RNFL over time. No clinical correlation with SPRS. Inconsistent findings [88].

4.1. Recommendations for Future Research

4.1.1. Choice of Outcome Measure

This scoping review is the first comprehensive reference of outcome measures available for HSP according to outcome/construct. Based on our findings, we propose that a COS include a CROM—SPRS, PROM—SF36, and an objective biomarker—DTI, gait analysis or serum/CSF NfL.

We recommend the SPRS as it was the most commonly used CROM in HSP-related studies, has undergone psychometric testing in a cohort of individuals with HSP [67], has demonstrated cross-cultural validity [68], has been tested for responsiveness in longitudinal and interventional studies [30,33,35,62,64,69,89], and is a disease-specific outcome measure. In addition, it includes an assessment of motor function, and therefore, a performance outcome measure (PerfOM) for motor function is not required. However, we note that SPRS was not designed for use in the pediatric population, and therefore, we identify a need for a validated pediatric HSP-specific CROM.

We recommend the SF-36 as the most suitable PROM as it was commonly used in HSP-related studies, is well-validated in healthy controls and other conditions [72], has cross-cultural validity in the HSP population [73,74], and correlates with disease severity as measured with the SPRS [75,90,91] and gait analysis [34]. However, we note that the SF-36 is a generic QoL measure and may not be sensitive to smaller changes in HSP-specific symptoms, such as spasticity and bladder function, that are likely key targets for intervention in future HSP trials [92]. Therefore, we identify a need for an HSP-specific QoL scale to address the need for more sensitive tools, particularly when evaluating small changes in response to treatment in a slowly progressive condition. These findings are echoed by a recent study of CROMs and PROMs in HSP, identifying the SPRS as a suitable CROM to measure disease progression and the need for an HSP-specific PROM [89].

Diffusion tensor imaging, gait analysis, and neurofilament light chain levels are promising objective biomarkers likely to be suitable for use in clinical trials. However, further studies are needed to establish the sensitivity and specificity of these biomarkers in HSP, including longitudinal studies. Biomarkers identified through knowledge of disease pathways in specific HSP genotypes can aid in the diagnosis and measurement of treatment response. Abnormal plasma and CSF 25- and 27-OH levels are seen in individuals with HSP-*CYPB1* (SPG5) and responded to treatment with statins [35,44]. More recently, two groups used different approaches to differentiate individuals with HSP-*SPAST* (SPG4) from healthy controls by analyzing peripheral blood mononuclear cells. Our group demonstrated reduced levels of acetylated α-tubulin seen on flow cytometry [93], while another group showed increased distance between cell and nucleus centroids on automated image analysis [94]. Both studies used surrogate markers of microtubule dysfunction based on the known role of spastin in regulating microtubule dynamics in HSP-*SPAST*. These findings highlight the need for further research to identify genotype-specific biomarkers that are more likely to be sensitive and specific for particular HSP genotypes. For the evaluation of existing outcome measures or the development of new outcome measures, we recommend referring to published guidelines [19] to ensure validity and reliability. Development of a COS for HSP should be informed by published recommendations, COS-STAD, to ensure that the included outcome measures meet minimum standards [95].

4.1.2. Recommendations for Trial Design

Recommendations for HSP clinical trial design will improve the quality and consistency of reporting of evidence to strengthen conclusions drawn from clinical trials. We identify a need for larger sample sizes in HSP studies, a particular challenge when studying rare conditions. International, multi-center collaborations are a potential solution to sample size challenges and have the added benefit of cross-cultural applicability of study results. It is important for outcome measures chosen to be available in different languages, relevant in different populations, and feasible for use in low-resource settings. Biomarkers that can be collected at recruitment sites and analyzed in a central facility are ideal for multi-center trials to improve the standardization of data. Although studies of single HSP genotypes are ideal for consistency within target populations, this may not be feasible when trying to attain large sample sizes. Therefore, a grouping of HSP genotypes or other neurodegenerative disorders according to similar underlying disease pathophysiology may allow for more efficient research of targeted therapeutic agents. Clinical trials for HSP may require longer periods of treatment and follow-up to demonstrate significant treatment effects, particularly in the more slowly progressive genotypes, such as HSP-*SPAST*. Using a combination of multiple outcome measures and biomarkers can account for phenotypic variability, even within the same genotype. It may improve the chance of detecting a treatment effect signal in heterogeneous patient cohorts.

4.1.3. Study Limitations

A limitation of this study was the use of the search terms "outcome measure" and "biomarker", which inevitably missed relevant studies [19]. When planning our study design, we considered the volume of search results from individual searches for each outcome, e.g., spasticity, mobility, neuroimaging, etc. and deemed that approach unfeasible.

Systematic reviews on specific outcome measures have been published previously [81,86,96,97] and provide information on the utility of these outcome measures. Due to the nature of a scoping review, we were unable to perform qualitative analysis of each outcome measure to assess validity and reliability. Therefore, a systematic review specific to an outcome measure is required to answer this question. Our review identifies outcome measures that require further validation and provides an overview of the currently available literature surrounding the identified outcome measures.

5. Conclusions

In this scoping review, we present a critical assessment of currently available outcome measures for use in HSP clinical trials. We discuss the benefits and limitations of commonly used outcome measures and propose areas for further research. Given the emergence of multiple candidate HSP therapies in recent years [35,44,98], there is an urgent need for further development of a core set of validated and standardized outcome measures for use in HSP clinical trials to test the efficacy of these therapies.

Supplementary Materials: The following supporting information can be downloaded at https://www.mdpi.com/article/10.3390/genes14091756/s1; Supplementary Material S1: List of the included studies in scoping review; Supplementary Material S2: Table of HSP outcome measures and biomarkers; Supplementary Material S3: Comparison of Randomized controlled trials in HSP.

Author Contributions: Conceptualization, S.-F.S., G.W., K.R.K. and C.M.S.; methodology, S.-F.S.; results screening, S.-F.S., D.Y. and K.R.K.; data extraction, S.-F.S., D.Y., L.I.R. and F.J.; data analysis, S.-F.S., D.Y. and G.W.; writing—original draft preparation, S.-F.S.; writing—review and editing, all authors. All authors have read and agreed to the published version of the manuscript.

Funding: S.-F.S. is the recipient of the National Health and Medical Research Council and Australian and New Zealand Association of Neurologists Postgraduate Scholarships APP113610.

Institutional Review Board Statement: Not applicable.

Informed Consent Statement: Not applicable.

Data Availability Statement: All data supporting reported materials are available in the manuscript and Supplementary Materials.

Acknowledgments: The authors would like to thank Tess Aitken, Academic Liaison Librarian at the University of Sydney, for guidance in searching the databases. K.R.K. is supported by the Paul Ainsworth Family Foundation and the Medical Research Future Fund.

Conflicts of Interest: The authors declare no conflict of interest.

Appendix A

All database searches performed on 9 August 2022.

Embase search
Embase Classic <1947 to 1973>
Embase <1974 to 9 August 2022>

1. spastic paraplegia/or spastic para*.mp. 11219
2. hereditary motor sensory neuropathy/or hereditary spastic parap*.mp. 13564
3. 1 or 2 21406
4. pharmacological biomarker/or biomarker*.mp. or biological marker/ 660089
5. patient-reported outcome/or outcome assessment/or "outcome measure*".mp. or "quality of life"/ 1469067
6. 4 or 5 2095172
7. 3 and 6 1267
8. remove duplicates from 7 1256

Medline search
Ovid MEDLINE(R) ALL <1946 to 9 August 2022>

1. Spastic Paraplegia, Hereditary/or "spastic para*".mp. 6979
2. Spastic Paraplegia, Hereditary/or "hereditary spastic para*".mp. or Paraparesis, Spastic/ 2875
3. biomarker*.mp. or Biomarkers, Pharmacological/or Biomarkers/ 719216
4. "outcome measure".mp. or Outcome Assessment, Health Care/ 147938
5. "patient reported outcome measure".mp. or Patient Reported Outcome Measures/ 13303
6. 1 or 2 7101
7. 3 or 4 or 5 873824
8. 6 and 7 199

Scopus search
TITLE-ABS-KEY (("hereditary spastic parap*" OR "spastic parap*") AND (biomarker* OR "outcome measure*"))
210 results
Web of Science search
("hereditary spastic parap*" OR "spastic parap*") AND (biomarker* OR "outcome measure*")
230 results
Central Cochrane Database
EBM Reviews—Cochrane Central Register of Controlled Trials <July 2022>

1. "hereditary spastic paraplegia".mp. or Spastic Paraplegia, Hereditary/ 35

References

1. Ruano, L.; Melo, C.; Silva, M.C.; Coutinho, P. The Global Epidemiology of Hereditary Ataxia and Spastic Paraplegia: A Systematic Review of Prevalence Studies. *Neuroepidemiology* **2014**, *42*, 174–183. [CrossRef]
2. Erichsen, A.K.; Koht, J.; Stray-Pedersen, A.; Abdelnoor, M.; Tallaksen, C.M.E. Prevalence of hereditary ataxia and spastic paraplegia in southeast Norway: A population-based study. *Brain* **2009**, *132*, 1577–1588. [CrossRef] [PubMed]
3. Coutinho, P.; Ruano, L.; Loureiro, J.L.; Cruz, V.T.; Barros, J.; Tuna, A.; Barbot, C.; Guimarães, J.; Alonso, I.; Silveira, I.; et al. Hereditary ataxia and spastic paraplegia in Portugal: A population-based prevalence study. *JAMA Neurol.* **2013**, *70*, 746–755. [CrossRef]
4. Vander Stichele, G.; Durr, A.; Yoon, G.; Schüle, R.; Blackstone, C.; Esposito, G.; Buffel, C.; Oliveira, I.; Freitag, C.; van Rooijen, S.; et al. An integrated modelling methodology for estimating global incidence and prevalence of hereditary spastic paraplegia subtypes SPG4, SPG7, SPG11, and SPG15. *BMC Neurol.* **2022**, *22*, 115. [CrossRef]
5. Kerstens, H.C.J.W.; Lith, B.J.H.v.; Nijkrake, M.J.; Swart, B.J.M.d.; Bemd, L.A.C.v.d.; Smeets, R.; Klemens, F.; Warrenburg, B.P.C.v.d.; Wees, P.J.v.d.; Geurts, A.C.H. Healthcare needs, expectations, utilization, and experienced treatment effects in patients with hereditary spastic paraplegia: A web-based survey in the Netherlands. *Orphanet J. Rare Dis.* **2021**, *16*, 1–283. [CrossRef]
6. Shribman, S.; Reid, E.; Crosby, A.H.; Houlden, H.; Warner, T.T. Hereditary spastic paraplegia: From diagnosis to emerging therapeutic approaches. *Lancet Neurol.* **2019**, *18*, 1136–1146. [CrossRef] [PubMed]
7. Meyyazhagan, A.; Orlacchio, A. Hereditary Spastic Paraplegia: An Update. *Int. J. Mol. Sci.* **2022**, *23*, 1697. [CrossRef]
8. Saputra, L.; Kumar, K.R. Challenges and Controversies in the Genetic Diagnosis of Hereditary Spastic Paraplegia. *Curr. Neurol. Neurosci. Rep.* **2021**, *21*, 15. [CrossRef]
9. Harding, A.E. Classification of the hereditary ataxias and paraplegias. *Lancet* **1983**, *1*, 1151–1155. [CrossRef]
10. Méreaux, J.-L.; Banneau, G.; Papin, M.; Coarelli, G.; Valter, R.; Raymond, L.; Kol, B.; Ariste, O.; Parodi, L.; Tissier, L.; et al. Clinical and genetic spectra of 1550 index patients with hereditary spastic paraplegia. *Brain* **2022**, *145*, 1029–1037. [CrossRef]
11. Wali, G.; Sutharsan, R.; Fan, Y.; Stewart, R.; Tello Velasquez, J.; Sue, C.M.; Crane, D.I.; Mackay-Sim, A. Mechanism of impaired microtubule-dependent peroxisome trafficking and oxidative stress in SPAST-mutated cells from patients with Hereditary Spastic Paraplegia. *Sci. Rep.* **2016**, *6*, 27004. [CrossRef]
12. Denton, K.R.; Lei, L.; Grenier, J.; Rodionov, V.; Blackstone, C.; Li, X.J. Loss of Spastin Function Results in Disease-Specific Axonal Defects in Human Pluripotent Stem Cell-Based Models of Hereditary Spastic Paraplegia. *Stem Cells* **2014**, *32*, 414–423. [CrossRef]
13. Fassier, C.; Tarrade, A.; Peris, L.; Courageot, S.; Mailly, P.; Dalard, C.; Delga, S.; Roblot, N.; Lefèvre, J.; Job, D.; et al. Microtubule-targeting drugs rescue axonal swellings in cortical neurons from spastin knockout mice. *Dis. Model. Mech.* **2013**, *6*, 72–83. [CrossRef]
14. Wali, G.; Liyanage, E.; Blair, N.F.; Sutharsan, R.; Park, J.S.; Mackay-Sim, A.; Sue, C.M. Oxidative Stress-Induced Axon Fragmentation Is a Consequence of Reduced Axonal Transport in Hereditary Spastic Paraplegia SPAST Patient Neurons. *Front. Neurosci.* **2020**, *14*, 401. [CrossRef]
15. Fan, Y.; Wali, G.; Sutharsan, R.; Bellette, B.; Crane, D.I.; Sue, C.M.; Mackay-Sim, A. Low dose tubulin-binding drugs rescue peroxisome trafficking deficit in patient-derived stem cells in Hereditary Spastic Paraplegia. *Biol. Open* **2014**, *3*, 494–502. [CrossRef]
16. Halley, M.C.; Smith, H.S.; Ashley, E.A.; Goldenberg, A.J.; Tabor, H.K. A call for an integrated approach to improve efficiency, equity and sustainability in rare disease research in the United States. *Nat. Genet.* **2022**, *54*, 219–222. [CrossRef] [PubMed]
17. Bellofatto, M.; De Michele, G.; Iovino, A.; Filla, A.; Santorelli, F.M. Management of Hereditary Spastic Paraplegia: A Systematic Review of the Literature. *Front. Neurol.* **2019**, *10*, 3. [CrossRef]
18. Chan, A.-W.; Tetzlaff, J.M.; Gøtzsche, P.C.; Altman, D.G.; Mann, H.; Berlin, J.A.; Dickersin, K.; Hróbjartsson, A.; Schulz, K.F.; Parulekar, W.R.; et al. SPIRIT 2013 explanation and elaboration: Guidance for protocols of clinical trials. *BMJ* **2013**, *346*, e7586. [CrossRef]
19. Prinsen, C.A.C.; Vohra, S.; Rose, M.R.; Boers, M.; Tugwell, P.; Clarke, M.; Williamson, P.R.; Terwee, C.B. How to select outcome measurement instruments for outcomes included in a "Core Outcome Set"—A practical guideline. *Curr. Control. Trials Cardiovasc. Med.* **2016**, *17*, 449. [CrossRef]
20. Trummer, B.; Haubenberger, D.; Blackstone, C. Clinical Trial Designs and Measures in Hereditary Spastic Paraplegias. *Front. Neurol.* **2018**, *9*, 1017. [CrossRef]
21. Williamson, P.R.; Altman, D.G.; Blazeby, J.M.; Clarke, M.; Devane, D.; Gargon, E.; Tugwell, P. Developing core outcome sets for clinical trials: Issues to consider. *Curr. Control. Trials Cardiovasc. Med.* **2012**, *13*, 132. [CrossRef] [PubMed]
22. Peters, M.D.J.; Godfrey, C.M.; Khalil, H.; McInerney, P.; Parker, D.; Soares, C.B. Guidance for conducting systematic scoping reviews. *Int. J. Evid. Based Healthc.* **2015**, *13*, 141–146. [CrossRef]
23. Peters, M.D.J.; Marnie, C.; Tricco, A.C.; Pollock, D.; Munn, Z.; Alexander, L.; McInerney, P.; Godfrey, C.M.; Khalil, H. Updated methodological guidance for the conduct of scoping reviews. *JBI Evid. Synth.* **2020**, *18*, 2119–2126. [CrossRef] [PubMed]
24. Veritas Health Innovation, Melbourne Australia. Covidence Systematic Review Software. Available online: www.covidence.org. (accessed on 12 July 2023).
25. U.S. Food and Drug Administration. Clinical Outcome Assessment (COA): Frequently Asked Questions. Available online: https://www.fda.gov/about-fda/clinical-outcome-assessment-coa-frequently-asked-questions#COADefinition (accessed on 2 June 2023).

26. Group OCfE-BMLoEW. The Oxford 2011 Levels of Evidence. Available online: http://www.cebm.net/index.aspx?o=56532011 (accessed on 12 July 2023).
27. Page, M.J.; McKenzie, J.E.; Bossuyt, P.M.; Boutron, I.; Hoffmann, T.C.; Mulrow, C.D.; Shamseer, L.; Tetzlaff, J.M.; Akl, E.A.; Brennan, S.E.; et al. The PRISMA 2020 statement: An updated guideline for reporting systematic reviews. *BMJ* **2021**, *372*, n71. [CrossRef] [PubMed]
28. Lassmann, C.; Ilg, W.; Schneider, M.; Volker, M.; Haeufle, D.F.B.; Schule, R.; Giese, M.; Schols, L.; Rattay, T.W. Specific gait changes in prodromal hereditary spastic paraplegia type 4—preSPG4 study. *Medrxiv* **2022**, *21*. [CrossRef]
29. Rattay, T.W.; Volker, M.; Rautenberg, M.; Kessler, C.; Wurster, I.; Winter, N.; Haack, T.B.; Lindig, T.; Hengel, H.; Synofzik, M.; et al. The prodromal phase of hereditary spastic paraplegia type 4: The preSPG4 cohort study. *Brain* **2023**, *146*, 1093–1102. [CrossRef] [PubMed]
30. Cubillos-Arcila, D.M.; Machado, G.D.; Sehnem, L.; Burguez, D.; Zanardi, A.P.J.; Martins, V.F.; Peyre-Tartaruga, L.A.; Saute, J.A.M. Progression of Functional Gait in Hereditary Spastic Paraplegias. *Cerebellum* **2022**, *21*, 350–357. [CrossRef] [PubMed]
31. Faber, I.; Martinez, A.R.M.; de Rezende, T.J.R.; Martins, C.R.; Martins, M.P.; Lourenço, C.M.; Marques, W.; Montecchiani, C.; Orlacchio, A.; Pedroso, J.L.; et al. SPG11 mutations cause widespread white matter and basal ganglia abnormalities, but restricted cortical damage. *NeuroImage Clin.* **2018**, *19*, 848–857. [CrossRef]
32. Kessler, C.; Serna-Higuita, L.M.; Wilke, C.; Rattay, T.W.; Hengel, H.; Reichbauer, J.; Stransky, E.; Leyva-Gutierrez, A.; Mengel, D.; Synofzik, M.; et al. Characteristics of serum neurofilament light chain as a biomarker in hereditary spastic paraplegia type 4. *Ann. Clin. Transl. Neurol.* **2022**, *9*, 326–338. [CrossRef]
33. Musacchio, T.; Nohl, K.; Boelmans, K.; Maltese, V.; Zeller, D.; Isaias, I.; Volkmann, J.; Klebe, S. Biological course and natural history of hereditary spastic paraplegia type 11 (SPG11). *Mov. Disord.* **2018**, *33* (Suppl. 2), S225.
34. Regensburger, M.; Spatz, I.T.; Ollenschlager, M.; Martindale, C.F.; Lindeburg, P.; Kohl, Z.; Eskofier, B.; Klucken, J.; Schule, R.; Klebe, S.; et al. Inertial Gait Sensors to Measure Mobility and Functioning in Hereditary Spastic Paraplegia: A Cross-Sectional Multicenter Clinical Study. *Neurology* **2022**, *99*, e1079–e1089. [CrossRef] [PubMed]
35. Schols, L.; Rattay, T.W.; Martus, P.; Meisner, C.; Baets, J.; Fischer, I.; Jagle, C.; Fraidakis, M.J.; Martinuzzi, A.; Saute, J.A.; et al. Hereditary spastic paraplegia type 5: Natural history, biomarkers and a randomized controlled trial. *Brain* **2017**, *140*, 3112–3127. [CrossRef] [PubMed]
36. Paparella, G.; Vavla, M.; Bernardi, L.; Girardi, G.; Stefan, C.; Martinuzzi, A. Efficacy of a Combined Treatment of Botulinum Toxin and Intensive Physiotherapy in Hereditary Spastic Paraplegia. *Front. Neurosci.* **2020**, *14*, 111. [CrossRef] [PubMed]
37. Scheuer, K.H.; Svenstrup, K.; Jennum, P.; Rogvi-Hansen, B.; Werdelin, L.; Fenger, K.; Nielsen, J.E. Double-blind crossover trial of gabapentin in SPG4-linked hereditary spastic paraplegia. *Eur. J. Neurol.* **2007**, *14*, 663–666. [CrossRef]
38. Du Montcel, S.T.; Charles, P.; Ribai, P.; Goizet, C.; Le Bayon, A.; Labauge, P.; Guyant-Marechal, L.; Forlani, S.; Jauffret, C.; Vandenberghe, N.; et al. Composite cerebellar functional severity score: Validation of a quantitative score of cerebellar impairment. *Brain* **2008**, *131*, 1352–1361. [CrossRef]
39. Du Montcel, S.T.; Charles, P.; Goizet, C.; Marelli, C.; Ribai, P.; Vincitorio, C.; Anheim, M.; Guyant-Marechal, L.; Le Bayon, A.; Vandenberghe, N.; et al. Factors influencing disease progression in autosomal dominant cerebellar ataxia and spastic paraplegia. *Arch. Neurol.* **2012**, *69*, 500–508. [CrossRef]
40. Simonini, C.; Zucchi, E.; Bedin, R.; Martinelli, I.; Gianferrari, G.; Fini, N.; Sorau, G.; Liguori, R.; Vacchiano, V.; Mandrioli, J. CSF Heavy Neurofilament May Discriminate and Predict Motor Neuron Diseases with Upper Motor Neuron Involvement. *Biomedicines* **2021**, *9*, 1623. [CrossRef]
41. Zucchi, E.; Bedin, R.; Fasano, A.; Fini, N.; Gessani, A.; Vinceti, M.; Mandrioli, J. Cerebrospinal Fluid Neurofilaments May Discriminate Upper Motor Neuron Syndromes: A Pilot Study. *Neurodegener. Dis.* **2018**, *18*, 255–261. [CrossRef]
42. Wilke, C.; Rattay, T.W.; Hengel, H.; Zimmermann, M.; Brockmann, K.; Schols, L.; Jens, K.A.; Schule, R.; Synofzik, M. Serum neurofilament light chain is increased in hereditary spastic paraplegias. *Ann. Clin. Transl. Neurol.* **2018**, *5*, 876–882. [CrossRef]
43. Lin, Q.; Liu, Y.; Ye, Z.; Hu, J.; Cai, W.; Weng, Q.; Chen, W.J.; Wang, N.; Cao, D.; Lin, Y.; et al. Potential markers for sample size estimations in hereditary spastic paraplegia type 5. *Orphanet J. Rare Dis.* **2021**, *16*, 1–9. [CrossRef]
44. Marelli, C.; Lamari, F.; Rainteau, D.; Lafourcade, A.; Banneau, G.; Humbert, L.; Monin, M.L.; Petit, E.; Debs, R.; Castelnovo, G.; et al. Plasma oxysterols: Biomarkers for diagnosis and treatment in spastic paraplegia type 5. *Brain* **2018**, *141*, 72–84. [CrossRef] [PubMed]
45. Prestsaeter, S.; Koht, J.; Lamari, F.; Tallaksen, C.M.E.; Hoven, S.T.J.; Vigeland, M.D.; Selmer, K.K.; Rydning, S.L. Elevated hydroxycholesterols in Norwegian patients with hereditary spastic paraplegia SPG5. *J. Neurol. Sci.* **2020**, *419*, 117211. [CrossRef] [PubMed]
46. Sarret, C.; Lemaire, J.J.; Tonduti, D.; Sontheimer, A.; Coste, J.; Pereira, B.; Feschet, F.; Roche, B.; Boespflug-Tanguy, O. Time-course of myelination and atrophy on cerebral imaging in 35 patients with PLP1-related disorders. *Dev. Med. Child Neurol.* **2016**, *58*, 706–713. [CrossRef] [PubMed]
47. Bonnefoy-Mazure, A.; Turcot, K.; Kaelin, A.; De Coulon, G.; Armand, S. Full body gait analysis may improve diagnostic discrimination between hereditary spastic paraplegia and spastic diplegia: A preliminary study. *Res. Dev. Disabil.* **2013**, *34*, 495–504. [CrossRef]
48. Adair, B.; Rodda, J.; McGinley, J.L.; Graham, H.K.; Morris, M.E. Kinematic gait deficits at the trunk and pelvis: Characteristic features in children with hereditary spastic paraplegia. *Dev. Med. Child Neurol.* **2016**, *58*, 829–835. [CrossRef]

49. Zhang, Y.; Roxburgh, R.; Huang, L.; Parsons, J.; Davies, T.C. The effect of hydrotherapy treatment on gait characteristics of hereditary spastic paraparesis patients. *Gait Posture* **2014**, *39*, 1074–1079. [CrossRef]
50. Lith, B.J.H.v.; Boer, J.J.d.; Warrenburg, B.P.C.v.d.; Weerdesteyn, V.G.M.; Geurts, A.C.H. Functional effects of botulinum toxin type a in the hip adductors and subsequent stretching in patients with hereditary spastic paraplegia. *J. Rehabil. Med.* **2019**, *51*, 434–441. [CrossRef]
51. Pointon, R.; Whelan, H.; Raza, R.; Peacock, S.; Wilsmore, C.; Mulkeen, A.; Goodden, J.; Lodh, R. The use of intrathecal baclofen for management of spasticity in hereditary spastic paraparesis: A case series. *Eur. J. Paediatr. Neurol.* **2022**, *36*, 14–18. [CrossRef]
52. Henderson, G.; De Gheldere, A.; Gilman, J.; Hewart, P.; Clements, P.; Bliss, W.; Henman, P. A service evaluation of gait correction surgery for patients with hereditary spastic paraplegia. *Gait Posture* **2019**, *73* (Suppl. 1), 251–252. [CrossRef]
53. Milenkovic, I.; Klotz, S.; Zulehner, G.; Sycha, T.; Wiest, G. Slowed vertical saccades as a hallmark of hereditary spastic paraplegia type 7. *Ann. Clin. Transl. Neurol.* **2019**, *6*, 2127–2132. [CrossRef]
54. Olchik, M.; Scudeiro, L.; Dos Santos, V.; Ayres, A.; Rech, R.; Burguez, D.; Machado, G.; Gonzalez-Salazar, C.; Padovani, M.; Franca, M.; et al. Speech biomarkers in hereditary spastic paraplegia. *Mov. Disord.* **2021**, *36* (Suppl. 1), S490. [CrossRef]
55. Marelli, C.; Badiou, S.; Genestet, S.; Larrieu, L.; Damier, P.; Camu, W.; Planes, M.; Koenig, M.; Guissart, C. Autosomal dominant SPG9: Intrafamilial variability and onset during pregnancy. *Neurol. Sci.* **2020**, *41*, 1931–1933. [CrossRef] [PubMed]
56. Coutelier, M.; Goizet, C.; Durr, A.; Habarou, F.; Morais, S.; Dionne-Laporte, A.; Tao, F.; Konop, J.; Stoll, M.; Charles, P.; et al. Alteration of ornithine metabolism leads to dominant and recessive hereditary spastic paraplegia. *Brain* **2015**, *138*, 2191–2205. [CrossRef]
57. Vaz, F.M.; McDermott, J.H.; Alders, M.; Wortmann, S.B.; Kolker, S.; Pras-Raves, M.L.; Vervaart, M.A.T.; van Lenthe, H.; Luyf, A.C.M.; Elfrink, H.L.; et al. Mutations in PCYT2 disrupt etherlipid biosynthesis and cause a complex hereditary spastic paraplegia. *Brain* **2019**, *142*, 3382–3397. [CrossRef] [PubMed]
58. Kloth, K.; Cozma, C.; Bester, M.; Gerloff, C.; Biskup, S.; Zittel, S. Dystonia as initial presentation of compound heterozygous GBA2 mutations: Expanding the phenotypic spectrum of SPG46. *Eur. J. Med. Genet.* **2020**, *63*, 5. [CrossRef]
59. Ebrahimi-Fakhari, D.; Alecu, J.E.; Brechmann, B.; Ziegler, M.; Eberhardt, K.; Jumo, H.; D'Amore, A.; Habibzadeh, P.; Faghihi, M.A.; De Bleecker, J.L.; et al. High-throughput imaging of ATG9A distribution as a diagnostic functional assay for adaptor protein complex 4-associated hereditary spastic paraplegia. *Brain Commun.* **2021**, *3*, fcab221. [CrossRef]
60. De la Casa-Fages, B.; Fernandez-Eulate, G.; Gamez, J.; Barahona-Hernando, R.; Moris, G.; Garcia-Barcina, M.; Infante, J.; Zulaica, M.; Fernandez-Pelayo, U.; Munoz-Oreja, M.; et al. Parkinsonism and spastic paraplegia type 7: Expanding the spectrum of mitochondrial Parkinsonism. *Mov. Disord.* **2019**, *34*, 1547–1561. [CrossRef]
61. Rattay, T.W.; Lindig, T.; Baets, J.; Smets, K.; Deconinck, T.; Sohn, A.S.; Hortnagel, K.; Eckstein, K.N.; Wiethoff, S.; Reichbauer, J.; et al. FAHN/SPG35: A narrow phenotypic spectrum across disease classifications. *Brain* **2019**, *142*, 1561–1572. [CrossRef]
62. Ardolino, G.; Bocci, T.; Nigro, M.; Vergari, M.; Di Fonzo, A.; Bonato, S.; Cogiamanian, F.; Cortese, F.; Cova, I.; Barbieri, S.; et al. Spinal direct current stimulation (tsDCS) in hereditary spastic paraplegias (HSP): A sham-controlled crossover study. *J. Spinal Cord. Med.* **2021**, *44*, 46–53. [CrossRef]
63. Bastani, P.B.; Kordjazi, M.; Oveisgharan, S.; Abdi, S. A Randomized Controlled Trial of the Effect of Repetitive Transcranial Magnetic Stimulation of the Motor Cortex on Lower Extremity Spasticity in Hereditary Spastic Paraplegia. *J. Clin. Neurophysiol. Off. Publ. Am. Electroencephalogr. Soc.* **2021**, *40*, 173–179. [CrossRef]
64. de Lima, F.D.; Faber, I.; Servelhere, K.R.; Bittar, M.F.R.; Martinez, A.R.M.; Piovesana, L.G.; Martins, M.P.; Martins, C.R.; Benaglia, T.; Carvalho, B.D.; et al. Randomized Trial of Botulinum Toxin Type A in Hereditary Spastic Paraplegia—The SPASTOX Trial. *Mov. Disord.* **2021**, *36*, 1654–1663. [CrossRef]
65. Mitchell, J.W.; Noble, A.; Baker, G.; Batchelor, R.; Brigo, F.; Christensen, J.; French, J.; Gil-Nagel, A.; Guekht, A.; Jette, N.; et al. Protocol for the development of an international Core Outcome Set for treatment trials in adults with epilepsy: The EPilepsy outcome Set for Effectiveness Trials Project (EPSET). *Trials* **2022**, *23*, 943. [CrossRef] [PubMed]
66. Mokkink, L.B.; de Vet, H.C.W.; Prinsen, C.A.C.; Patrick, D.L.; Alonso, J.; Bouter, L.M.; Terwee, C.B. COSMIN Risk of Bias checklist for systematic reviews of Patient-Reported Outcome Measures. *Qual. Life Res.* **2018**, *27*, 1171–1179. [CrossRef] [PubMed]
67. SchÜLe, R.; Holland-Letz, T.; Klimpe, S.; Kassubek, J.; Klopstock, T.; Mall, V.; Otto, S.; Winner, B.; Schols, L. The Spastic Paraplegia Rating Scale (SPRS): A reliable and valid measure of disease severity. *Neurology* **2006**, *67*, 430–434. [CrossRef]
68. Servelhere, K.R.; Faber, I.; Coan, A.C.; França, J.M. Translation and validation into Brazilian Portuguese of the Spastic Paraplegia Rating Scale (SPRS). *Arq. De Neuro-Psiquiatr.* **2016**, *74*, 489–494. [CrossRef]
69. Montanaro, D.; Vavla, M.; Frijia, F.; Aghakhanyan, G.; Baratto, A.; Coi, A.; Stefan, C. Multimodal MRI Longitudinal Assessment of White and Gray Matter in Different SPG Types of Hereditary Spastic Paraparesis. *Front. Neurosci.* **2020**, *14*, 325. [CrossRef]
70. Meseguer-Henarejos, A.B.; Sánchez-Meca, J.; López-Pina, J.A.; Carles-Hernández, R. Inter- and intra-rater reliability of the Modified Ashworth Scale: A systematic review and meta-analysis. *Eur. J. Phys. Rehabil. Med.* **2018**, *54*, 576–590. [CrossRef]
71. ATS statement: Guidelines for the six-minute walk test. *Am. J. Respir. Crit. Care Med.* **2002**, *166*, 111–117. [CrossRef]
72. Stewart, A.; Sherbourne, C.D.; Ware, J.E.; Hays, R.D.; Wells, K.B.; Berry, S.H.; Kamberg, C.; Nelson, E.C.; Davies, A.R.; Rogers, W.H.; et al. *Measuring Functioning and Well Being: The Medical Outcomes Study Approach*; Duke University Press: Durham, UK; London, UK, 1992.

73. Lins, L.; Carvalho, F.M. SF-36 total score as a single measure of health-related quality of life: Scoping review. *SAGE Open Med.* **2016**, *4*, 2050312116671725. [CrossRef]
74. Braschinsky, M.; Rannikmae, K.; Krikmann, U.; Luus, S.M.; Raidvee, A.; Gross-Paju, K.; Haldre, S. Health-related quality of life in patients with hereditary spastic paraplegia in Estonia. *Spinal Cord.* **2011**, *49*, 175–181. [CrossRef]
75. Gassner, H.; List, J.; Martindale, C.F.; Regensburger, M.; Klucken, J.; Winkler, J.; Kohl, Z. Functional gait measures correlate to fear of falling, and quality of life in patients with Hereditary Spastic Paraplegia: A cross-sectional study. *Clin. Neurol. Neurosurg.* **2021**, *209*, 106888. [CrossRef] [PubMed]
76. Rabin, R.; de Charro, F. EQ-5D: A measure of health status from the EuroQol Group. *Ann. Med.* **2001**, *33*, 337–343. [CrossRef] [PubMed]
77. Kessler, C.; Serna-Higuita, L.M.; Rattay, T.W.; Maetzler, W.; Wurster, I.; Hayer, S.; Wilke, C.; Hengel, H.; Reichbauer, J.; Armbruster, M.; et al. Neurofilament light chain is a cerebrospinal fluid biomarker in hereditary spastic paraplegia. *Ann. Clin. Transl. Neurol.* **2021**, *8*, 1122–1131. [CrossRef]
78. Alecu, J.E.; Saffari, A.; Ziegler, M.; Jordan, C.; Tam, A.; Kim, S.; Leung, E.; Szczaluba, K.; Mierzewska, H.; King, S.D.; et al. Plasma Neurofilament Light Chain Is Elevated in Adaptor Protein Complex 4-Related Hereditary Spastic Paraplegia. *Mov. Disord.* **2023**. *online ahead of print*. [CrossRef] [PubMed]
79. Gaiottino, J.; Norgren, N.; Dobson, R.; Topping, J.; Nissim, A.; Malaspina, A.; Bestwick, J.P.; Monsch, A.U.; Regeniter, A.; Lindberg, R.L.; et al. Increased neurofilament light chain blood levels in neurodegenerative neurological diseases. *PLoS ONE* **2013**, *8*, e75091. [CrossRef]
80. Navas-Sanchez, F.J.; Fernandez-Pena, A.; de Blas, D.M.; Aleman-Gomez, Y.; Marcos-Vidal, L.; Guzman-de-Villoria, J.A.; Fernandez-Garcia, P.; Romero, J.; Catalina, I.; Lillo, L.; et al. Thalamic atrophy in patients with pure hereditary spastic paraplegia type 4. *J. Neurol.* **2021**, *268*, 2429–2440. [CrossRef]
81. Mishra, V.; Guo, X.; Delgado, M.R.; Huang, H. Toward tract-specific fractional anisotropy (TSFA) at crossing-fiber regions with clinical diffusion MRI. *Magn. Reson. Med.* **2015**, *74*, 1768–1779. [CrossRef]
82. Rezende, T.J.R.; de Albuquerque, M.; Lamas, G.M.; Martinez, A.R.M.; Campos, B.M.; Casseb, R.F.; Silva, C.B.; Branco, L.M.T.; D'Abreu, A.; Lopes-Cendes, I.; et al. Multimodal MRI-based study in patients with SPG4 mutations. *PLoS ONE* **2015**, *10*, e0117666. [CrossRef]
83. Martinuzzi, A.; Montanaro, D.; Vavla, M.; Paparella, G.; Bonanni, P.; Musumeci, O.; Brighina, E.; Hlavata, H.; Rossi, G.; Aghakhanyan, G.; et al. Clinical and Paraclinical Indicators of Motor System Impairment in Hereditary Spastic Paraplegia: A Pilot Study. *PLoS ONE* **2016**, *11*, e0153283. [CrossRef]
84. Groppa, S.; Oliviero, A.; Eisen, A.; Quartarone, A.; Cohen, L.G.; Mall, V.; Kaelin-Lang, A.; Mima, T.; Rossi, S.; Thickbroom, G.W.; et al. A practical guide to diagnostic transcranial magnetic stimulation: Report of an IFCN committee. *Clin. Neurophysiol.* **2012**, *123*, 858–882. [CrossRef]
85. Siow, S.-F.; Smail, R.C.; Ng, K.; Kumar, K.R.; Sue, C.M. Motor Evoked Potentials in Hereditary Spastic Paraplegia—A Systematic Review. *Front. Neurol.* **2019**, *10*, 967. [CrossRef]
86. Loris, E.; Ollenschläger, M.; Greinwalder, T.; Eskofier, B.; Winkler, J.; Gaßner, H.; Regensburger, M. Mobile digital gait analysis objectively measures progression in hereditary spastic paraplegia. *Ann. Clin. Transl. Neurol.* **2023**, *10*, 447–452. [CrossRef]
87. Vavla, M.; Paparella, G.; Papayannis, A.; Pascuzzo, R.; Girardi, G.; Pellegrini, F.; Capello, G.; Prosdocimo, G.; Martinuzzi, A. Optical Coherence Tomography in a Cohort of Genetically Defined Hereditary Spastic Paraplegia: A Brief Research Report. *Front. Neurol.* **2019**, *10*, 1193. [CrossRef] [PubMed]
88. Amprosi, M.; Indelicato, E.; Eigentler, A.; Fritz, J.; Nachbauer, W.; Boesch, S. Toward the Definition of Patient-Reported Outcome Measurements in Hereditary Spastic Paraplegia. *Neurol. Genet.* **2023**, *9*, e200052. [CrossRef] [PubMed]
89. Klimpe, S.; Schule, R.; Kassubek, J.; Otto, S.; Kohl, Z.; Klebe, S.; Klopstock, T.; Ratzka, S.; Karle, K.; Schols, L. Disease severity affects quality of life of hereditary spastic paraplegia patients. *Eur. J. Neurol.* **2012**, *19*, 168–171. [CrossRef] [PubMed]
90. Orsucci, D.; Petrucci, L.; Caldarazzo Ienco, E.; Chico, L.; Simi, P.; Fogli, A.; Baldinotti, F.; Simoncini, C.; Logerfo, A.; Carlesi, C.; et al. Hereditary spastic paraparesis in adults. A clinical and genetic perspective from Tuscany. *Clin. Neurol. Neurosurg.* **2014**, *120*, 14–19. [CrossRef] [PubMed]
91. Patrick, D.L.; Deyo, R.A. Generic and Disease-Specific Measures in Assessing Health Status and Quality of Life. *Med. Care* **1989**, *27*, S217–S232. [CrossRef]
92. Wali, G.; Siow, S.F.; Liyanage, E.; Kumar, K.R.; Mackay-Sim, A.; Sue, C.M. Reduced acetylated α-tubulin in SPAST hereditary spastic paraplegia patient PBMCs. *Front Neurosci.* **2023**, *17*, 1073516. [CrossRef]
93. Sardina, F.; Valente, D.; Fattorini, G.; Cioffi, E.; Zanna, G.D.; Tessa, A.; Trisciuoglio, D.; Soddu, S.; Santorelli, F.M.; Casali, C.; et al. New cellular imaging-based method to distinguish the SPG4 subtype of hereditary spastic paraplegia. *Eur. J. Neurol.* **2023**, *30*, 1734–1744. [CrossRef]
94. Kirkham, J.J.; Davis, K.; Altman, D.G.; Blazeby, J.M.; Clarke, M.; Tunis, S.; Williamson, P.R. Core Outcome Set-STAndards for Development: The COS-STAD recommendations. *PLoS Med.* **2017**, *14*, e1002447. [CrossRef]
95. da Graça, F.F.; de Rezende, T.J.R.; Vasconcellos, L.F.R.; Pedroso, J.L.; Barsottini, O.G.P.; França, J.M.C. Neuroimaging in Hereditary Spastic Paraplegias: Current Use and Future Perspectives. *Front. Neurol.* **2019**, *9*, 1117. [CrossRef]

96. Vavla, M.; Montanaro, D.; Pizzighello, S.; Frijia, F.; Arrigoni, F.; Baratto, A.; Piccoli, G.; Paparella, G.; Martinuzzi, A. Brain Magnetic Spectroscopy Imaging and Hereditary Spastic Paraplegia: A Focused Systematic Review on Current Landmarks and Future Perspectives. *Front. Neurol.* **2020**, *11*, 515. [CrossRef]
97. de Freitas, J.L.; Rezende Filho, F.M.; Sallum, J.M.F.; França, M.C.; Pedroso, J.L.; Barsottini, O.G.P. Ophthalmological changes in hereditary spastic paraplegia and other genetic diseases with spastic paraplegia. *J. Neurol. Sci.* **2020**, *409*, 116620. [CrossRef] [PubMed]
98. Wali, G.; Berkovsky, S.; Whiten, D.R.; Mackay-Sim, A.; Sue, C.M. Single cell morphology distinguishes genotype and drug effect in Hereditary Spastic Paraplegia. *Sci. Rep.* **2021**, *11*, 16635. [CrossRef] [PubMed]

Disclaimer/Publisher's Note: The statements, opinions and data contained in all publications are solely those of the individual author(s) and contributor(s) and not of MDPI and/or the editor(s). MDPI and/or the editor(s) disclaim responsibility for any injury to people or property resulting from any ideas, methods, instructions or products referred to in the content.

genes

Review

Distribution, Function, and Expression of the Apelinergic System in the Healthy and Diseased Mammalian Brain

Martin N. Ivanov [1,2,*], Dimo S. Stoyanov [1], Stoyan P. Pavlov [1] and Anton. B. Tonchev [1,2]

1 Department of Anatomy and Cell Biology, Medical University-Varna, 9000 Varna, Bulgaria
2 Department of Stem Cell Biology, Research Institute, Medical University-Varna, 9000 Varna, Bulgaria
* Correspondence: martin.ivanov@mu-varna.bg

Abstract: Apelin, a peptide initially isolated from bovine stomach extract, is an endogenous ligand for the Apelin Receptor (APLNR). Subsequently, a second peptide, ELABELA, that can bind to the receptor has been identified. The Apelin receptor and its endogenous ligands are widely distributed in mammalian organs. A growing body of evidence suggests that this system participates in various signaling cascades that can regulate cell proliferation, blood pressure, fluid homeostasis, feeding behavior, and pituitary hormone release. Additional research has been done to elucidate the system's potential role in neurogenesis, the pathophysiology of Glioblastoma multiforme, and the protective effects of apelin peptides on some neurological and psychiatric disorders-ischemic stroke, epilepsy, Parkinson's, and Alzheimer's disease. This review discusses the current knowledge on the apelinergic system's involvement in brain physiology in health and disease.

Keywords: APLNR; APJ; Apelin; ELABELA; CNS; CNS-associated diseases; neurogenesis; gene expression

Citation: Ivanov, M.N.; Stoyanov, D.S.; Pavlov, S.P.; Tonchev, A.B. Distribution, Function, and Expression of the Apelinergic System in the Healthy and Diseased Mammalian Brain. *Genes* **2022**, *13*, 2172. https://doi.org/10.3390/genes13112172

Academic Editor: Claudia Ricci

Received: 25 October 2022
Accepted: 14 November 2022
Published: 21 November 2022

Publisher's Note: MDPI stays neutral with regard to jurisdictional claims in published maps and institutional affiliations.

Copyright: © 2022 by the authors. Licensee MDPI, Basel, Switzerland. This article is an open access article distributed under the terms and conditions of the Creative Commons Attribution (CC BY) license (https://creativecommons.org/licenses/by/4.0/).

1. Introduction

Apelin is a small, secreted protein first isolated from bovine stomach extracts and later identified as an endogenous ligand for the Apelin receptor [1]. It is secreted as a 77 aa precursor protein which is then cleaved into several mature forms that bind to the apelin receptor with different affinity and specificity.

Apelin receptors and their ligands are widely expressed in the mammalian brain [2–7]. Increasing interest in the apelinergic system has led to numerous discoveries and shed light on this system's complex functional diversity. It is not only involved in the physiological processes in the brain, but it also plays a role in the pathophysiology of neuropsychiatric disorders. The system is a suitable pharmacological target: recent discoveries of novel synthetic Apelin-receptor ligands with increased half/life (LIT01-196, l-homoarginine aks l-hArg, and non-canonical amino acids l-cyclohexylalanine aka l-Cha, etc.) and antagonists have paved the way towards clinical trials [8–10]. For a comprehensive review of the newly synthesized APLNR ligands, refer to Fischer et al. [11].

2. Structure and Function of Apelin Receptor (APLNR; APJ)

Research on the apelinergic system history dates back to 1993 when the Apelin receptor (APLNR, also knowns as APJ) was first described and cloned from a human genomic library due to its significant structural similarity (~40% identity between amino acid sequences) to angiotensin II receptor (AT2R). However, because of its inability to bind Angiotensin II and because no endogenous ligand was known to interact with it at the time, APLNR was considered an orphan receptor [12]. This notion was updated in 1998 when a ligand named apelin was first isolated from bovine and human stomach extracts [1].

The human APLNR is a G protein-coupled receptor of class A (Rhodopsin-like receptors subclass A3) containing 380 amino acid residues with characteristics of the GCPR proteins family containing 7 TM α-helical segments [13,14]. Significant insight into its structure was recently described utilizing X-ray crystallography with a synthetic 17-residue mimetic [13]. The gene coding for APLNR is located on the long arm of chromosome 11 at locus 11q12.1.

APLNR expression has been described in other vertebrates, including rhesus macaque, mouse, rat, and cattle. In addition, lower vertebrates such as chicken, red-eared slider, zebrafish, and tilapia possess variants of APLNR receptors (APLNR1, 2, 2a, 2b, and 3a) with variable homology to the human receptor [3,6,15–19].

3. Structure and Function of Apelin Receptor Ligands

3.1. Apelin (APLN)

The first described endogenous ligand of the APLNR is apelin, generated from a 77 amino acids long inactive pre-proprotein named pre-proapelin. [1] It is encoded by a gene located in the X chromosome in rats (Xq35), mice (XA3.2), and humans (Xq25-26.1) [20]. The pre-proapelin gene contains three exons, with exons 1 and 2 being the coding regions [21].

The discovery of apelin was made by Tatemoto et al. in 1998 through extraction from bovine stomach tissue and was shown to act as a ligand for an orphaned at that time G-coupled protein receptor APLNR. The name Apelin comes from APJ Endogenous Ligand. Interestingly the homology of the protein is conserved across humans, mice, rats, and bovine [1,20].

Pre-proapelin has a secretory N-terminal signaling peptide (22 amino acid residues) and a C-terminal binding site. The last 55 amino acid residues in the binding peptide are highly conserved. Additionally, the many basic amino acid residues represent multiple cleaving sites for peptidases [1,22]. Pro-preapelin is subsequently cleaved, generating a range of peptides that bind to APLNR. However, each ligand can exert different tissue distribution, binding affinity, half-life, and functions (Figure 1A).

Upon cleavage, pre-proapelin is transformed into several fragments with sizes ranging from 55 to 13 residues (Figure 1A). The 55-residue is formed by removing the 22-amino acid N-Signaling peptide of the pre-proapelin. Apelin-55 is then further processed to generate shorter active isoforms like apelin-36, 17, and 13 through protease-mediated cleavage. Indeed, this is shown to be the case with converting proapelin to apelin-13 with the help of PCSK3 (FURIN) [23].

At first, the apelin ligand was isolated from bovine stomach extracts as a 36-amino-acid peptide capable of binding to CHO-cells expressing APLNR [1]. Additional experiments by the same group showed that shorter peptides (apelin-17, 13, and [Pyr^1] apelin-13) exert stronger binding affinity to APLNR than apelin-36. Post-translational modifications of apelin-13 ([Pyr^1] apelin-13) prevent ligand degradation by exopeptidases [22]. Forskolin-induced cAMP inhibition assays using CHO transfected with APLNR cDNA also showed that [Pyr^1] apelin-13 prompted the most potent response. However, in another study, competition binding analysis in the same cell line (HEK293) established that apelin-13 has a higher affinity to the receptor than [Pyr^1]apelin-13 [24]. Zhen et al. (2013) found that [Pyr^1]apelin-13 is the most common ligand isoform in the blood [25]. For a long time, it was accepted that apelin-55 serves only as a precursor, but current evidence suggests that it may bind to APLNR with similar potency as apelin-17 and apelin-13 [26].

Several proteases (CD10 aka Neprilysin, angiotensin-converting enzyme 2 aka ACE2, and Kallikrein aka KLKB1) can regulate apelin by modification or inactivation [27–29]. The physiologically relevant action of metalloprotease CD10 is the cleavage of [Pyr^1]apelin-13 between Arg4 and Leu5 (RPRL motif) and between Leu5 and Ser6, creating fragments 5–13 and 6–13, respectively [30]. Cleavage of the RPRL motif inactivates the protein and prevents it from binding to APLNR [27].

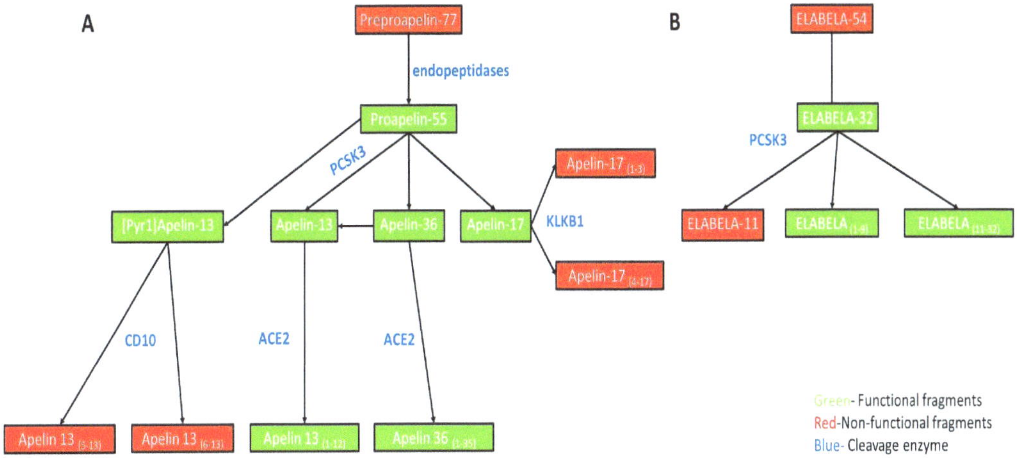

Figure 1. Enzymatic processing of Apelin/ELABELA. (**A**). The inactive pre-proapelin (77aa) is processed via endogenous endonucleases to Apelin-55, which can bind to APLNR. Apelin-55 can be additionally processed, generating four isoforms: Apelin-13, Apelin-17, Apelin-36, and [Pyr[1]] apelin-13. The generation of Apelin-13 is achieved by PCSK3 (FURIN). All four isoforms are capable of binding to the APLNR. CD10 (Neprilysin) is capable of inactivating [Pyr[1]] apelin-13, creating two inactive Apelin-13 isoforms (5–13aa and 6–13aa). On the other hand, Angiotensin-converting enzyme 2 (ACE2) converts Apelin-13 and Apelin-36 to active forms, Apelin-$13_{(1-12)}$ and Apelin-$32_{(1-35)}$. (**B**). ELABELA gene codes for a non-functional 54aa-long pro-protein, which generates an active 32aa-long protein upon processing. ELABELA-32 can generate three fragments, ELABELA-11, an inactive form, generated with the help of PCSK3, and two functional ones, ELABELA$_{(1-9)}$ and $_{(11-32)}$, with the activity of unknown proteases. Abbreviations: PCSK3, proprotein convertases subtilisin/kexin type; CD10, Neprilysin; ACE2, Angiotensin Converting enzyme-2; KLKB1, Kallikrein.

Angiotensin-converting enzyme 2 (ACE2) is a zinc-metalloprotease with carboxypeptidase activity that removes the C-terminal phenylalanine and produces apelin isoforms with modified action [28]. Cleaving of apelin-13 and apelin-36 by ACE2 produces two fragments, apelin-$13_{(1-12)}$ or apelin-$36_{(1-35)}$. Apelin-$13_{(1-12)}$ binds to the APLNR and activates downstream pathways [31]. It is worth mentioning that ACE2 was recently identified as the main SARS-CoV-2 receptors. Saravi and Beer proposed that apelin peptides can be used as a potential drug for improving the outcome of COVID-19 lung and cardio-vascular injuries [32]. Apelin-17 was shown to be indirectly related to the COVID-19 severity [33].

Kallikrein is a serine protease that cleaves apelin-17 between Arg3 and Arg4 into two isoforms, apelin-$17_{(1-3)}$ and apelin-$17_{(4-17)}$ fragments. Those fragments bind to the receptor with high affinity but cannot activate the Ca^{2+} ions mobilization [29].

3.2. *Apela/ELA (Apelin Receptor Early Endogenous Ligand/Elabela/Toddler)*

The human ELABELA gene is located on chromosome 4 and contains three exons. The transcript, previously believed to be a non-coding RNA, contains an open reading frame (ORF) and encodes a pre-proprotein with a length of 54 amino acids, containing a 22 amino acids secretory signaling terminus and a 31 amino acids mature portion. This protein is highly conserved across the species with nearly perfect homology of the C-terminus's last 13 amino acid residues [34–36].

ELABELA (Toddler) was discovered and described as an essential regulator of heart development [34,35]. In embryos, ELABELA acts as an early developmental signal required for the migration of mesendodermal cells. ELABELA$^{-/-}$ knock-out mice show cardiac agenesia or form only a rudimentary heart. This phenotype resembles the effects observed in APLNR knock-out mice [34,35]. Apelin and ELABELA have sequence similarity of 25%, isoelectric points above 12, and are rich in basic amino acid residues. These observations led to the experimental evidence and conclusion that ELABELA can also activate the APLNR [34,35,37].

Additional experiments utilizing HEK293 cells transfected with lentivirus containing *APLNR* enchased by GFP and treated with ELABELA show that after binding to the APLNR, ELABELA causes transient internalization of the receptor. Similarly, the binding of ELABELA to APLNR-expressing cells acts on Gi-coupled proteins leading to stimulation of ERK 1/2 phosphorylation, an increase in calcium mobilization, and a decrease in cAMP production [38,39]. Interestingly, competition binding analysis showed that ELABELA binds with a higher affinity to APLNR than apelin [39].

Chng et al. have speculated on the cleavage sites of the ELABELA peptide because of the presence of dibasic amino acids: Arg9–Arg10 and Arg20–Arg21 [34]. Indeed, incubation of ELA-32 in the presence of PCSK3 (FURIN) leads to the generation of fragment ELA-$11_{(22-32)}$ [40] (Figure 1B). In addition, incubation of ELABELA in rat plasma induces the formation of two other bioactive fragments- $ELA_{(1-9)}$ and ELA-$22_{(11-32)}$ with a cleavage site Arg9-Arg10 and Arg10-Lys11 [41].

Some of those smaller fragments can bind effectively to the receptor and activate the Gαi1 pathway, recruiting β-arrestins, leading to the internalization of the receptor with efficacy similar to that of ELABELA (parent form) and Apelin-13 [40,41]. ELA-11 is a notable exception, with reduced receptor binding affinity and β-arrestin recruitment but with the capability to inhibit cAMP production [42]. The plasma half-life of ELA is very short ($t_{1/2}$ = 2 min), similar to Apelin [41]. The plasma concentration of apelin-13 is in the range of 0.13 ± 0.05 ng/mL [43].

4. Transcriptional Regulation of *APLNR* and *Apelin*

Multiple transcriptional factors are involved in the regulation of the apelin receptor gene expression, including Sp1 (stimulating protein-1), glucocorticoid, estrogen receptors, and CCAAT enhancer-binding protein (C/EBP) [44]. Additionally, insulin can also affect the expression of both apelin and APLNR [45]. Upon knock-out of *Apelin*, the expression of APLNR was reduced, suggesting that *Apelin* regulates the expression of APLNR [46].

Based on the data so far, transcriptional factors regulating apelin gene expression are upstream transcription factor 1/ upstream transcription factor 2 (USF1/USF2), signal transducer and activator of transcription 3 (STAT3), hypoxia-inducible factor 1-α (HIF1-α), TNF-α (tumor necrosis factor-α) collaborative binding interaction between RARα (retinoic acid receptor α), KLF5 (*Krüppel-like* factor 5) and Sp1 (stimulating protein-1) [47–51]. Insulin is also shown to increase apelin expression. [45] Down-regulation of apelin is associated with the transcriptional factors ATF4 and its binding partner C/EBP-b via the p38 MAPK pro-apoptotic signal pathway [52].

Moreover, various substances such as resveratrol metabolites, andrographolide, and insulin can increase the expression of mRNA of *Apelin*, while the application of corticosteroids can downregulate it [53–55].

5. Signaling Pathways Associated with Activation of APLNR

APLNR is coupled with heterogenous guanine nucleotide-binding proteins (Gαi/o, Gαq/11, Gαs, and Gα12/13), which upon activation of the receptor, can regulate diverse intracellular events. (Figure 2) It is possible that the APLNR is also coupled with inhibitory G-coupled protein (Gi) [1].

APLNR is associated with different G-coupled proteins Gαi/o, Gαq/11, Gαs, and Gα12/13. Activation of the receptor and recruitment of Gαs activates Adenylyl Cyclase leading to the production of PKA. Adenylyl cyclase can be inhibited by Gαi/o. Gαi/o and, on the other hand, is responsible for downstream reaction associated with cell cycle progression, inhibition of autophagy, and cell survival and response to injury through activation of PI3K/Akt/mTOR pathways. mTOR can phosphorylate P70S6K kinase, leading to cell migration and proliferation. Additionally, mTOR can be phosphorylated (Ser2448), leading to apoptosis and autophagy inhibition. Gαi/o can also activate MEK1/2/ERK1/2 and phospholipase Cβ leading to an increase in cell survival due to inhibition of apoptosis. In addition, Gαq/11 and Gq also can activate phospholipase Cβ. Gα12/13 mediates cytoskeleton remodeling by activating RhoGEF/Rhoa pathway. Apelin is also involved in inducing self-renewal in cancer stem cells. A possible mechanism involving the apelinergic system is associated with increased gene suppression and enhanced stem cell self-renewal. GSK3b increases the stability of KDM1A via phosphorylation at s683, which can interact with USP22. KDM1A is responsible for the demethylation of histone H3K4 downregulating genes (BMP2, CDKN1A, and GATA6) associated with stem cell self-renewal and tumorigenesis. When bound to APLNR, Apelin-13 recruits GRK2 and β-arrestins leading to internalization of the receptor via transferrin and rapid recycling back to the cell membrane. Apelin-36, on the other hand, also recruits GRK2 and β-arrestins, but in this case, the APLNR is targeted for lysosomal degradation. Apelin is also callable of attenuating the mitochondrion permeabilization caused by NMDAR activation, decreasing the generation of ROS, Cytochrome C, and Caspase-3, thus, inhibiting apoptosis. Another attenuation mechanism is the $Ca2^{+}$-dependent Casein kinase-2 (CK2) phosphorylation of NR2B subunit at S1480, leading to decreased activity of NMDAR. Additionally, the administration of apelin-13 can inhibit the effect of the GRP78/CHOP pathway and caspase-12 cascade associated with ER stress. Abbreviations: NMDAr, N-methyl-D-aspartate receptor; NR2B, The N-methyl-D-aspartate receptor subunit 2B; CK2, Casein kinase 2; cAMP- cyclic Adenosine monophosphate; ATP, Adenosine triphosphate; PKA, protein kinase A; PI3K, Phosphoinositide 3-kinase; Akt, Protein kinase B; mTOR, mammalian target of rapamycin; MEK1/2, Mitogen-activated protein kinase kinase 1/2; ERK 1/2, extracellular signal-regulated kinase; RhoGEF, Guanine nucleotide exchange factor for Rho/Rac/Cdc42-like GTPases; Rhoa, Ras homolog family member A; PLC β2, Phospholipase C β2; IP3, Inositol trisphosphate, PIP2, Phosphatidylinositol 4,5-bisphosphate; DAG, diacylglycerol; PKC, Protein kinase C; GRK2, G Protein-Coupled Receptor Kinase 2; BAX, Bcl-2-associated X protein; BID, BH3 interacting-domain death agonist; BIK, Bcl-2-interacting killer; ROS, Reactive oxygen species; CHOP, C/EBP Homologous Protein; GRP78, Glucose regulated protein-78; GSK3β, Glycogen synthase kinase-3 β, KDM1A, Lysine-specific histone demethylase 1A; USP22, Ubiquitin Specific Peptidase 22.

Those G-proteins are associated with distinct cellular events like activation or inhibition of adenylate cyclase by Gαs or Gαi/o, respectively, increase in intracellular Ca^{2+} with activation of phospholipase Cβ by Gαq, actin cytoskeleton remodeling by Rho activated by Gα12/13 and others. Additional pathways involving EGFR and -β-arrestin signaling are also relevant to downstream reactions associated with APLNR activation.

Coupling to Gαi/o inactivates the Adenylyl Cyclase-reducing protein kinase A (PKA) [2,5,22,24,56]. APLNR/Apelin interactions are shown to enhance downstream effects related to the kinases Akt and ERK1/2, which are functionally related to cell survival and injury protection [57]. Apelin-13 enhances the phosphorylation of Akt while reducing phosphorylated ERK1/2, demonstrating a potent neuroprotective effect and promoting cell survival in a serum deprivation (SD) neuronal model of apoptosis in cultured cortical neurons. [58] Interestingly, upon exposure to apelin ligands, neurons demonstrate an increase in intracellular calcium [59]. On the other hand, Gαs can activate Adenylyl Cyclase, which increases cAMP, inducing PKA activation [60,61].

Activation of Gα12/13 leads to activation of Phospholipase C, which hydrolyses phosphatidylinositol 4,5-bisphosphate (PIP_2) into diacylglycerol (DAG) and inositol 1,4,5-trisphosphate (IP_3). Increased Ca^{2+} concentration, a result of IP_3 binding to the endoplasmic reticulum, is required in addition to DAG to activate PKA [60,61]. Apelin can also regulate cell cycle progression by activating p70S6K through PI3K/Akt/mTOR pathway [62]. mTOR can be phosphorylated through PI3K/Akt at the Ser2448 position leading to the inhibition of autophagy and caspase-3 activation associated with apoptosis [63].

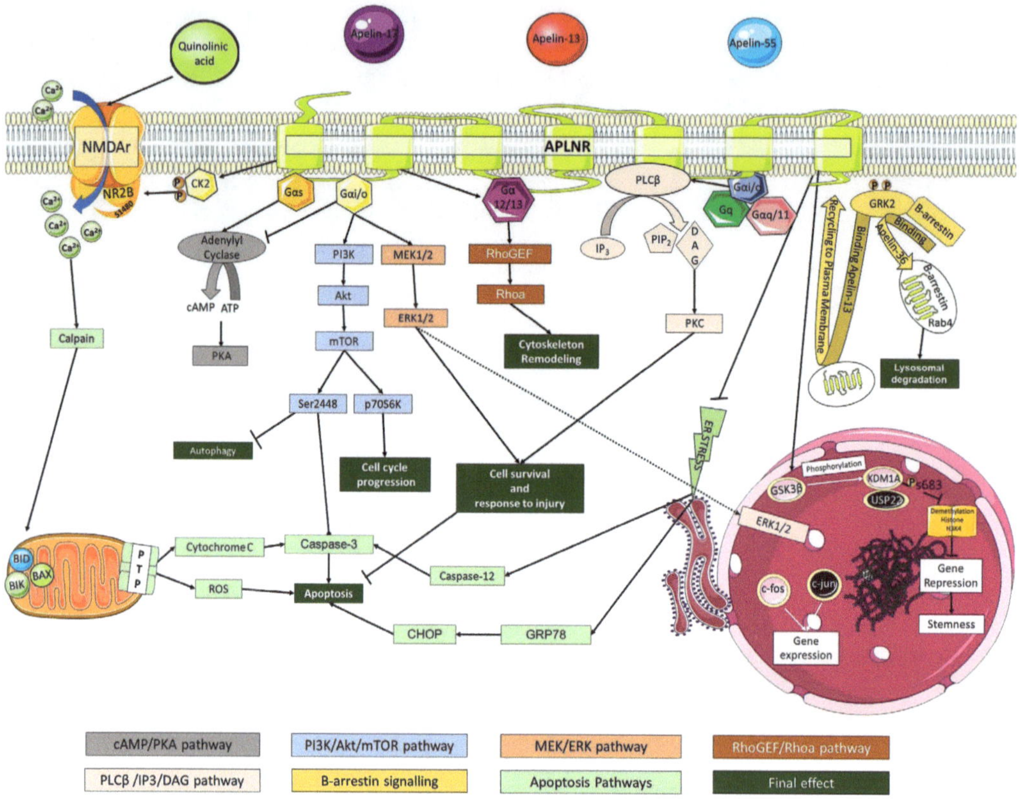

Figure 2. Signaling pathways associated with activation of APLNR.

Different ligands exhibit different overall effects on downstream processes like receptor phosphorylation, recruitment of GRK2, and β-arrestins [60,64].

Of note, when internalized, both apelin-13 and 36 colocalize with transferrin. Interestingly, the binding of apelin-36 to the receptor recruits β-arrestin-1, which leads to the internalization of the receptor via early Rab4 endosomes. On the contrary, the apelin-13 binding leads to transient internalization of the receptor followed by rapid recycling to the cell surface. The internalization process is not associated with β-arrestin complex formation, as observed during apelin-36 internalization [65,66]. This suggests ligand-depending trafficking mechanisms and downstream effects of the system.

Studies have shown that activating pro-survival pathways (ERK1/2 and AKT) by pre-treating hippocampal cultures with apelin ligands rescues the neurons from N-methyl-D-aspartic acid (NMDAr) receptor-mediated excitotoxicity injury [57]. Mitochondrial depolarization and increased permeabilization can increase ROS and cytochrome C, which are the main starting points for both apoptosis and necrosis. Interestingly, infusion of apelin-13 in serum-deprived cortical neuron cultures can attenuate ROS generation by stabilizing the mitochondrial membrane [58]. The underlying mechanism of this process involves IP3, PKC, MEK1/2, and ERK1/2 Signaling pathways.

Since NMDA receptors include glycine binding subunits-NR1 and glutamate binding subunits-NR2 (NR2A–NR2D), a potential Casein kinase-2 (CK2) or Ca^{2+}-dependent phosphorylation of NR2B S1480 by the APLNR pathway provides additional mechanisms of protection against excitotoxicity [67]. Moreover, apelin ligands inhibit HIV-associated neurotoxicity and apoptosis in neurons [57].

6. APLNR, Apelin, and ELABELA Expression in the Normal Brain

Brain distribution of APLNR and its endogenous ligands has been extensively studied utilizing different molecular and histo-anatomical techniques. APLNR and Apelin are present in the mammalian central and peripheral nervous systems. The topographical localization of APLNR and Apelin in the brain suggests multiple roles of this system in neurogenesis, pituitary hormone release, body fluid homeostasis, regulation of blood pressure, feeding behaviors, etc. [5,16,56,68]. Regionally, APLNR is expressed in restricted areas of the cerebral cortex like frontal, temporal, occipital, piriform, and entorhinal cortices (Figure 3, Table 1) [2,5,12,20,57,68–70]. Compared to APLNR, the overall levels of *Apelin* mRNA measured with RT-PCR in the cerebrum is higher [70].

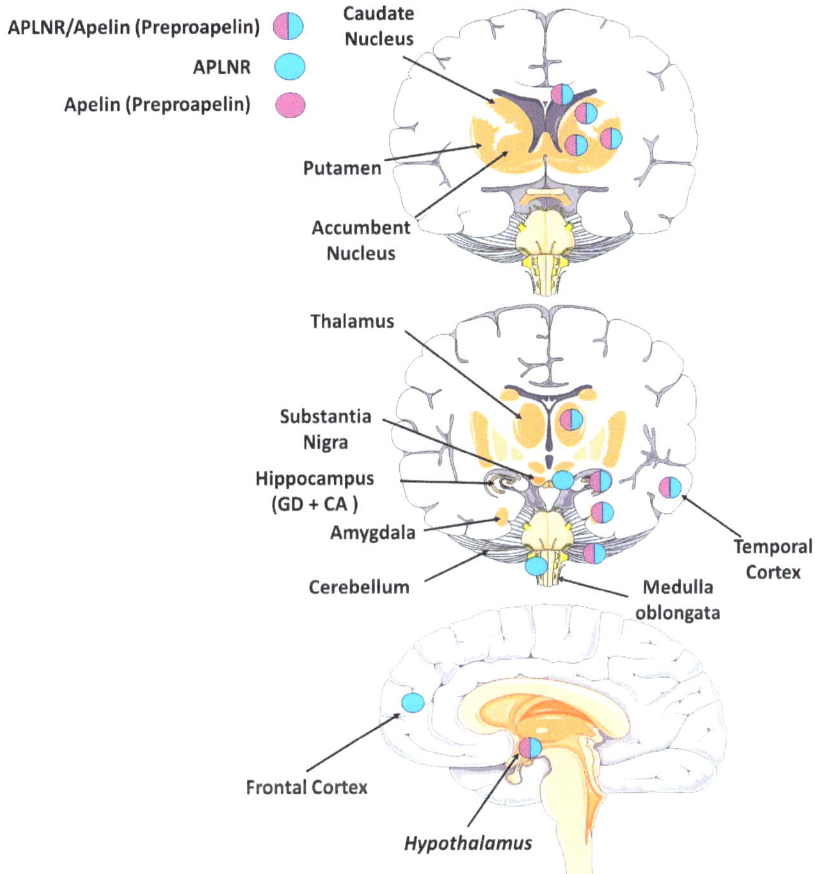

Figure 3. Expression of APLNR and Apelin ligand in the normal human brain. Schematic representation depicting the adult human brain at different levels showing the structures in which APLNR and Apelin ligand are expressed.

Using a variety of molecular biology techniques (qPCR, RT-PCR, WB, NB) for detecting APLNR and Apelin, multiple studies have shown that both are present in the dentate gyrus and the hippocampus proper (Cornu Ammonis, CA), Apelin having a higher expression level as compared to the APLNR [2,5,12,19,20,57,68–70]. The subgranular zone of the dentate gyrus is one of the main niches for postnatal neurogenesis, and it will be interesting to investigate if the APLNR/Apelinsystem can modulate its cellular output. Our recent data in primates show that APLNR is expressed in the other major neurogenic niche: the anterior subventricular zone along the cerebral lateral ventricle (SVZa) [16]. Probably, the cells expressing APLNR in SVZa represent the neural progenitor subpopulation [16]. Interestingly, APLNR shows a strong expression in the caudate nucleus [16], another region of adult neurogenesis in primates and humans [71]. APLNR is expressed in the thalamus [2,69]. In the hypothalamus, both apelin and APLNR have been localized [2,3,5,20,70]. Their expression was restricted to the supraoptic (SON) and the paraventricular nucleus, both magnocellular and parvocellular parts, contributing to the maintenance of fluid homeostasis [3,5,6,20,68,72]. The apelinergic system has also been detected in other brain areas, including the substantia nigra, cerebellum, preoptic area, pituitary gland, medulla oblongata, pons, and the spinal cord [2,5,19,24,69,70,72,73]. At the cellular level, APLNR has been detected in neurons, oligodendrocytes, and astrocytes but not in microglia, while Apelin has been detected in neurons but not astrocytes and microglia [2,57–59,74].

Table 1. Table with different brain regions in the human brain with known expression of APLNR and Apelin ligand.

	Human		
Tissue	**Apelin (Preproapelin)**	**APLNR**	**References**
Whole Brain	Unknown	+	[7]
Frontal Cortex	Unknown	+	[2,69]
Temporal Cortex	+	+	[2,75]
Striatum (Overall)	Unknown	+	[69]
Putamen	+	+	[2]
Caudate nucleus	+	+	[2]
Accumbent nucleus	+	+	[2]
Corpus callosum	+	+	[2,69]
Hippocampus (GD + CA)	+	+	[2,69,75]
Amygdala	+	+	[2,69]
Thalamus	+	+	[2,69]
Hypothalamus (Overall)	+	+	[2]
Substantia nigra	+	+	[2,69]
Cerebellum	+	+	[2]
Medulla oblongata	Unknown	+	[69]
Spinal cord	+	+	[2,69]

7. A Role of the Apelinergic System in Brain Diseases

Mounting evidence suggests that the apelinergic system is a prominent player in the pathogenesis of different neuronal and mental diseases, such as stroke, epilepsy, Alzheimer's disease, and Parkinson's, among others.

7.1. Apelinergic System Involvement in Ischemic Stroke

Ischemic stroke is the most common cause of disability and death worldwide [76]. Damage caused by cerebral blood vessel occlusion leads to the regional increase in Ca^{2+} (via NMDAR activation), depolarization of the mitochondrial membrane, caspase activation, neuronal cell death, and cerebral edema. Infusion of apelin-13 in mice reduces the infarct zone volume [77], cerebral edema, and caspase-3 activation but does not alter the neurological deficits [78]. Apelin-36 in lower concentrations can also reduce the infarct volume, but unlike apelin-13, it also improves neurological function after ischemia/reperfusion injury. LY294002, a potent inhibitor of PI3K, reduced the phosphorylation of Akt, thus, lowering the activity of the PI3K/Akt pathway activated by the APLNR ligands. Applying this substance to the ischemic stroke model treated with apelin-13 or 36 elevates the pro-apoptotic proteins caspase-3 and BAX, confirming that the antiapoptotic effect of apelin-36 is induced by PI3K/Akt pathway [79].

Apelin-13 treatment significantly reduced the levels of neutrophil infiltration in the ischemic penumbra and the levels of the pro-inflammatory mediators IL-1β, TNF-α, and ICAM-1. Moreover, it can also lower the number of cells activated in the penumbral region, thus, inducing a neuroprotective effect by blocking or suppressing neuroinflammation [77,80]. Intranasal administration of Apelin-13 effectively reduced the number of apoptotic cells and of activated microglial cells, increasing the expression of antiapoptotic factors (Bcl-2). It could also reduce the pro-inflammatory cytokines and chemokines TNF-a, IL-1b, MIP-1a, and MCP-1 and increase the anti-inflammatory cytokine IL-10. Angiogenesis in the peri-infarct region can be explained by the enhanced activity of pro-angiogenic factors VEGF and MMP9, which were also elevated after treatment with apelin-13. Because of the enhanced angiogenesis after treatment, better recovery was reported compared to non-treated animals [81]. Upon treatment with apelin, an upregulation of the expression of VEGF and VEGF-2 can be observed. This elevation is associated with the protective effects of apelin, mediated by ERK and PI3K/Akt pathways, which can be blocked by intraventricular injection with an anti-VEGF antibody [80].

Following cerebral ischemia in primates, *APLNR* and *Apelin mRNA* was strongly induced in monkey SVZa and caudate nucleus [16].

7.2. Apelinergic System Involvement in Epilepsy

Neurons in the mammalian neocortex are either excitatory, glutamatergic projecting neurons or inhibitory, GABAergic interneurons that branch in the local circuits. A disbalance in the excitation levels leads to pathological hyperexcitability manifested by spontaneous and recurrent seizures [82,83].

Extended epileptic periods and poorly managed or drug-resistant epilepsy can cause neuronal loss either by apoptosis or necrosis. The observed overexpression of Apelin in patients with drug-resistant temporal lobe epilepsy and rats with lithium–pilocarpine-induced epilepsy may be a compensatory mechanism [75]. Apelin can salvage the hippocampal neurons from the effects of excitotoxicity by downregulating metabotropic Glutamate Receptor-1 (mGluR1), increasing phosphorylation of Akt, and upregulating Bcl2, thus, reducing caspase-3 activation [84]. Treatment with brain-specific micro-RNA-182 (miR-182) that blocks Apelin leads to increased apoptosis in epilepsy models. Blocking miR-182 can increase the effects on Apelin, lower pro-apoptotic proteins (Bax; caspase-3), and increase the antiapoptotic ones (Bcl-2) [84].

Treatment with apelin-13 in an experimental rat epilepsy model prevented the induction of seizures and neuronal loss. This effect is lost when F13A, an APLNR receptor antagonist, is applied [85]. Apelin can exert a level of neuroprotection in the PTZ model of epilepsy thanks to its ability to maintain mitochondrial potentials, reduce intracellular Ca^{2+}, and inhibit ROS generation and COX2 (Cyclooxygenase 2) [86].

7.3. Apelinergic System Involvement in Neurogenesis and Glioblastoma Multiforme

Glioblastomas are brain tumors showing high invasiveness, angiogenesis, and an unusual tumor environment. There is substantial evidence showing that Glioblastoma multiforme is derived from SVZa stem cells [87].

Apelin is secreted from the endothelial cell near Glioblastoma stem-like cells (GSCs). It mediates self-renewal, but it is not associated with proliferation. Apelin protein expression is also correlated with the levels of vascularization of GBM [88].

Silencing the apelin Signaling pathway either by knocking down or blocking the APLNR reduces tumor volume, vascularization, and proliferation [89]. GSC are in a quiescent state maintained by the vascular niche in the tumor, which is the main reason for the inefficiency of chemotherapies [90]. Interestingly, applying an antagonist of APLNR in combination with chemotherapies improves the response and decreases the GSC numbers. This effect is possibly mediated by activation of GSK3β (Glycogen synthase kinase-3 pathway) [88,89,91]. Nuclear GSK3β phosphorylates KDM1A at s683, which can interact with USP22, thus, increasing the stability of KDM1A. KDM1A is responsible for the demethylation of histone H3K4 leading to the downregulation of genes (*BMP2*, *CDKN1A*, and *GATA6*) associated with stem cell self-renewal [92].

ELABELA was also shown to be expressed in GSCs. Moreover, brain tumor datasets have shown that expression levels of ELABELA are linked to tumor grading and patient survival [93].

Current therapies relying on anti-VEGF mAb usually target tumor angiogenesis. Unfortunately, such therapies have not increased patient survival [94]. These treatments have been shown to decrease the apelin expression inside the tumor, thus, increasing its invasiveness [95]. Interestingly, using a partial agonist for APLNR (apelin-F13A) combined with anti-VEGF therapy lessens the invasiveness and angiogenesis properties of GBM [95,96].

7.4. Apelinergic System Involvement in Alzheimer's Disease (AD)

Alzheimer's disease is a progressive neurodegenerative disorder characterized by the deposition of intracellular senile plaques composed of insoluble neurofibrillary tangles and extracellular amyloid β (Aβ) peptides. Neuronal loss in the hippocampus and neocortex leads to memory loss and cognitive impairments [97,98].

In newly discovered AD patients, the levels of Apelin-13 were lower compared to healthy individuals [43].

Apelin-13 can reduce memory deficits in a mouse model of Alzheimer's disease. [63,99] Aβ deposition in neurons induces apoptosis and autophagy, which can be attenuated by Apelin-13 treatment. The molecular basis of these neuroprotective effects in AD models is: (i). Decreased autophagy pathway (e.g., LC3II/I), (ii). Increase of autophagic clearance (HDAC6), (iii). Decreased apoptosis (caspase-3), and (iv). Increasing survival of neurons through the mTOR pathway [63].

Neuroinflammation plays a critical role in the pathophysiology of Alzheimer's disease. Important components of the neuroinflammation response, including microglial and astroglial activation and pro-inflammatory cytokine (e.g., IL-1β and TNF-α) production are attenuated from Apelin-13 [99].

Apelin can also increase the expression of hippocampal neurotrophins/neurotrophin receptors, such as Brain-Derived Neurotrophic Factor (BDNF) and Tropomyosin receptor kinase B (TrkB), which are typically at low levels in Alzheimer's mouse models. Blocking the TrkB receptor with an apelin antagonist, K252a, blocked the apelin-13 effects, showing that the beneficial effects of apelin in the hippocampus are mediated by activation of the BDNF/TrkB Signaling pathway. Synaptophysin (SYP) generaly used for evaluating synaptic transmission plasticity is downregulated in AD and restores its normal levels upon reapplication of apelin-13 [99]. Tissue necrosis is also initiated in AD by activation of the proteins RIP1 and RIP3, controlled by TNF-α. Reduction of RIP1, RIP3, and TNF-α is observed when apelin is applied [100]. Wan et al. have provided an in-depth review of the role of apelin in AD and its mechanism of neuroprotection [101].

7.5. Apelinergic System Involvement in Parkinson's Disease (PD)

Parkinson's disease (PD) is a neurodegenerative disorder affecting the dopaminergic neurons in the substantia nigra. It manifests with motor dysfunctions, including muscle rigidity, tremor, slow movement, and cognitive impairments, including depression, anxiety, and in later stages, dementia. The main histological hallmark of the disease is the aggregation of a misfolded protein called α-synuclein, which accumulates and becomes cytotoxic [102,103]. Additionally, factors such as mitochondrial dysfunction, inflammation, oxidative stress, and synaptic dysfunction can play a crucial role in the pathophysiology of the disease. To show the role of apelin-13, Pouresmaeili-Babaki et al. used SH-SY5Y cells treated with 6-hydroxydopamine (6-OHDA), which is a widely used cell model for PD. Upon treatment with 6-OHDA, dopaminergic cell death can be observed. Application of Apelin-13 is capable of inhibiting cytochrome-3 release and activation of caspase-3, effects through activation of APLNR/PI3K/Akt Signaling pathway [104]. The same group was able to show also that Apelin can improve memory and cognitive deficits in a Parkinson's disease model treated with 6-hydroxydopamine (6-OHDA) [105].

Another study, utilizing the same SH-SY5Y cell line but induced cell damage by applications of 1-methyl-4-phenyl-pyridine (MPP+) showed that apelin-13 could attenuate the neurotoxicity and the Endoplasmic Reticulum Stress (ER stress), the level of GRP78, CHOP and cleaved caspase-12 and significantly increase the levels of phosphorylated ERK1/2, thus, preventing the apoptosis [106]. Similarly, another study using 1-methyl-4-phenyl-1,2,3,6-tetrahydropyridin (MPTP) to induce Parkinson-like damage has shown that apelin-13 significantly increases autophagy by upregulation of LC3B and Beclin1 and down-regulation of p62. Apelin-13 was also capable of inhibiting the effect of the GRP78/ IRE1α/XBP1s/CHOP pathway associated with ER stress [107].

The neuroprotective effect of apelin was also shown in the methamphetamine PC12 cell model. Applying methamphetamine increased the generation of ROS, autophagy, and apoptosis, which were reduced by apelin [108]. Furthermore, some evidence suggests that it can also alleviate motor deficits [107] and prevent pathological alterations to the synaptic elements in the striatum and substantia nigra [109].

8. Conclusions and Future Directions

The components of the apelinergic system are widely expressed in the adult brain, and interaction between APLNR and Apelin isoforms or ELABELA are responsible for numerous physiological functions. Apelin and ELABELA are modified and processed into multiple functional and non-functional isoforms. Although the homology between apelin isoforms is similar, and they can all activate APLNR, the activated downstream signaling cascades may vary. The studies we discussed in this review show the role of the apelinergic system in the physiology of the brain and the pathophysiology of brain-related diseases. In animal models, the apelinergic system can activate critical signaling pathways related to cell survival, response to injury, reduction of apoptosis, cell cycle regulation, and stem cell biology.

Future work will need to address the system's function in relation to gene regulation, stem cell biology, neuronal development, and the related implications in those areas.

The expression of Apelin and APLNR in the hippocampus and the SVZ along the lateral wall of the cerebral lateral ventricle suggests that the apelinergic system may affect neural regeneration.

Finally, the identification and characterization of novel analogs and ligands with an increased half-life, specificity, and binding strength will advance the quest for novel therapeutic approaches in treating neuropsychiatric disorders and will increase our understanding of how the components of the system interact.

Author Contributions: All authors took part in the conceptualizing of the study; M.N.I. performed the initial literature review; D.S.S., A.B.T. and S.P.P. performed a review on disease and mechanisms; D.S.S., M.N.I., A.B.T. and S.P.P. conceptualized and produced the figures, M.N.I. wrote the draft manuscript; A.B.T., S.P.P. and D.S.S. performed critical revisions; A.B.T. revised and approved the final version of the manuscript for submitting. All authors have read and agreed to the published version of the manuscript.

Funding: This work was supported by an intramural grant of the Medical University—Varna (19028/2019) and the Bulgarian Medical Association grants for research projects (#NO-6/30.08.2018) M.I. is supported by the European Commission Horizon 2020 Framework Program (Project 856871—TRANSTEM).

Institutional Review Board Statement: Not applicable.

Informed Consent Statement: Not applicable.

Data Availability Statement: Not applicable.

Acknowledgments: The figures are prepared using Servier Medical Art Collection (https://smart.servier.com/, accessed on 24 October 2022).

Conflicts of Interest: The authors declare no conflict of interest.

References

1. Tatemoto, K.; Hosoya, M.; Habata, Y.; Fujii, R.; Kakegawa, T.; Zou, M.-X.; Kawamata, Y.; Fukusumi, S.; Hinuma, S.; Kitada, C.; et al. Isolation and Characterization of a Novel Endogenous Peptide Ligand for the Human APJ Receptor. *Biochem. Biophys. Res. Commun.* **1998**, *251*, 471–476. [CrossRef]
2. Medhurst, A.D.; Jennings, C.A.; Robbins, M.J.; Davis, R.P.; Ellis, C.; Winborn, K.Y.; Lawrie, K.W.M.; Hervieu, G.; Riley, G.; Bolaky, J.E.; et al. Pharmacological and Immunohistochemical Characterization of the APJ Receptor and Its Endogenous Ligand Apelin. *J. Neurochem.* **2003**, *84*, 1162–1172. [CrossRef]
3. Pope, G.R.; Roberts, E.M.; Lolait, S.J.; O'Carroll, A.M. Central and Peripheral Apelin Receptor Distribution in the Mouse: Species Differences with Rat. *Peptides* **2012**, *33*, 139–148. [CrossRef] [PubMed]
4. De Mota, N.; Lenkei, Z.; Llorens-Cortès, C. Cloning, Pharmacological Characterization and Brain Distribution of the Rat Apelin Receptor. *Neuroendocrinology* **2000**, *72*, 400–407. [CrossRef] [PubMed]
5. Reaux, A.; De Mota, N.; Skultetyova, I.; Lenkei, Z.; El Messari, S.; Gallatz, K.; Corvol, P.; Palkovits, M.; Llorens-Cortès, C. Physiological Role of a Novel Neuropeptide, Apelin, and Its Receptor in the Rat Brain. *J. Neurochem.* **2001**, *77*, 1085–1096. [CrossRef]
6. O'Carroll, A.M.; Selby, T.L.; Palkovits, M.; Lolait, S.J. Distribution of MRNA Encoding B78/Apj, the Rat Homologue of the Human APJ Receptor, and Its Endogenous Ligand Apelin in Brain and Peripheral Tissues. *Biochim. Biophys. Acta Gene Struct. Expr.* **2000**, *1492*, 72–80. [CrossRef]
7. De Falco, M.; De Luca, L.; Onori, N.; Cavallotti, I.; Artigiano, F.; Esposito, V.; De Luca, B.; Laforgia, V.; Groeger, A.M.; De Luca, A. Apelin Expression in Normal Human Tissues. *Vivo* **2002**, *16*, 333–336.
8. Flahault, A.; Keck, M.; Girault-Sotias, P.E.; Esteoulle, L.; De Mota, N.; Bonnet, D.; Llorens-Cortes, C. LIT01-196, a Metabolically Stable Apelin-17 Analog, Normalizes Blood Pressure in Hypertensive DOCA-Salt Rats via a NO Synthase-Dependent Mechanism. *Front. Pharmacol.* **2021**, *12*, 1913. [CrossRef] [PubMed]
9. Fernandez, K.X.; Fischer, C.; Vu, J.; Gheblawi, M.; Wang, W.; Gottschalk, S.; Iturrioz, X.; Llorens-Cortés, C.; Oudit, G.Y.; Vederas, J.C. Metabolically Stable Apelin-Analogues, Incorporating Cyclohexylalanine and Homoarginine, as Potent Apelin Receptor Activators. *RSC Med. Chem.* **2021**, *12*, 1402–1413. [CrossRef] [PubMed]
10. Trân, K.; Van Den Hauwe, R.; Sainsily, X.; Couvineau, P.; Côté, J.; Simard, L.; Echevarria, M.; Murza, A.; Serre, A.; Théroux, L.; et al. Constraining the Side Chain of C-Terminal Amino Acids in Apelin-13 Greatly Increases Affinity, Modulates Signaling, and Improves the Pharmacokinetic Profile. *J. Med. Chem.* **2021**, *64*, 5345–5364. [CrossRef] [PubMed]
11. Fischer, C. A Patent Review of Apelin Receptor (APJR) Modulators (2014–2019). *Expert Opin. Ther. Pat.* **2020**, *30*, 251–261. [CrossRef] [PubMed]
12. O'Dowd, B.F.; Heiber, M.; Chan, A.; Heng, H.H.Q.; Tsui, L.-C.; Kennedy, J.L.; Shi, X.; Petronis, A.; George, S.R.; Nguyen, T. A Human Gene That Shows Identity with the Gene Encoding the Angiotensin Receptor Is Located on Chromosome 11. *Gene* **1993**, *136*, 355–360. [CrossRef]
13. Ma, Y.; Yue, Y.; Ma, Y.; Zhang, Q.; Zhou, Q.; Song, Y.; Shen, Y.; Li, X.; Ma, X.; Li, C.; et al. Structural Basis for Apelin Control of the Human Apelin Receptor. *Structure* **2017**, *25*, 858–866.e4. [CrossRef] [PubMed]
14. Pitkin, S.L.; Maguire, J.J.; Bonner, T.I.; Davenport, A.P. Apelin Receptor Nomenclature, Distribution, Pharmacology, and Function; International Union of Basic and Clinical Pharmacology; LXXIV. *Pharmacol. Rev.* **2010**, *62*, 331–342. [CrossRef] [PubMed]
15. Zhang, J.; Zhou, Y.; Wu, C.; Wan, Y.; Fang, C.; Li, J.; Fang, W.; Yi, R.; Zhu, G.; Li, J.; et al. Characterization of the Apelin/Elabela Receptors (APLNR) in Chickens, Turtles, and Zebrafish: Identification of a Novel Apelin-Specific Receptor in Teleosts. *Front. Endocrinol.* **2018**, *9*, 756. [CrossRef] [PubMed]
16. Chongtham, M.C.; Wang, H.; Thaller, C.; Hsiao, N.-H.; Vachkov, I.H.; Pavlov, S.P.; Williamson, L.H.; Yamashima, T.; Stoykova, A.; Yan, J.; et al. Transcriptome Response and Spatial Pattern of Gene Expression in the Primate Subventricular Zone Neurogenic Niche After Cerebral Ischemia. *Front. Cell Dev. Biol.* **2020**, *8*, 584314. [CrossRef] [PubMed]
17. Scott, I.C.; Masri, B.; D'Amico, L.A.; Jin, S.W.; Jungblut, B.; Wehman, A.M.; Baier, H.; Audigier, Y.; Stainier, D.Y.R. The G Protein-Coupled Receptor Agtrl1b Regulates Early Development of Myocardial Progenitors. *Dev. Cell* **2007**, *12*, 403–413. [CrossRef] [PubMed]
18. Tucker, B.; Hepperle, C.; Kortschak, D.; Rainbird, B.; Wells, S.; Oates, A.C.; Lardelli, M. Zebrafish Angiotensin II Receptor-like 1a (Agtrl1a) Is Expressed in Migrating Hypoblast, Vasculature, and in Multiple Embryonic Epithelia. *Gene Expr. Patterns* **2007**, *7*, 258–265. [CrossRef]
19. Hosoya, M.; Kawamata, Y.; Fukusumi, S.; Fujii, R.; Habata, Y.; Hinuma, S.; Kitada, C.; Honda, S.; Kurokawa, T.; Onda, H.; et al. Molecular and Functional Characteristics of APJ. *J. Biol. Chem.* **2000**, *275*, 21061–21067. [CrossRef]
20. Lee, D.K.; Cheng, R.; Nguyen, T.; Fan, T.; Kariyawasam, A.P.; Liu, Y.; Osmond, D.H.; George, S.R.; O'Dowd, B.F. Characterization of Apelin, the Ligand for the APJ Receptor. *J. Neurochem.* **2001**, *74*, 34–41. [CrossRef]
21. Flahault, A.; Couvineau, P.; Alvear-Perez, R.; Iturrioz, X.; Llorens-Cortes, C. Role of the Vasopressin/Apelin Balance and Potential Use of Metabolically Stable Apelin Analogs in Water Metabolism Disorders. *Front. Endocrinol.* **2017**, *8*, 120. [CrossRef] [PubMed]

22. Habata, Y.; Fujii, R.; Hosoya, M.; Fukusumi, S.; Kawamata, Y.; Hinuma, S.; Kitada, C.; Nishizawa, N.; Murosaki, S.; Kurokawa, T.; et al. Apelin, the Natural Ligand of the Orphan Receptor APJ, Is Abundantly Secreted in the Colostrum. *Biochim. Biophys. Acta Mol. Cell Res.* **1999**, *1452*, 25–35. [CrossRef]
23. Shin, K.; Pandey, A.; Liu, X.Q.; Anini, Y.; Rainey, J.K. Preferential Apelin-13 Production by the Proprotein Convertase PCSK3 Is Implicated in Obesity. *FEBS Open Bio* **2013**, *3*, 328–333. [CrossRef]
24. Kawamata, Y.; Habata, Y.; Fukusumi, S.; Hosoya, M.; Fujii, R.; Hinuma, S.; Nishizawa, N.; Kitada, C.; Onda, H.; Nishimura, O.; et al. Molecular Properties of Apelin: Tissue Distribution and Receptor Binding. *Biochim. Biophys. Acta Mol. Cell Res.* **2001**, *1538*, 162–171. [CrossRef]
25. Zhen, E.Y.; Higgs, R.E.; Gutierrez, J.A. Pyroglutamyl Apelin-13 Identified as the Major Apelin Isoform in Human Plasma. *Anal. Biochem.* **2013**, *442*, 1–9. [CrossRef]
26. Shin, K.; Chapman, N.A.; Sarker, M.; Kenward, C.; Huang, S.K.; Weatherbee-Martin, N.; Pandey, A.; Dupré, D.J.; Rainey, J.K. Bioactivity of the Putative Apelin Proprotein Expands the Repertoire of Apelin Receptor Ligands. *Biochim. Biophys. Acta Gen. Subj.* **2017**, *1861*, 1901–1912. [CrossRef] [PubMed]
27. McKinnie, S.M.K.; Fischer, C.; Tran, K.M.H.; Wang, W.; Mosquera, F.; Oudit, G.Y.; Vederas, J.C. The Metalloprotease Neprilysin Degrades and Inactivates Apelin Peptides. *ChemBioChem* **2016**, *17*, 1495–1498. [CrossRef] [PubMed]
28. Vickers, C.; Hales, P.; Kaushik, V.; Dick, L.; Gavin, J.; Tang, J.; Godbout, K.; Parsons, T.; Baronas, E.; Hsieh, F.; et al. Hydrolysis of Biological Peptides by Human Angiotensin-Converting Enzyme-Related Carboxypeptidase. *J. Biol. Chem.* **2002**, *277*, 14838–14843. [CrossRef]
29. Fischer, C.; Lamer, T.; Wang, W.; McKinnie, S.M.K.; Iturrioz, X.; Llorens-Cortes, C.; Oudit, G.Y.; Vederas, J.C. Plasma Kallikrein Cleaves and Inactivates Apelin-17: Palmitoyl- and PEG-Extended Apelin-17 Analogs as Metabolically Stable Blood Pressure-Lowering Agents. *Eur. J. Med. Chem.* **2019**, *166*, 119–124. [CrossRef] [PubMed]
30. Murza, A.; Belleville, K.; Longpré, J.M.; Sarret, P.; Marsault, É. Stability and Degradation Patterns of Chemically Modified Analogs of Apelin-13 in Plasma and Cerebrospinal Fluid. *Biopolym. Pept. Sci. Sect.* **2014**, *102*, 297–303. [CrossRef] [PubMed]
31. Yang, P.; Kuc, R.E.; Brame, A.L.; Dyson, A.; Singer, M.; Glen, R.C.; Cheriyan, J.; Wilkinson, I.B.; Davenport, A.P.; Maguire, J.J. [Pyr1]Apelin-13(1-12) Is a Biologically Active ACE2 Metabolite of the Endogenous Cardiovascular Peptide [Pyr1]Apelin-13. *Front. Neurosci.* **2017**, *11*, 92. [CrossRef] [PubMed]
32. Saeedi Saravi, S.S.; Beer, J.H. Apelin-Potential Therapy for COVID-19? *J. Mol. Cell. Cardiol.* **2020**, *145*, 84–87. [CrossRef] [PubMed]
33. Fekri, M.S.; Barfzade, E.; Farokhnia, M.; Bajgani, S.M.H.; Shafahi, A.; Shafiepour, M.; Movahedinia, S.; Dabiri, S.; Yousefi, M. Investigating COVID-19 Severity Based on Serum Apelin-17 Levels and Inflammatory Mediators. *Curr. Respir. Med. Rev.* **2022**, *18*, 65–71. [CrossRef]
34. Chng, S.C.; Ho, L.; Tian, J.; Reversade, B. ELABELA: A Hormone Essential for Heart Development Signals via the Apelin Receptor. *Dev. Cell* **2013**, *27*, 672–680. [CrossRef] [PubMed]
35. Pauli, A.; Norris, M.L.; Valen, E.; Chew, G.L.; Gagnon, J.A.; Zimmerman, S.; Mitchell, A.; Ma, J.; Dubrulle, J.; Reyon, D.; et al. Toddler: An Embryonic Signal That Promotes Cell Movement via Apelin Receptors. *Science* **2014**, *343*, 1248636. [CrossRef]
36. Pauli, A.; Valen, E.; Schier, A.F. Identifying (Non-)Coding RNAs and Small Peptides: Challenges and Opportunities. *BioEssays* **2015**, *37*, 103–112. [CrossRef] [PubMed]
37. Xie, F.; Lv, D.; Chen, L. ELABELA: A Novel Hormone in Cardiac Development Acting as a New Endogenous Ligand for the APJ Receptor. *Acta Biochim. Biophys. Sin.* **2014**, *46*, 620–622. [CrossRef]
38. Wang, Z.; Yu, D.; Wang, M.; Wang, Q.; Kouznetsova, J.; Yang, R.; Qian, K.; Wu, W.; Shuldiner, A.; Sztalryd, C.; et al. Elabela-Apelin Receptor Signaling Pathway Is Functional in Mammalian Systems. *Sci. Rep.* **2015**, *5*, 8170. [CrossRef]
39. Deng, C.; Chen, H.; Yang, N.; Feng, Y.; Hsueh, A.J.W. Apela Regulates Fluid Homeostasis by Binding to the APJ Receptor to Activate Gi Signaling. *J. Biol. Chem.* **2015**, *290*, 18261–18268. [CrossRef]
40. Chen, H.; Wang, L.; Wang, W.; Cheng, C.; Zhang, Y.; Zhou, Y.; Wang, C.; Miao, X.; Wang, J.; Wang, C.; et al. ELABELA and an ELABELA Fragment Protect against AKI. *J. Am. Soc. Nephrol.* **2017**, *28*, 2694–2707. [CrossRef]
41. Murza, A.; Sainsily, X.; Coquerel, D.; Côté, J.; Marx, P.; Besserer-Offroy, É.; Longpré, J.M.; Lainé, J.; Reversade, B.; Salvail, D.; et al. Discovery and Structure-Activity Relationship of a Bioactive Fragment of ELABELA That Modulates Vascular and Cardiac Functions. *J. Med. Chem.* **2016**, *59*, 2962–2972. [CrossRef] [PubMed]
42. Yang, P.; Read, C.; Kuc, R.E.; Buonincontri, G.; Southwood, M.; Torella, R.; Upton, P.D.; Crosby, A.; Sawiak, S.J.; Carpenter, T.A.; et al. Elabela/Toddler Is an Endogenous Agonist of the Apelin APJ Receptor in the Adult Cardiovascular System, and Exogenous Administration of the Peptide Compensates for the Downregulation of Its Expression in Pulmonary Arterial Hypertension. *Circulation* **2017**, *135*, 1160–1173. [CrossRef] [PubMed]
43. Yildiz, Z.; Eren, N.; Orcun, A.; Münevver Gokyigit, F.; Turgay, F.; Gündogdu Celebi, L. Serum Apelin-13 Levels and Total Oxidant/Antioxidant Status of Patients with Alzheimer's Disease. *Aging Med.* **2021**, *4*, 201–205. [CrossRef] [PubMed]
44. O'Caroll, A.M.; Lolait, S.J.; Howell, G.M. Transcriptional Regulation of the Rat Apelin Receptor Gene: Promoter Cloning and Identification of an Sp1 Site Necessary for Promoter Activity. *J. Mol. Endocrinol.* **2006**, *36*, 221–235. [CrossRef]
45. Dray, C.; Debard, C.; Jager, J.; Disse, E.; Daviaud, D.; Martin, P.; Attané, C.; Wanecq, E.; Guigné, C.; Bost, F.; et al. Apelin and APJ Regulation in Adipose Tissue and Skeletal Muscle of Type 2 Diabetic Mice and Humans. *Am. J. Physiol. Endocrinol. Metab.* **2010**, *298*, E1161–E1169. [CrossRef]

46. Wang, G.; Kundu, R.; Han, S.; Qi, X.; Englander, E.W.; Quertermous, T.; Greeley, G.H. Ontogeny of Apelin and Its Receptor in the Rodent Gastrointestinal Tract. *Regul. Pept.* **2009**, *158*, 32–39. [CrossRef]
47. Han, S.; Wang, G.; Qi, X.; Englander, E.W.; Greeley, G.H. Involvement of a Stat3 Binding Site in Inflammation-Induced Enteric Apelin Expression. *Am. J. Physiol. Gastrointest. Liver Physiol.* **2008**, *295*, 1068–1078. [CrossRef]
48. Wang, G.; Qi, X.; Wei, W.; Englander, E.W.; Greeley, G.H.; Wang, G.; Qi, X.; Wei, W.; Englander, E.W.; Greeley, G.H. Characterization of the 5′-regulatory Regions of the Rat and Human Apelin Genes and Regulation of Breast Apelin by USF. *FASEB J.* **2006**, *20*, 2639–2641. [CrossRef]
49. Lv, X.R.; Zheng, B.; Li, S.Y.; Han, A.L.; Wang, C.; Shi, J.H.; Zhang, X.H.; Liu, Y.; Li, Y.H.; Wen, J.K. Synthetic Retinoid Am80 Up-Regulates Apelin Expression by Promoting Interaction of RARα with KLF5 and Sp1 in Vascular Smooth Muscle Cells. *Biochem. J.* **2013**, *456*, 35–46. [CrossRef]
50. Ronkainen, V.; Ronkainen, J.J.; Hänninen, S.L.; Leskinen, H.; Ruas, J.L.; Pereira, T.; Poellinger, L.; Vuolteenaho, O.; Tavi, P. Hypoxia Inducible Factor Regulates the Cardiac Expression and Secretion of Apelin. *FASEB J.* **2007**, *21*, 1821–1830. [CrossRef]
51. Daviaud, D.; Boucher, J.; Gesta, S.; Dray, C.; Guigne, C.; Quilliot, D.; Ayav, A.; Ziegler, O.; Carpene, C.; Saulnier-Blache, J.-S.; et al. TNFα Up-regulates Apelin Expression in Human and Mouse Adipose Tissue. *FASEB J.* **2006**, *20*, 1528–1530. [CrossRef] [PubMed]
52. Jeong, K.; Oh, Y.; Kim, S.J.; Kim, H.; Park, K.C.; Kim, S.S.; Ha, J.; Kang, I.; Choe, W. Apelin Is Transcriptionally Regulated by ER Stress-Induced ATF4 Expression via a P38 MAPK-Dependent Pathway. *Apoptosis* **2014**, *19*, 1399–1410. [CrossRef] [PubMed]
53. Alipanah-Moghadam, R.; Mehri, A.; Manafi, F.; Malekzadeh, V.; Nemati, A.; Aghamohammadi, V.; Mazani, M.; Cain, C.T.C.; Mohammadzadeh-Vardin, M. Andrographolide, a Novel Inducer of Apelin Gene Expression. *J. Ethnopharmacol.* **2021**, *280*, 114487. [CrossRef]
54. Wei, L.; Hou, X.; Tatemoto, K. Regulation of Apelin MRNA Expression by Insulin and Glucocorticoids in Mouse 3T3-L1 Adipocytes. *Regul. Pept.* **2005**, *132*, 27–32. [CrossRef] [PubMed]
55. Eseberri, I.; Lasa, A.; Churruca, I.; Portillo, M.P. Resveratrol Metabolites Modify Adipokine Expression and Secretion in 3T3-L1 Pre-Adipocytes and Mature Adipocytes. *PLoS ONE* **2013**, *8*, 2–9. [CrossRef]
56. Reaux, A.; Gallatz, K.; Palkovits, M.; Llorens-Cortes, C. Distribution of Apelin-Synthesizing Neurons in the Adult Rat Brain. *Neuroscience* **2002**, *113*, 653–662. [CrossRef]
57. O'Donnell, L.A.; Agrawal, A.; Sabnekar, P.; Dichter, M.A.; Lynch, D.R.; Kolson, D.L. Apelin, an Endogenous Neuronal Peptide, Protects Hippocampal Neurons against Excitotoxic Injury. *J. Neurochem.* **2007**, *102*, 1905–1917. [CrossRef]
58. Zeng, X.J.; Yu, S.P.; Zhang, L.; Wei, L. Neuroprotective Effect of the Endogenous Neural Peptide Apelin in Cultured Mouse Cortical Neurons. *Exp. Cell Res.* **2010**, *316*, 1773–1783. [CrossRef]
59. Choe, W.; Albright, A.; Sulcove, J.; Jaffer, S.; Hesselgesser, J.; Lavi, E.; Crino, P.; Kolson, D.L. Functional Expression of the Seven-Transmembrane HIV-1 Co-Receptor APJ in Neural Cells. *J. Neurovirol.* **2000**, *6*, 61–69.
60. Zhang, Y.; Wang, Y.; Lou, Y.; Luo, M.; Lu, Y.; Li, Z.; Wang, Y.; Miao, L. *Elabela, a Newly Discovered APJ Ligand: Similarities and Differences with Apelin*; Elsevier Inc.: Amsterdam, The Netherlands, 2018; Volume 109, ISBN 8643181136520.
61. Chapman, N.A.; Dupré, D.J.; Rainey, J.K. The Apelin Receptor: Physiology, Pathology, Cell Signalling, and Ligand Modulation of a Peptide-Activated Class A GPCR. *Biochem. Cell Biol.* **2014**, *92*, 431–440. [CrossRef]
62. Eyries, M.; Siegfried, G.; Ciumas, M.; Montagne, K.; Agrapart, M.; Lebrin, F.; Soubrier, F. Hypoxia-Induced Apelin Expression Regulates Endothelial Cell Proliferation and Regenerative Angiogenesis. *Circ. Res.* **2008**, *103*, 432–440. [CrossRef] [PubMed]
63. Aminyavari, S.; Zahmatkesh, M.; Farahmandfar, M.; Khodagholi, F.; Dargahi, L.; Zarrindast, M.R. Protective Role of Apelin-13 on Amyloid Β25–35-Induced Memory Deficit; Involvement of Autophagy and Apoptosis Process. *Prog. Neuro-Psychopharmacol. Biol. Psychiatry* **2019**, *89*, 322–334. [CrossRef] [PubMed]
64. Chen, J.; Chen, X.; Li, S.; Jiang, Y.; Mao, H.; Zhang, R.; Ji, B.; Yan, M.; Cai, X.; Wang, C. Individual Phosphorylation Sites at the C-Terminus of the Apelin Receptor Play Different Roles in Signal Transduction. *Redox Biol.* **2020**, *36*, 101629. [CrossRef] [PubMed]
65. Lee, D.K.; Ferguson, S.S.G.; George, S.R.; O'Dowd, B.F. The Fate of the Internalized Apelin Receptor Is Determined by Different Isoforms of Apelin Mediating Differential Interaction with β-Arrestin. *Biochem. Biophys. Res. Commun.* **2010**, *395*, 185–189. [CrossRef] [PubMed]
66. Zhou, N.; Fan, X.; Mukhtar, M.; Fang, J.; Patel, C.A.; DuBois, G.C.; Pomerantz, R.J. Cell-Cell Fusion and Internalization of the CNS-Based, HIV-1 Co-Receptor, APJ. *Virology* **2003**, *307*, 22–36. [CrossRef]
67. Cook, D.R.; Gleichman, A.J.; Cross, S.A.; Doshi, S.; Ho, W.; Jordan-Sciutto, K.L.; Lynch, D.R.; Kolson, D.L. NMDA Receptor Modulation by the Neuropeptide Apelin: Implications for Excitotoxic Injury. *J. Neurochem.* **2011**, *118*, 1113–1123. [CrossRef] [PubMed]
68. De Mota, N.; Reaux-Le Goazigo, A.; El Messari, S.; Chartrel, N.; Roesch, D.; Dujardin, C.; Kordon, C.; Vaudry, H.; Moos, F.; Llorens-Cortes, C. Apelin, a Potent Diuretic Neuropeptide Counteracting Vasopressin Actions through Inhibition of Vasopressin Neuron Activity and Vasopressin Release. *Proc. Natl. Acad. Sci. USA* **2004**, *101*, 10464–10469. [CrossRef] [PubMed]
69. Matsumoto, M.; Hidaka, K.; Akiho, H.; Tada, S.; Okada, M.; Yamaguchi, T. Low Stringency Hybridization Study of the Dopamine D4 Receptor Revealed D4-like MRNA Distribution of the Orphan Seven-Transmembrane Receptor, APJ, in Human Brain. *Neurosci. Lett.* **1996**, *219*, 119–122. [CrossRef]
70. Kasai, A.; Kinjo, T.; Ishihara, R.; Sakai, I.; Ishimaru, Y.; Yoshioka, Y.; Yamamuro, A.; Ishige, K.; Ito, Y.; Maeda, S. Apelin Deficiency Accelerates the Progression of Amyotrophic Lateral Sclerosis. *PLoS ONE* **2011**, *6*, e23968. [CrossRef]
71. Ernst, A.; Alkass, K.; Bernard, S.; Salehpour, M.; Perl, S.; Tisdale, J.; Possnert, G.; Druid, H.; Frisén, J. Neurogenesis in the Striatum of the Adult Human Brain. *Cell* **2014**, *156*, 1072–1083. [CrossRef]

72. Brailoiu, G.C.; Dun, S.L.; Yang, J.; Ohsawa, M.; Chang, J.K.; Dun, N.J. Apelin-Immunoreactivity in the Rat Hypothalamus and Pituitary. *Neurosci. Lett.* **2002**, *327*, 193–197. [CrossRef]
73. Xiong, Q.; He, W.; Wang, H.; Zhou, J.; Zhang, Y.; He, J.; Yang, C.; Zhang, B. Effect of the Spinal Apelin-APJ System on the Pathogenesis of Chronic Constriction Injury-Induced Neuropathic Pain in Rats. *Mol. Med. Rep.* **2017**, *16*, 1223–1231. [CrossRef]
74. Puffer, B.A.; Sharron, M.; Coughlan, C.M.; Baribaud, F.; McManus, C.M.; Lee, B.; David, J.; Price, K.; Horuk, R.; Tsang, M.; et al. Expression and Coreceptor Function of APJ for Primate Immunodeficiency Viruses. *Virology* **2000**, *276*, 435–444. [CrossRef]
75. Zhang, X.; Peng, X.; Fang, M.; Zhou, C.; Zhao, F.; Zhang, Y.; Xu, Y.; Zhu, Q.; Luo, J.; Chen, G.; et al. Up-Regulation of Apelin in Brain Tissue of Patients with Epilepsy and an Epileptic Rat Model. *Peptides* **2011**, *32*, 1793–1799. [CrossRef]
76. Donnan, D.A.; Fisher, M.; Macleod, M.S.M.D. Lancet Emergency and Comprehensive Care for Stroke Needed. *Lancet* **2009**, *373*, 1496. [CrossRef]
77. Xin, Q.; Cheng, B.; Pan, Y.; Liu, H.; Yang, C.; Chen, J.; Bai, B. Neuroprotective Effects of Apelin-13 on Experimental Ischemic Stroke through Suppression of Inflammation. *Peptides* **2015**, *63*, 55–62. [CrossRef]
78. Khaksari, M.; Aboutaleb, N.; Nasirinezhad, F.; Vakili, A.; Madjd, Z. Apelin-13 Protects the Brain against Ischemic Reperfusion Injury and Cerebral Edema in a Transient Model of Focal Cerebral Ischemia. *J. Mol. Neurosci.* **2012**, *48*, 201–208. [CrossRef] [PubMed]
79. Gu, Q.; Zhai, L.; Feng, X.; Chen, J.; Miao, Z.; Ren, L.; Qian, X.; Yu, J.; Li, Y.; Xu, X.; et al. Apelin-36, a Potent Peptide, Protects against Ischemic Brain Injury by Activating the PI3K/Akt Pathway. *Neurochem. Int.* **2013**, *63*, 535–540. [CrossRef]
80. Huang, C.; Dai, C.; Gong, K.; Zuo, H.; Chu, H. Apelin-13 Protects Neurovascular Unit against Ischemic Injuries through the Effects of Vascular Endothelial Growth Factor. *Neuropeptides* **2016**, *60*, 67–74. [CrossRef] [PubMed]
81. Chen, D.; Lee, J.; Gu, X.; Wei, L.; Yu, S.P. Intranasal Delivery of Apelin-13 Is Neuroprotective and Promotes Angiogenesis after Ischemic Stroke in Mice. *ASN Neuro* **2015**, *7*, 1759091415605114. [CrossRef]
82. Takano, T.; Sawai, C. Interneuron Dysfunction in Epilepsy: An Experimental Approach Using Immature Brain Insults to Induce Neuronal Migration Disorders. *Epilepsy Res.* **2019**, *156*, 106185. [CrossRef] [PubMed]
83. Liu, Y.Q.; Yu, F.; Liu, W.H.; He, X.H.; Peng, B.W. Dysfunction of Hippocampal Interneurons in Epilepsy. *Neurosci. Bull.* **2014**, *30*, 985–998. [CrossRef] [PubMed]
84. Dong, H.; Dong, B.; Zhang, N.; Liu, S.; Zhao, H. MicroRNA-182 Negatively Influences the Neuroprotective Effect of Apelin against Neuronal Injury in Epilepsy. *Neuropsychiatr. Dis. Treat.* **2020**, *16*, 327–338. [CrossRef] [PubMed]
85. Kalantaripour, T.P.; Esmaeili-Mahani, S.; Sheibani, V.; Asadi-Shekaari, M.; Pasban-Aliabadi, H. Anticonvulsant and Neuroprotective Effects of Apelin-13 on Pentylenetetrazole-Induced Seizures in Male Rats. *Biomed. Pharmacother.* **2016**, *84*, 258–263. [CrossRef] [PubMed]
86. Kalantaripour, T.P.; Esmaeili-Mahani, S.; Sheibani, V.; Najafipour, H.; Asadi-Shekaari, M.M. Apelin-13 Protects Rat Primary Cortical Glia-Neuron Co-Culture against Pentylenetetrazole-Induced Toxicity. *Biomed. Pharmacother.* **2017**, *87*, 661–668. [CrossRef]
87. Zhang, G.L.; Wang, C.F.; Qian, C.; Ji, Y.X.; Wang, Y.Z. Role and Mechanism of Neural Stem Cells of the Subventricular Zone in Glioblastoma. *World J. Stem Cells* **2021**, *13*, 877–893. [CrossRef] [PubMed]
88. Harford-Wright, E.; Andre-Gregoire, G.; Jacobs, K.A.; Treps, L.; Le Gonidec, S.; Leclair, H.M.; Gonzalez-Diest, S.; Roux, Q.; Guillonneau, F.; Loussouarn, D.; et al. Pharmacological Targeting of Apelin Impairs Glioblastoma Growth. *Brain* **2017**, *140*, 2939–2954. [CrossRef] [PubMed]
89. Frisch, A.; Kälin, S.; Monk, R.; Radke, J.; Heppner, F.L.; Kälin, R.E. Apelin Controls Angiogenesis-Dependent Glioblastoma Growth. *Int. J. Mol. Sci.* **2020**, *21*, 4179. [CrossRef]
90. Sun, R.; Kim, A.H. The Multifaceted Mechanisms of Malignant Glioblastoma Progression and Clinical Implications. *Cancer Metastasis Rev.* **2022**, 1–28. [CrossRef] [PubMed]
91. Harford-Wright, E.; Gavard, J. Apelin, the Devil Inside Brain Tumors. *J. Exp. Neurosci.* **2018**, *12*, 3–5. [CrossRef]
92. Zhou, A.; Lin, K.; Zhang, S.; Chen, Y.; Zhang, N.; Xue, J.; Wang, Z.; Aldape, K.D.; Xie, K.; Woodgett, J.R.; et al. Nuclear GSK3β Promotes Tumorigenesis by Phosphorylating KDM1A and Inducing Its Deubiquitylation by USP22. *Nat. Cell Biol.* **2016**, *18*, 954–966. [CrossRef] [PubMed]
93. Ganguly, D.; Cai, C.; Sims, M.M.; Yang, C.H.; Thomas, M.; Cheng, J.; Saad, A.G.; Pfeffer, L.M. APELA Expression in Glioma, and Its Association with Patient Survival and Tumor Grade. *Pharmaceuticals* **2019**, *12*, 45. [CrossRef] [PubMed]
94. Mosteiro, A.; Pedrosa, L.; Ferrés, A.; Diao, D.; Sierra, À.; González, J.J. The Vascular Microenvironment in Glioblastoma: A Comprehensive Review. *Biomedicines* **2022**, *10*, 1285. [CrossRef] [PubMed]
95. Amoozgar, Z.; Jain, R.K.; Duda, D.G. Role of Apelin in Glioblastoma Vascularization and Invasion after Anti-VEGF Therapy: What Is the Impact on the Immune System? *Cancer Res.* **2019**, *79*, 2104–2106. [CrossRef] [PubMed]
96. Mastrella, G.; Hou, M.; Li, M.; Stoecklein, V.M.; Zdouc, N.; Volmar, M.N.M.; Miletic, H.; Reinhard, S.; Herold-Mende, C.C.; Kleber, S.; et al. Targeting APLN/APLNR Improves Antiangiogenic Efficiency and Blunts Proinvasive Side Effects of VEGFA/VEGFR2 Blockade in Glioblastoma. *Cancer Res.* **2019**, *79*, 2298–2313. [CrossRef] [PubMed]
97. Verma, A.; Kumar Waiker, D.; Bhardwaj, B.; Saraf, P.; Shrivastava, S.K. The Molecular Mechanism, Targets, and Novel Molecules in the Treatment of Alzheimer's Disease. *Bioorg. Chem.* **2022**, *119*, 105562. [CrossRef]
98. García-Morales, V.; González-Acedo, A.; Melguizo-Rodríguez, L.; Pardo-Moreno, T.; Costela-Ruiz, V.J.; Montiel-Troya, M.; Ramos-Rodríguez, J.J. Current Understanding of the Physiopathology, Diagnosis and Therapeutic Approach to Alzheimer's Disease. *Biomedicines* **2021**, *9*, 1910. [CrossRef]

99. Luo, H.; Xiang, Y.; Qu, X.; Liu, H.; Liu, C.; Li, G.; Han, L.; Qin, X. Apelin-13 Suppresses Neuroinflammation against Cognitive Deficit in a Streptozotocin-Induced Rat Model of Alzheimer's Disease through Activation of BDNF-TrkB Signaling Pathway. *Front. Pharmacol.* **2019**, *10*, 395. [CrossRef]
100. Nasseri, B.; Zareian, P.; Alizade, H. Apelin Attenuates Streptozotocin-Induced Learning and Memory Impairment by Modulating Necroptosis Signaling Pathway. *Int. Immunopharmacol.* **2020**, *84*, 106546. [CrossRef]
101. Wan, T.; Fu, M.; Jiang, Y.; Jiang, W.; Li, P.; Zhou, S. Research Progress on Mechanism of Neuroprotective Roles of Apelin-13 in Prevention and Treatment of Alzheimer's Disease. *Neurochem. Res.* **2021**, *47*, 205–217. [CrossRef] [PubMed]
102. Yoo, J.M.; Lin, Y.; Heo, Y.; Lee, Y.H. Polymorphism in Alpha-Synuclein Oligomers and Its Implications in Toxicity under Disease Conditions. *Front. Mol. Biosci.* **2022**, *9*, 959425. [CrossRef] [PubMed]
103. Delenclos, M.; Burgess, J.D.; Lamprokostopoulou, A.; Outeiro, T.F.; Vekrellis, K.; McLean, P.J. Cellular Models of Alpha-Synuclein Toxicity and Aggregation. *J. Neurochem.* **2019**, *150*, 566–576. [CrossRef] [PubMed]
104. Pouresmaeili-Babaki, E.; Esmaeili-Mahani, S.; Abbasnejad, M.; Ravan, H. Protective Effect of Neuropeptide Apelin-13 on 6-Hydroxydopamine-Induced Neurotoxicity in SH-SY5Y Dopaminergic Cells: Involvement of Its Antioxidant and Antiapoptotic Properties. *Rejuvenation Res.* **2018**, *21*, 162–167. [CrossRef] [PubMed]
105. Haghparast, E.; Esmaeili-Mahani, S.; Abbasnejad, M.; Sheibani, V. Apelin-13 Ameliorates Cognitive Impairments in 6-Hydroxydopamine-Induced Substantia Nigra Lesion in Rats. *Neuropeptides* **2018**, *68*, 28–35. [CrossRef]
106. Jiang, Y.; Liu, H.; Ji, B.; Wang, Z.; Wang, C.; Yang, C.; Pan, Y.; Chen, J.; Cheng, B.; Bai, B. Apelin-13 Attenuates ER Stress-associated Apoptosis Induced by MPP+ in SH-SY5Y Cells. *Int. J. Mol. Med.* **2018**, *42*, 1732–1740. [CrossRef] [PubMed]
107. Zhu, J.; Dou, S.; Jiang, Y.; Chen, J.; Wang, C.; Cheng, B. Apelin-13 Protects Dopaminergic Neurons in MPTP-Induced Parkinson's Disease Model Mice through Inhibiting Endoplasmic Reticulum Stress and Promoting Autophagy. *Brain Res.* **2019**, *1715*, 203–212. [CrossRef] [PubMed]
108. Foroughi, K.; Khaksari, M.; Rahmati, M.; Bitaraf, F.S.; Shayannia, A. Apelin-13 Protects PC12 Cells against Methamphetamine-Induced Oxidative Stress, Autophagy and Apoptosis. *Neurochem. Res.* **2019**, *44*, 2103–2112. [CrossRef]
109. Haghparast, E.; Sheibani, V.; Abbasnejad, M.; Esmaeili-Mahani, S. Apelin-13 Attenuates Motor Impairments and Prevents the Changes in Synaptic Plasticity-Related Molecules in the Striatum of Parkinsonism Rats. *Peptides* **2019**, *117*, 170091. [CrossRef] [PubMed]

MDPI AG
Grosspeteranlage 5
4052 Basel
Switzerland
Tel.: +41 61 683 77 34

Genes Editorial Office
E-mail: genes@mdpi.com
www.mdpi.com/journal/genes

Disclaimer/Publisher's Note: The statements, opinions and data contained in all publications are solely those of the individual author(s) and contributor(s) and not of MDPI and/or the editor(s). MDPI and/or the editor(s) disclaim responsibility for any injury to people or property resulting from any ideas, methods, instructions or products referred to in the content.

www.ingramcontent.com/pod-product-compliance
Lightning Source LLC
Chambersburg PA
CBHW041352210926
51850CB00013B/3058

* 9 7 8 3 7 2 5 8 2 5 3 1 8 *